THE ACUTE-PHASE RESPONSE TO INJURY AND INFECTION

The roles of interleukin 1 and other mediators

Research monographs in cell and tissue physiology

Volume 10

General Editors

J.T. DINGLE and J.L. GORDON

Cambridge

ELSEVIER
AMSTERDAM · NEW YORK · OXFORD

The acute-phase response to injury and infection

The roles of interleukin 1 and other mediators

Editors

A.H. GORDON

London

and

A. KOJ

Cracow

1985

ELSEVIER
AMSTERDAM · NEW YORK · OXFORD

ISBN 0-444-80648-2 (volume)
ISBN 0-444-80234-7 (series)

Published by:

Elsevier Science Publishers B.V. (Biomedical Division)
P.O. Box 211, 1000 AE Amsterdam
The Netherlands

Sole distributors for the USA and Canada:

Elsevier Science Publishing Company, Inc.
52 Vanderbilt Avenue
New York, NY 10017
USA

Library of Congress Cataloging in Publication Data
Main entry under title:

The Acute-phase response to injury and infection.

(Research monographs in cell and tissue physiology ; v. 10)
Includes bibliographies and index.
1. Inflammation. 2. Interleukins. 3. Wounds and injuries. 4. Infection. I. Gordon, A. H. II. Koj, A. III. Series. [DNLM: 1. Infection--immunology. 2. Interleukin 1--metabolism. 3. Plasma Proteins--metabolism. 4. Wounds and Injuries--immunology. W1 RE232GL v.10 / QW 504 A189]
RB131.A28 1985 616'.0473 85-12968
ISBN 0-444-80648-2 (U.S.)

PRINTED IN THE NETHERLANDS

Preface

It has long been known that such disparate conditions as bacterial infection, myocardial infarction, surgery, burns or neoplastic disease all lead to the appearance in the blood of certain liver-produced plasma proteins collectively described as acute-phase reactants. In such conditions the rates of synthesis of these proteins increase with the result that certain of them, which may originally have been almost undetectable (such as C-reactive protein in man or α_2-macroglobin in rat), appear as important constituents of the plasma. The mechanism of induced synthesis of acute-phase proteins has been the subject of numerous studies but only recently has a clearer picture begun to emerge due to the combined efforts of biochemists, immunologists and clinicians.

It appears that macrophages are of primary importance in initiating the acute-phase response. When accumulated at the site of injury, these phagocytic cells elaborate and secrete numerous active factors, the so-called monokines, which in the past have been given different names depending on the method of their assay and principal biological activity: leucocyte endogenous mediator, endogenous pyrogen, lymphocyte activating factor or serum amyloid A inducer. Recent evidence has shown that this group of closely related molecules can bring about most, if not all, of the symptoms of the acute-phase response.

Interleukin 1 proteins are hormone-like substances, most of which with molecular weights near to 15,000. Interleukin 1 has a multiplicity of pleomorphic amplifying effects on immunological and inflammatory reactions. Depending on the target cell, it can bring about fever, activation of thymocytes to form interleukin 2, release of granulocytes from bone marrow, decreased plasma iron and zinc concentration, increased breakdown of muscle tissue and transfer of amino acids to the liver, and finally, increased synthesis of acute-phase proteins. Thus a better understanding of the mechanisms of the acute-phase response is important for both experimental biology and clinical medicine. In human patients measurements of plasma levels of acute-phase proteins provide useful diagnostic and prognostic indicators, especially in cases of tissue necrosis, such as myocardial infarction, in various inflammatory diseases or during tumour proliferation.

Since the introduction of the name interleukin 1 in 1982, understanding of the relationship of the acute-phase response, considered narrowly as increases in concentration of certain plasma proteins and the other changes also brought about by interleukin 1, has increased greatly. Numerous reports of these studies are available in biochemical, immunological and clinical journals. The present monograph attempts to bring together the more important of these findings with particular emphasis on the underlying mechanisms.

During the period of its writing, it has become apparent that interleukin 1 is a group of similar but chemically distinct small proteins. A survey of the field at this time seems especially appropriate because it indicates not only the multifarious roles of interleukin 1 but also that some biological effects are brought about by distinct species of interleukin 1.

Acknowledgements

I am very grateful to Ms Anne Richardson for her excellent secretarial help and to Ms Fiona Forrester who prepared many of the figures.

A.H. Gordon

List of contributors

A. FLECK
Department of Chemical Pathology, Charing Cross Hospital Medical School, University of London, Fulham Palace Road, London W6 8RF, U.K.

A.H. GORDON
National Institute for Biological Standards and Control, Holly Hill, Hampstead, London NW3 6RB, U.K.

A. KOJ
Institute of Molecular Biology, Jagiellonian University, Al. Mickiewicza 3, 31-120 Cracow, Poland

M.A. MYERS
Department of Chemical Pathology, Charing Cross Hospital Medical School, University of London, Fulham Palace Road, London W6 8RF, U.K.

J.D. SIPE
Boston University School of Medicine, 71 East Concord Street, Boston, MA 02118, U.S.A.

Contents

Part I. The inflammatory reaction and the production of monokines

Part II. Extrahepatic actions of monokines and injury-derived mediators

Part III. Liver response to inflammation and synthesis of acute-phase plasma proteins

Part IV. Clinical aspects of the acute-phase response

Part V. Assays of monokines

ABBREVIATIONS

AEF	Amyloid enhancing factor
APC	Amyloid P component
ASC	Synovial cells
BAF	B cell activating factor
Bb	Cleavage product of complement component B
CETAF	Corneal epidermal cell-derived thymocyte activating factor
CK	Creatine kinase
Con A	Concanavalin A
CPS	Pneumococcal C-polysaccharide
DMA	Drug metabolising activity
EAP	Endotoxin-associated protein
EC	Extracellular
EGF	Epidermal-growth factor
ELISA	Enzyme-linked immunosorbent assay
EP	Endogenous pyrogen
ESR	Erythrocyte sedimentation rate
ETAF	Epidermal cell-derived thymocyte activating factor
FCS	Foetal calf serum
FCR	Fractional catabolic rate
FIM	Monocytosis inducing factor
FMF	Familial mediterranian fever
GAF	Glucocorticoid antagonising factor
GVHD	Graft versus host disease
HMP	Hexose monophosphate pathway
HPBM-SN	Human peripheral blood monocyte supernatant
HSF	Hepatocyte stimulating factor
IC	Intracellular
ICV	Intracerebroventricular
LAF	Lymphocyte activating factor
LDCF	Lymphocyte-derived chemotatic factor
LDH	Lactic dehydrogenase
LEM	Leucocyte endogenous mediator
LNC	Lymphatic node cell
LP	Leucocyte pyrogen
LPS	Lipopolysaccharide
MAF	Macrophage activating factor
MCF	Mononuclear cell factor
MCSF	Macrophage cell stimulating factor
MDCF	Macrophage-derived competence factor
MEM	Modified Eagle's medium
MGF	Macrophage growth factor
MI	Myocardial infarction
MØ	Macrophage
NADH	Nicotinamide adenine dinucleotide (reduced form)
NAD(P)H	Nicotinamide adenine dinucleotide phosphate (reduced form)
NBT	Nitro blue tetrazolium
OVA	Ovalbumin

PA	Plasminogen activator
PBM	Peripheral blood monocytes
PDGF	Platelet-derived growth factor
PEC	Peritoneal exudate cells
PFC	Plaque-forming cell
PHA	Phytohaemagglutinin
PI	Proteinase inhibitor
PIF	Proteolysis inducing factor
PMA	Phorbolmyristate acetate
PO/AH	Preoptic area of the anterior hypothalamus
RBC	Red blood cell
RIA	Radioimmunoassay
RPD	Rabbit pyrogen dose
SDS	Sodium dodecyl sulphate
SLE	Systemic lupus erythromatosis
SLO	Soluble streptolysin antigen
SOD	Superoxide dismutase
SpA	*S-aureus* protein A
TER	Transcapillary escape rate
TRF	T cell replacing factor

Acute-phase proteins

α-CPI	α-Cysteine proteinase inhibitor
α_1-Ach	α_1-Antichymotrypsin
α_1-AGP	α_1-Acid glycoprotein
α_1-AP globulin	α_1-Acute-phase globulin
α_1-AT	α_1-Antitrypsin
α_1-PI	α_1-Proteinase inhibitor
α_2-APl	α_2-Antiplasmin
α_2-M	α_2-Macroglobulin
α_2-PAG	Pregnancy-associated glycoprotein
β_1-AC	β_1-Anticollagenase
AFP	Alpha-foetoprotein
Alb	Albumin
AP	Acute phase
AT III	Antithrombin III
C3	C3 component of complement
C1-INA	C1-esterase inactivator
CRP	C-reactive protein
Hp	Haptoglobin
ITI	Inter-α-trypsin inhibitor
SAA	Serum amyloid A Protein
SAP	Serum amyloid P protein

Gordon/Koj (eds.) The acute-phase response to injury and infection.

Introduction

A. KOJ and A.H. GORDON

1. Definitions and historical background

The acute-phase response (AP-response) represents an early and unspecific but highly complex reaction of the animal organism to a variety of injuries such as bacterial or parasitic infection, mechanical or thermal trauma, malignant growth or ischaemic necrosis (Fig. 1). It includes not only the local reaction but also neurological, endocrine and metabolic alterations which are expressed as fever, leucocytosis, increased secretion of certain hormones (among which cortisol is particularly important), changes in the concentration of some heavy metals in blood and liver, activation of the clotting, complement, kinin-forming and fibrinolytic pathways, negative nitrogen balance, transfer of amino acids from muscles to the liver followed by a drastic rearrangement of plasma protein synthesis (for reviews see Kampschmidt, 1981, 1983; Sipe and Rosenstreich, 1981; Kaplan et al., 1982; Kushner, 1982; Koj, 1983; Pepys and Baltz, 1983; Sundsmo and Fair,1983; Movat, 1985). The effects of the AP-response are generally thought to be beneficial to the injured organism helping to restore disturbed homeostasis by checking bleeding, by demarcation and resorption of necrotic tissues, by binding and removal of excessive amounts of proteases and exogenous substances, and by preparing conditions for reparative processes and wound healing. The AP-response usually subsides quickly after removal of the noxious stimulus and is often followed by a specific reaction of the immune system although in rare cases it can continue during chronic inflammation.

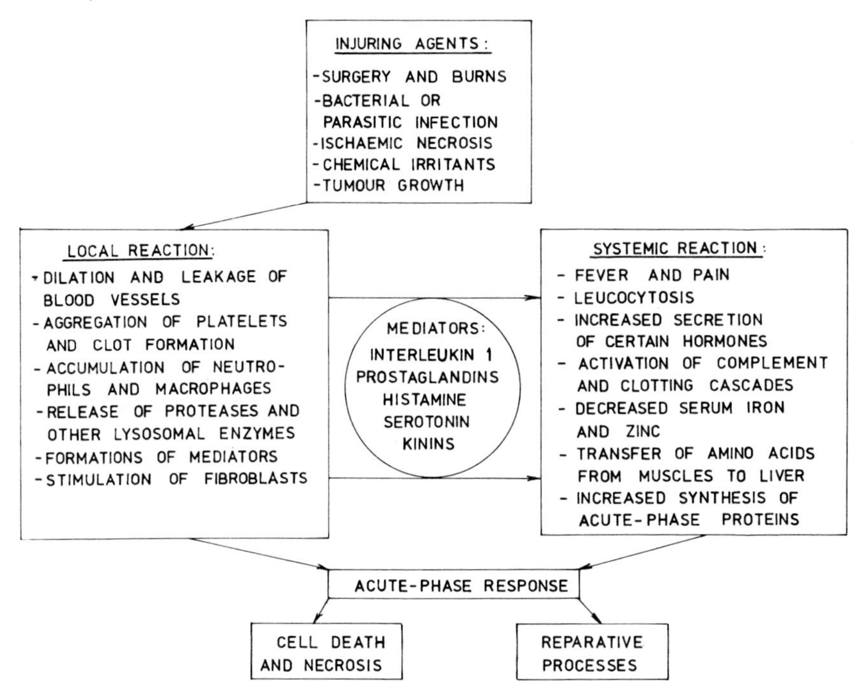

Fig. 1. Development and effects of the AP-response.

The term 'acute-phase' was introduced in 1941 by Avery and co-workers (Abernethy and Avery, 1941) to describe the properties of sera from patients with febrile infectious diseases. They were aware that the sera contained the specific protein (C-reactive protein, CRP), which in the presence of calcium precipitates with the C-polysaccharide fraction of pneumococci, a phenomenon described in 1930 by Tillet and Francis (Pepys, 1981). Subsequent studies demonstrated the occurrence of CRP in the blood of injured rabbits (Anderson and McCarty, 1951), monkeys and dogs (Riley and Coleman, 1970), while injured rats produced different 'acute-phase globulins' present either in α_1- or α_2-fractions (Darcy, 1964; Heim and Lane, 1964; Gordon and Darcy, 1967; Weimer and Humelbaugh, 1967; Gordon and Koj, 1968). As a result of these investigations it became clear that despite considerable species-dependent variability almost all types of injuries lead to increased concentration of a number of plasma proteins collectively named 'acute-phase reactants' (Owen, 1967) and defined as 'trauma-inducible liver-produced plasma glycoproteins' (Koj, 1970a, 1974). Since, however, the term acute-phase reactants is also occasionally used for mediators of inflammation detectable in the plasma early after injury (Sipe and Rosenstreich, 1981), or for major clinical symptoms or systemic inflammatory reaction, such as leucocytosis or increased ESR (Owen, 1967), the more precise name 'acute-phase proteins' (AP-proteins) is rather to be recommended for those plasma proteins which are affected by injury.

Between 1930, when Tillet and Francis noticed the formation of precipitates in the plasma from acutely ill patients with pneumonia and named the protein responsible C-reactive protein, and 1960 when Darcy started to investigate a parallel phenomenon in rats, very little attention was paid to plasma protein changes consequent upon injury or inflammation. Unlike CRP, the presence of which revealed itself by precipitation, the protein investigated by Darcy, an α_1-glycoprotein, remained soluble and required an immunological procedure for its estimation. Inevitably, without the refined immunological and electrophoretic methods, which only became available in the post-war period, for the estimation of changed concentrations of individual plasma proteins, this aspect of the AP-response which involves at least 20 of the plasma proteins (cf. Chapter 12, Table 2) would not have been revealed.

The main importance of Darcy's work was that he was able to show that surprisingly mild intervention, such as subcutaneous injection of water, brings about an increased concentration in the plasma of a protein now recognized as a member of the acute-phase group of plasma proteins. Since his work, the considerable number of plasma proteins subject to changes in concentration as a result of trauma, inflammation and infection has become apparent and it has become possible to divide them according to their physiological functions. As suggested by Geisow and Gordon (1978), the most general classification of all plasma proteins is, into those concerned with defense and repair, transport and finally modulation of the immune system. As described later, AP-proteins exist as members of all three groups (cf. Chapter 13).

The tissue origin of AP-proteins and the regulatory mechanisms involved in their synthesis have been the subjects of speculations and studies during the last 50 years (see Pepys, 1981; Kushner, 1982; Pepys and Baltz, 1983). For a long time the injury-elicited increase in the plasma level of glycoproteins, and especially protein-bound carbohydrates soluble in perchloric acid (seromucoid fraction; Winzler, 1955), was linked to tissue destruction or to healing and proliferative processes (Keyser, 1952; Heppleston and Keyser, 1959). The foundation for the modern view was laid in 1951 by Miller and co-workers who convincingly demonstrated that the perfused rat liver is the source of the majority of plasma proteins. Despite this, many authors continued to believe that the seromucoid fraction (comprising among others α_1-acid glycoprotein, α_1-antitrypsin and haptoglobin) arises from the ground substance of the connective tissue. Even the site of the synthesis of CRP remained unknown until Hurlimann et al. (1966) demonstrated incorporation of radioactive amino acids into immunologically identified CRP after incubation of liver preparations from rabbits and monkeys which has been subjected to inflammatory stimuli. The widespread use of labelled amino acids, immunochemical techniques and the perfused liver from control and injured rats and rabbits finally proved that practically all AP-proteins are synthesized by hepatocytes and that in many cases this synthesis requires the presence of glucocorticoids (Barnhart and Forman, 1963; Sarcione, 1963; Krauss and Sarcione, 1964; Gordon and Koj, 1968; John and Miller, 1969; Kushner and Feldmann, 1978).

Because AP-proteins like most plasma proteins, except the immunoglobulins, are synthesized by hepatocytes, and because injury to another part of the body leads to increased rates of synthesis of these proteins, the existence of hormone-like effector substances originating at the injury site was postulated a long time ago. Already in 1947 Chanutin and Ludewig suggested that the increase in plasma fibrinogen during inflammation must be the result of stimulation of the liver by products of tissue destruction. Even earlier Homburger (1945) had demonstrated that sterile pus from a turpentine abscess contains a thermolabile factor which causes the plasma fibrinogen level to rise when it is injected intramuscularly. Although Homburger may have been the first investigator to have experimented with crude interleukin 1 (IL-1), his observations, presumably because they were premature, were never properly recognized.

In the early 1970's two independent groups of investigators: Pekarek, Wannemacher, Powanda, Beisel and associates from the group at Fort Dietrick (Maryland) and Kampschmidt and co-workers in Ardmore (Oklahoma), drew attention to leucocytes as a possible source of important mediators of the AP-response. In a series of well documented papers both these groups demonstrated that cells from glycogen-induced rabbit peritoneal exudates release a factor, then referred to as leucocyte endogenous mediator (LEM) which when injected into rats or rabbits elevates body temperature, decreases serum iron and zinc, causes the release of neutrophils from the bone marrow, and stimulates a flux of amino acids into the liver which is then followed by enhanced synthesis of AP-proteins (Kampschmidt et al., 1973; Powanda et al., 1973; Wannemacher et al., 1975). LEM was partly purified and shown to consist of a protein or proteins of 15 kDa and p*I* 7.8 which required -SH groups for activity and which could be inactivated by heat or proteolytic enzymes (for references see Kampschmidt, 1978). Since all these studies on biological activities of LEM were carried out in vivo, considerable effort was required to avoid contamination by bacterial endotoxin which was known to be able to evoke almost identical effects. Only when the *Limulus amebocyte lysate* test, as well as endotoxin-tolerant rats and endotoxin-resistant strains of mice, became available, were the direct biological effects of LEM on various target tissues finally established (Pekarek et al., 1974; Bornstein and Walsh, 1978; Kampschmidt et al., 1980). By 1980 it had become clear that LEM is not a single molecular species but rather a family of proteins with p*I*s ranging from 5 to 7 with a close similarity to other mediators of the AP-response, such as endogenous pyrogen (EP) (Kampschmidt, 1980, 1981).

EP, known also as leucocyte pyrogen (LP), has been extensively investigated since 1948 when Beeson described a pyrogenic substance from polymorphonuclear leucocytes (for review see Dinarello and Wolff, 1982; Movat, 1985). Studies of Atkins, Bodel and co-workers (Atkins et al., 1967; Bodel, 1974), Wood, Murphy and associates (Hahn et al., 1967; Murphy et al., 1974; Cebula et al., 1979) and Bornstein and Walsh (1978), demonstrated that LP is a group of low molecular weight proteins showing charge and size heterogeneity and is produced by endotoxin-stimulated leucocytes and macrophages of various animal species. Serious doubts were cast on the ability of granulocytes to form either LEM or EP when Hanson et al. (1980) demonstrated that in the mixed cell population from rabbit peritoneal exudate the adherent mononuclear cells were the only source of EP. The picture was additionally complicated by the fact that stimulated macrophages or blood monocytes were known to release other cytokines including lymphocyte activating factor (LAF) (Gery et al., 1972; Calderon et al., 1975; Rosenwasser et al., 1979), and serum amyloid A (SAA-inducer) capable of switching on synthesis of acute-phase serum amyloid A protein in mice (Sipe et al., 1979; McAdam et al., 1982).

About the same time several papers were published pointing to similarity or even identity of LEM, EP, LAF and SAA-inducers (Kampschmidt et al., 1978; Murphy et al., 1980; Sztein et al., 1981; McAdam et al., 1982). In the light of this evidence, participants in the Second International Lymphokine Workshop at Ermatingen (Switzerland) proposed the name IL-1 for LAF and related factors (cf. Oppenheim and Gery, 1982). Although the name is somewhat misleading as IL-1 appears to be much more than a signal between various types of leucocytes, the name IL-1 has been generally accepted. Oppenheim and Gery (1982) define IL-1 as 'a macrophage derived, hormone-like factor of Mr 12,000 – 16,000 that has a multiplicity of pleomorphic amplifying effects on immunological and inflammatory reactions. It is a genetically unrestricted, immunologically non-specific factor which is active at low ($< 10^{-10}$M) concentrations'. As described in subsequent chapters of this monograph, it is still disputable whether all functions of LAF, EP, LEM and SAA-inducers are indeed subserved by a single factor but introduction of the name IL-1 helped to integrate the efforts of formerly separate research groups working on widely different aspects of the AP-response which were rather artificially divided between immunology, cell biology, biochemistry, genetics, pathology and clinical medicine. It should be remembered, however, that even if IL-1 is the most important mediator of the AP-response there are other agents whose reaction may or may not involve IL-1 as an intermediate (cf. Fig. 1 and subsequent chapters).

2. The inflammatory stimulus and experimental models of the AP-response

One of the difficulties of comparing the results of the studies of the AP-response carried out in various laboratories stems from the lack of a uniform animal model of injury and/or inflammation. In studies in vivo an incredible range of stresses has been employed including exposure to UV-light or X-rays, injury by the rotating Noble-Collip drum, burns, contact with dry ice, injections of parasites, brewer's yeasts, kaolin, silica and very many chemical irritants (for references see Glenn et al., 1968; Kellermeyer and Warren, 1970; Beck and Whitehouse, 1974). The types of injury most commonly used for eliciting the acute inflammatory reaction can be arbitrarily divided into three classes.

2.1. Local injection of chemical irritants or foreign materials leading to a limited aseptic inflammatory abscess or granuloma

Turpentine oil or turpentine spirit is probably the oldest and most common irritant. It is interesting to recall that before the era of sulphonamides and antibiotics, injections of turpentine were sometimes used in human therapy as a means for enhancing an unspecific mechanism by formation of 'fixed abscesses' (Mouray, 1966). When injected subcutaneously or intramuscularly turpentine gives not only a strong local inflammation but also a pronounced systemic response including fever and increased synthesis of AP-proteins. Although the composition of commercial preparations of turpentine may vary, and the reaction cannot be regarded as strictly local since part of the turpentine disappears from the site of injury and may act directly on other organs, a rather good logarithmic dose-dependent relationship was observed between the amount of turpentine injected and the increased level of haptoglobin (Mouray, 1966), or level of α_1-AP-globulin and local granuloma weight (Darcy, 1970). Less popular irritants include carrageenan (Robert et al., 1959; Schumer et al., 1967; Scherer and Ruhenstroth-Bauer, 1977; Vinegar et al., 1982), suspensions of talc or Celite (Gordon and Koj, 1968; Koj and Dubin, 1976), casein (McAdam and Sipe, 1976; Benson et al., 1977) or silver nitrate (Kisilevsky et al., 1979). These agents appear to be no more effective than turpentine in eliciting the prompt acute reaction and in certain cases they lead to prolonged inflammation complicated by amyloidosis and various secondary adaptive responses.

2.2. Mechanical, ischaemic, thermal or irradiation injury causing tissue necrosis

Although Noble-Collip drum trauma (Loegering et al., 1976) or X-irradiation (John and Miller, 1968; Koj, 1970b) allow the application of a standardized insult, the response is often erratic (drum trauma) or low and delayed (X-irradiation). Surgery, such as a laparotomy has been recommended by several authors (Neuhaus et al., 1966; Chandler and Neuhaus, 1968; van Gool and Ladiges,1969) but quantitation of injury is variable and surgery may often be complicated by secondary infections.

2.3. Intraperitoneal or intravenous injection of killed bacteria, endotoxins or purified bacterial lipopolysaccharides (LPS)

It has long been established that endotoxins exert various metabolic effects on mammalian cells including such typical symptoms of the AP-response as fever (Snell and Atkins, 1967), reduced plasma concentration of iron (Kampschmidt et al., 1969) and zinc (Kampschmidt and Upchurch, 1970), increased hepatic uptake of amino acids (Wannemacher et al., 1972), increased synthesis of fibrinogen (Regoeczi et al.,1963; Koj and McFarlane, 1968) of other AP-proteins (Pekarek et al., 1974; Abd-el-Fattah et al., 1976; Scherer and Ruhenstroth-Bauer, 1977; Gordon and Limaos, 1979). Although endotoxins affect many organs and body functions, tissue macrophages and blood monocytes are their primary targets (Moore et al., 1978; Morrison and Ulevitch, 1978). Thus a unique opportunity is available for comparing the same inflammatory stimulus both in vivo and in vitro; indeed LPS has been widely used in order to stimulate production of mediators of inflammation. An additional advantage of its use is that standardized preparations of endotoxins are commercially available. On the other hand, endotoxins are highly complex materials and their structure and composition may vary considerably depending on their origin and method of purification. Also susceptibility to LPS of various species and strains of laboratory animals ranges widely.

3. Effects of partial hepatectomy

In addition to the types of injury listed above, another procedure, partial hepatectomy, has found limited application. However, until recently the full significance of the results thus obtained was not fully apparent. As long ago as 1970, Chandler and Snider observed a bimodal curve for plasma and fibrinogen and seromucoid synthesis following partial hepatectomy in rats. The maximum response of these proteins occurred after 24 h and was followed by a second lower but broader peak between 4 and 8 days. On the other hand, in control rats which had been subjected only to laparotomy, the time of the initial peak of fibrinogen or seromucoid response was the same but the return to the pre-operational level was much more rapid. A comparison of stimuli which led to increased plasma concentration of α_2-macroglobulin (α_2-M) and α-foetoprotein (AFP) in normal and partially hepatectomized rats has indicated a probable explanation for the bimodal plasma fibrinogen and seromucoid response to hepatectomy. Concentrations of α_2-M and AFP in normal laparotomized and partially hepatectomized rats were investigated by Hudig et al. (1979). These proteins were chosen for investigation because both are known to be synthesized by foetal liver and yolk sac and are present in neonatal rats. In older animals only trace concentrations of both these proteins can be found in the plasma. After injury to normal rats AFP remains at a very low concentration whereas α_2-M concentrations show great increases. On the other hand, with chemically induced liver necrosis, the AFP plasma concentration is greatly increased with only a slight change in α_2-M. Thus the existence of different control mechanisms for the synthesis of these two plasma proteins must be recognized. The consequences of different control mechanism were most evident after partial hepatectomy, as the maximum plasma concentration of α_2-M was at a maximum after 48 h, whereas AFP showed a broad maximum between 3 and 7 days. As was found for fibrinogen the α_2-M concentration declined slowly after partial hepatectomy and was still elevated at 11 days.

The above results suggest two tentative conclusions. The first is the negative one that neither IL-1 nor any other injury-derived cytokine can be capable of stimulation of AFP synthesis. The second conclusion is that AFP formation occurs only in a few tissues in which rapid cell division is taking place. These include regenerating liver, hepatomas and certain embryonic tumours (for references see Smith and Kelleher, 1980). As demonstrated by Abelev (1979) the derepression of AFP synthesis is not governed by a humoral factor since in the regenerating liver AFP is produced solely by the dividing hepatocytes in the perinecrotic area. It is not yet clear whether the same cells are involved in enhanced production of fibrinogen or α_2-M after hepatectomy, or whether liver macrophages stimulated during liver regeneration produce IL-1 which may in turn activate all hepatocytes to increased synthesis of AP-proteins.

4. Concluding remarks

The acute-phase reaction manifests itself in several different ways (cf. Fig. 1), some of which take place only in the whole organism (e.g. fever, leucocytosis). Others may be studied both in vivo and in various isolated systems. Thus perfused rat, mouse or rabbit livers have been successfully used for studying both kinetics and the nature of the mediators which stimulate synthesis of AP-proteins (Sarcione, 1963, 1970; Gordon and Koj, 1968; John and Miller, 1968, 1969; Koj and Dubin, 1974; Metcalfe and Tavill, 1975; Kushner et al., 1980; Kirsch and Franks, 1982) and enhance uptake of amino acids (Wannemacher et al., 1975). Evaluation of the liver's response to inflammatory stimuli has also been achieved by using liver explants (Hurlimann et al., 1966) or liver slices (Jamieson et al., 1975, 1982; Koj, 1980) but considerable progress has been recently achieved by use of cultured hepatocytes (cf. Chapter 17).

Until now, advance towards fuller understanding of the various aspects of the AP-response has been delayed by the extremely limited amounts of even semi-purified IL-1 that have been available, and also because such IL-1 is already known to contain more than one active constituent. However, due to replacement of peritoneal cells and isolated monocytes as the main sources of IL-1, by established cell lines grown in culture with enormously greater rates of IL-1 production, the situation is changing. Two important objectives are now in sight. Firstly, the isolation of individual constituents of the IL-1 group of mediators, and secondly the identification of which of these may be responsible for one or more aspects of the AP-response.

References

Abd-el-Fattah, M., Scherer, R. and Ruhenstroth-Bauer, G. (1976) J. Molec. Med. *1*, 211 – 221.

Abelev, I. (1979) in Carcino-Embryonic Proteins. Chemistry. Biology, Clinical Applications (Lehman, F.G., Ed.) Vol. 1, pp. 99 – 100, Elsevier/North-Holland Biomedical Press, Amsterdam – New York – Oxford.

Abernethy, T.J. and Avery, O.T. (1941) J. Exp. Med. *73*, 173 – 182.

Anderson, H.C. and McCarty, M. (1951) J. Exp. Med. *93*, 25 – 36.

Atkins, E., Bodel, P. and Francis, L. (1967) J. Exp. Med. *126*, 357 – 386.

Barnhart, M.I. and Forman, W.B. (1963) Vox Sang. *8*, 461 – 473.

Beck, F.J. and Whitehouse, M.W. (1974) Proc. Soc. Exp. Biol. Med. *145*, 134 – 140.

Benson, M.D., Scheinberg, M.A., Shirahama, T., Cathcart, E.S. and Skinner, M. (1977) J. Clin. Invest. *59*, 412 – 417.

Bodel, P. (1974) J. Exp. Med. *140*, 954 – 965.

Bornstein, D.L. and Walsh, E.C. (1978) J. Lab. Clin. Med. *91*, 236 – 245.

Calderon, J., Kiely, J.H., Lefko, J.L. and Unanue, E.R. (1975) J. Exp. Med. *142*, 151 – 164.

Cebula, T.A., Hanson, D.F. Moore, D.M. and Murphy, P.A. (1979) J. Lab. Clin. Med. *94*, 95 – 105.

Chandler, A.M. and Neuhaus, O.M. (1968) Biochim. Biophys. Acta *166*, 186 – 194.

Chanutin, A. and Ludewig, S. (1947) J. Biol. Chem. *167*, 313 – 320.

Darcy, D.A. (1960) Br. J. Cancer *14*, 524 – 533; 534 – 546.

Darcy, D.A. (1964) Br. J. Exp. Pathol. *45*, 281 – 293.

Darcy, D.A. (1970) Br. J. Exp. Pathol. *51*, 59 – 72.

Dinarello, C.A. and Wolff, S.M. (1982) Am. J. Physiol. *72*, 799 – 819.

Geisow, M.J. and Gordon, A.H. (1978) Trends Biochem. Sci. *3*, 169 – 171.

Gery, I., Gershon, R.K. and Waksman, B.H. (1972) J. Exp. Med. *136* 128 – 142.

Gordon, A.H. and Darcy, D.A. (1967) Br. J. Exp. Pathol. *48*, 81 – 89.

Gordon, A.H. and Koj, A. (1968). Br. J. Exp. Pathol. *49*, 436 – 447.

Gordon, A.H. and Limaos, E.A. (1979) Br. J. Exp. Pathol. *60*, 441 – 446.

Glenn, E.M., Bowman, B.J. and Koslowske, T.C. (1968) Biochem. Pharmacol. *17*, Suppl., 27 – 49.

Hahn, H.H., Chan, D.C., Postel, W.B. and Wood, Jr., W.B. (1967) J. Exp. Med. *126*, 385 – 394.
Hanson, D.F., Murphy, P.A. and Windle, B.E. (1980) J. Exp. Med. *151*, 1360 – 1371.
Heim, W.G. and Lane, P.H. (1964). Nature (London) *203*, 1077 – 1078.
Heppleston, A.G. and Keyser, W.J. (1959) Br. J. Exp. Pathol. *40*, 263 – 272.
Homburger, F. (1945) J. Clin. Invest. *24*, 43 – 45.
Hudig, D., Sell, S., Newell, L. and Smuckler, E.A. (1979) Lab. Invest. *40*, 134 – 139.
Hurlimann, J., Thorbecke, G.J. and Hochwald, G.M. (1966) J. Exp. Med. *123*, 365 – 378.
Jamieson, J.C., Morrison, K.E., Molaski, D. and Turchen, B. (1975) Can. J. Biochem. *53*, 401 – 414.
Jamieson, J.C., Kutryk, M., Woloski, B.M.R.N.J. and Kaplan, H.A. (1982) Biochem. Med. *28*, 176 – 187.
John, D.W. and Miller, L.L. (1968) J. Biol. Chem. *243*, 268 – 273.
John, D.W. and Miller, L.L. (1969) J. Biol. Chem. *244*, 6134 – 6142.
Kampschmidt, R.F. (1978) J. Reticuloendothel. Soc. *23*, 287 – 297.
Kampschmidt, R.F. (1980) in Microbiology 1980 (Schlesinger, D., ed.) pp. 150 – 153, Washington D.C. American Society for Biology.
Kampschmidt, R.F. (1981) in Infection: The Physiologic and Metabolic Responses of the Host (Powanda, M.C. and Canonico, P.G., eds.) pp. 55 – 74, Elsevier/North-Holland Biomedical Press, Amsterdam.
Kampschmidt, R.F. (1983) Lymphokine Res. *2*, 97 – 102.
Kampschmidt, R.F. and Upchurch, H.F. (1970) Proc. Soc. Exp. Biol. Med. *134*, 1150 – 1152.
Kampschmidt, R.F., Upchurch, H.F. and Eddington, C.L. (1969) Proc. Soc. Exp. Biol. Med. *132*, 817 – 820.
Kampschmidt, R.F., Upchurch, H.F., Eddington, C.L. and Pulliam, L.A. (1973) Am. J. Physiol. *224*, 530 – 533.
Kampschmidt, R.F., Pulliam, L.A. and Merriman, C.R. (1978) Am. J. Physiol. *235*, C118 – 121.
Kampschmidt, R.F., Pulliam, L.A. and Upchurch, H.F. (1980) J. Lab. Clin. Med. *95*, 616 – 623.
Kaplan, A.P., Silverberg, M., Dunn, J.T. and Ghebrehiwet, B. (1982) Ann. N.Y. Acad. Sci. *389*, 25 – 38.
Kellermeyer, R.W. and Warren, K.S. (1970) J. Exp. Med. *131*, 21 – 30.
Keyser, J.W. (1952) J. Clin. Pathol. *5*, 194 – 198.
Kirsch, R.E. and Franks, J.J. (1982) Hepatology *2*, 205 – 208.
Kisilevsky, R., Benson, M.D., Axelrod, M.A. and Boudreau, L. (1979) Lab. Invest. *41*, 206 – 210.
Koj, A. (1970a) in Ciba Foundation Symposium on Energy Metabolism in Trauma (Porter, R.R. and Knight, J., eds.) pp. 72 – 79, Churchill, London.
Koj, A. (1970b) Folia Biol. (Krakow) *18*, 275 – 286.
Koj, A. (1974) in Structure and Function of Plasma Proteins (Allison, A.C., ed.), Vol. 1, pp. 73 – 131, Plenum Press, London and New York.
Koj, A. (1980) Br. J. Exp. Pathol. *61*, 332 – 338.
Koj, A. (1984) in: Pathophysiology of Plasma Protein Metabolism (Mariani, G., ed.) pp. 221 – 248, Macmillan, London.
Koj, A. and Dubin, A. (1974) Acta Biochim. Polon. *21*, 159 – 167.
Koj, A. and Dubin, A. (1976) Br. J. Exp. Pathol. *57*, 733 – 741.
Koj, A. and McFarlane, A.S. (1968) Biochem. J. *108*, 137 – 146.
Krauss, S. and Sarcione, E.J. (1964) Biochim. Biophys. Acta *90*, 301 – 308.
Kushner, I. (1982) Ann. N.Y. Acad. Sci. *389*, 39 – 48.
Kushner, I. and Feldmann, G. (1978) J. Exp. Med. *148*, 466 – 477.
Kushner, I., Ribitch, W.N. and Blair, J.B. (1980) J. Lab. Clin. Med. *96*, 1037 – 1045.
Loegering, D.J., Kaplan, J.E. and Saba, T.M. (1976) Proc. Soc. Exp. Biol. Med. *152*, 42 – 46.

McAdam, K.P.W.J. and Sipe, J.D. (1976) J. Exp. Med. *144*, 1121 – 1127.
McAdam, K.P.W.J., Li, J., Knowles, J., Foss, N.T., Dinarello, C.A., Rosenwasser, L.J. Selinger, M.J., Kaplan, M.M., Goodman, R., Herbert, P.N., Bausserman, L.L. and Nadler, L.M. (1982) Ann. N.Y. Acad. Sci. *389*, 126 – 136.
Metcalfe, J. and Tavill, A.S. (1975) Br. J. Exp. Pathol. *56*, 570 – 578.
Miller, L.L., Bly, C.G., Watson, M.L. and Bale, W.F. (1951) J. Exp. Med. *94*, 431 – 453.
Moore, R.N., Goodrum, K.J. and Berry, L.J. (1978) J. Reticuloendothel. Soc. *19*, 187 – 197.
Morrison, D.C. and Ulevitch, R.J. (1978) Am. J. Pathol. *93*, 527 – 617.
Mouray, H. (1966) Biosynthese de l'Haptoglobine Chez le Lapin, R. Foulon and Co., Paris.
Movat, H.Z. (1985) The Inflammatory Reaction, Elsevier/North-Holland Biomedical Press, Amsterdam, in press.
Murphy, P.A., Chesney, P.J. and Wood, Jr., W.B. (1974) J. Lab. Clin. Med. *83*, 310 – 322.
Murphy, P.A., Simon, P.I. and Willoughby, W.F. (1980) J. Immunol. *12*, 2498 – 2501.
Neuhaus, O.W., Balegno, H.F. and Chandler, A.M. (1966) Am. J. Physiol. *211*, 151 – 156.
Oppenheim, J.J. and Gery, I. (1982) Immunology Today. *3*, 113 – 129.
Owen, J.A. (1967) Adv. Clin. Chem. *9*, 1 – 41.
Pekarek, S., Wannemacher, Jr., R.W., Powanda, M.C., Abeles, F., Mosher, D., Dinterman, R. and Beisel, W.R. (1974) Life Sci. *14*, 1765 – 1776.
Pepys, M.B. (1981) Lancet *i*, 653 – 657.
Pepys, M.B. and Baltz, M.L. (1983) Adv. Immunol. *34*, 141 – 212.
Powanda, M.C., Cockerell, G.L. and Pekarek, R.S. (1973) Am. J. Physiol. *225*, 399 – 401.
Regoeczi, E., Henley, G.E., Holloway, R.C. and MacFarlane, A.S. (1963) Br. J. Exp. Pathol. *44*, 397 – 403.
Riley, R.F. and Coleman, M.K. (1970) Clin. Chim. Acta *30*, 483 – 496.
Robert, B., Robert, L. and Jayle, M.F. (1959) Experientia *15*, 385 – 387.
Rosenwasser, L.J., Dinarello, C.A. and Rosenthal, A.S. (1979) J. Exp. Med. *150*, 709 – 714.
Sarcione, E.J. (1963) Arch. Biochem. Biophys. *100*, 516 – 519.
Sarcione, E.J. (1970) Biochemistry *9*, 3059 – 3062.
Scherer, R. and Ruhenstroth-Bauer, G. (1977) Naturwissenschaften *64*, 471 – 478.
Schumer, W., Molnar, J. Dowling, J.N. and Winzler, R.J. (1967) Am. J. Physiol. *212*, 90.
Sipe, J.D. and Rosenstreich, D.L. (1981) in Cellular Functions in Immunity and Inflammation, (Oppenheim, J.J., Rosenstreich, D.L. and Potter, M., eds.) pp. 411 – 429, Elsevier/North-Holland Biomedical Press, Amsterdam.
Sipe, J.D., Vogel, S.N., Ryan, J.L., McAdam, K.P.W.J. and Rosenstreich, L. (1979) J. Exp. Med. *150*, 597 – 606.
Smith, C.J.P. and Kelleher, P.C. (1980) Biochim. Biophys. Acta *605*, 1 – 32.
Sundsmo, J.S. and Fair, D.S. (1983) Clin. Physiol. Biochem. *1*, 225 – 284.
Snell, E.S. and Atkins, E. (1967) Am. J. Physiol. *212*, 1103 – 1112.
Sztein, M.B., Vogel, S.N., Sipe, J.D., Murphy, P.A., Mizel, S.B., Oppenheim, J.J. and Rosenstreich, D.L. (1981) Cell. Immunol. *63*, 164 – 176.
Van Gool, J. and Ladiges, N.C.J.J. (1969) J. Pathol. *97*, 115 – 126.
Vinegar, R. Truax, J., Gelph, J.L. and Volkner, P.A. (1982) Fed. Proc. *41*, 2588 – 2591.
Wannemacher, Jr., R.W., Pekarek, R.S. and Beisel, W.R. (1972) Proc. Soc. Exp. Biol. Med. *139*, 128 – 132.
Wannemacher, Jr., R.W., Pekarek, R.S., Thompson, W.L., Curnow, R.T., Beall, F.A., Zenser, T.V., DeRubertis, F.R. and Beisel, W.R. (1975) Endocrinology *96*, 651 – 661.
Weimer, H.E. and Humelbaugh, C. (1967) Can. J. Physiol. Pharmacol. *45*, 241 – 247.
Winzler, R.J. (1955) in Methods of Biochem. Analysis (Glick, D., ed.) Vol. 2, pp. 279 – 311, Interscience Publishers, New York.

PART I

The inflammatory reaction and the production of monokines

Gordon/Koj (eds.) The acute-phase response to injury and infection.

CHAPTER 1

Cellular and humoral components of the early inflammatory reaction

J.D. SIPE

A mechanism for responding to injury has been observed at all levels of life. For single cell organisms, it is the acid of injury (Chambers and Chambers, 1961) while for multicellular organisms (Metchnikoff, 1905) and multiorgan animals (Oppenheim and Potter, 1981) there are cellular and humoral events. Inflammation, the inflammatory response or the inflammatory reaction (described in detail in the Introduction) is the organism's reaction to injury in defense of its existence. Inflammation occurs following injury as a series of alterations in vascularized tissues. The inflammatory response is an extensive network of interacting mediators, cells and mechanisms. While pain and swelling may occur during the process of removal of the inciting agent and repair of the injured site, inflammation is primarily a beneficial reaction of tissues to injury.

When tissue injury, which may or may not involve cell death, occurs the immediate response is a local one. The cellular responses which act to destroy foreign cells and damaged material are in themselves destructive and require control and modulation (Munck et al., 1984). Either the primary injury itself or the cellular response to it can initiate the systemic humoral response that affects various parts of the body: blood vessels, connective tissue, brain and liver. This chapter will examine the sequence of early events that sets into motion the acute-phase response (AP-response).

1. The inflammatory exudate and blood leukocytosis

Inflammation is a nonspecific response of tissues to diverse stimuli ranging from relatively mild local changes which occur without or before cell death to the pronounced tissue injury and cell necrosis associated with trauma and surgery. Thus, as indicated in the Introduction (Fig. 1), excess heat, cold, chemical stimulation, the presence of foreign agents, malignancy, infection, trauma, surgery, all initiate the acute inflammatory response. The mechanisms by which the inflammatory response proceeds can be expected to be uniform despite the variety of possible inciting agents. This is because the process is regulated by common groups of mediators and inhibitors (Larsen and Henson, 1983).

Cell necrosis sets in motion a series of reactions: first local, later systemic designed to repair injury, remove foreign material and debris and eliminate bacteria. Injured and dying cells release lysosomal enzymes, oxygen radicals, prostaglandins, chemotactic factors and histamine. Within minutes of cell necrosis, there are pronounced hemodynamic changes including dilatation of arterial blood vessels followed by an increased rate of blood flow through the microvasculature. If necrosis is associated with a blood vessel injury, blood flows into any openings and clots with fibrin formation. The increased blood flow serves to wash away toxic products and to bring large numbers of leucocytes, plasma proteins and nutrients to the injured area. Dilatation of the capillaries results in increased vascular permeability, thus leading to an exudation of fluid from the vessels to the tissue (Movat, 1985). Thus, the early inflammatory response is characterized by vasodilation, increased blood flow and increased vascular permeability. As a result of the hemodynamic changes just described, inflammatory cells are recruited to the site of injury.

A variety of chemoattractants has been described for neutrophils sufficient to provide insurance that the host can mount an inflammatory response under a variety of circumstances. Neutrophils make their way between endothelial cells to the site of tissue injury whereas their cytocidal activity is remarkably directed toward foreign or noxious cells. Later, the infiltrating neutrophils are replaced by mononuclear cells. A normal, circumscribed response is shown in Fig. 1 (Larsen and Henson, 1983). The response was initiated by instilling a fragment of the fifth component of complement, C5a des-Arg, into the airways of rabbits. Within 6 h, there was an accumulation of neutrophils in the airspaces (Fig. 1B). By 24 – 48 h the neutrophilic infiltration was replaced by mononuclear cells (Fig. 1C) and over several days the alveoli were cleared of cells and fluid and the alveolar walls returned to normal thickness (Fig. 1A). The early and local inflammatory reaction has been described in detail by Movat (1985).

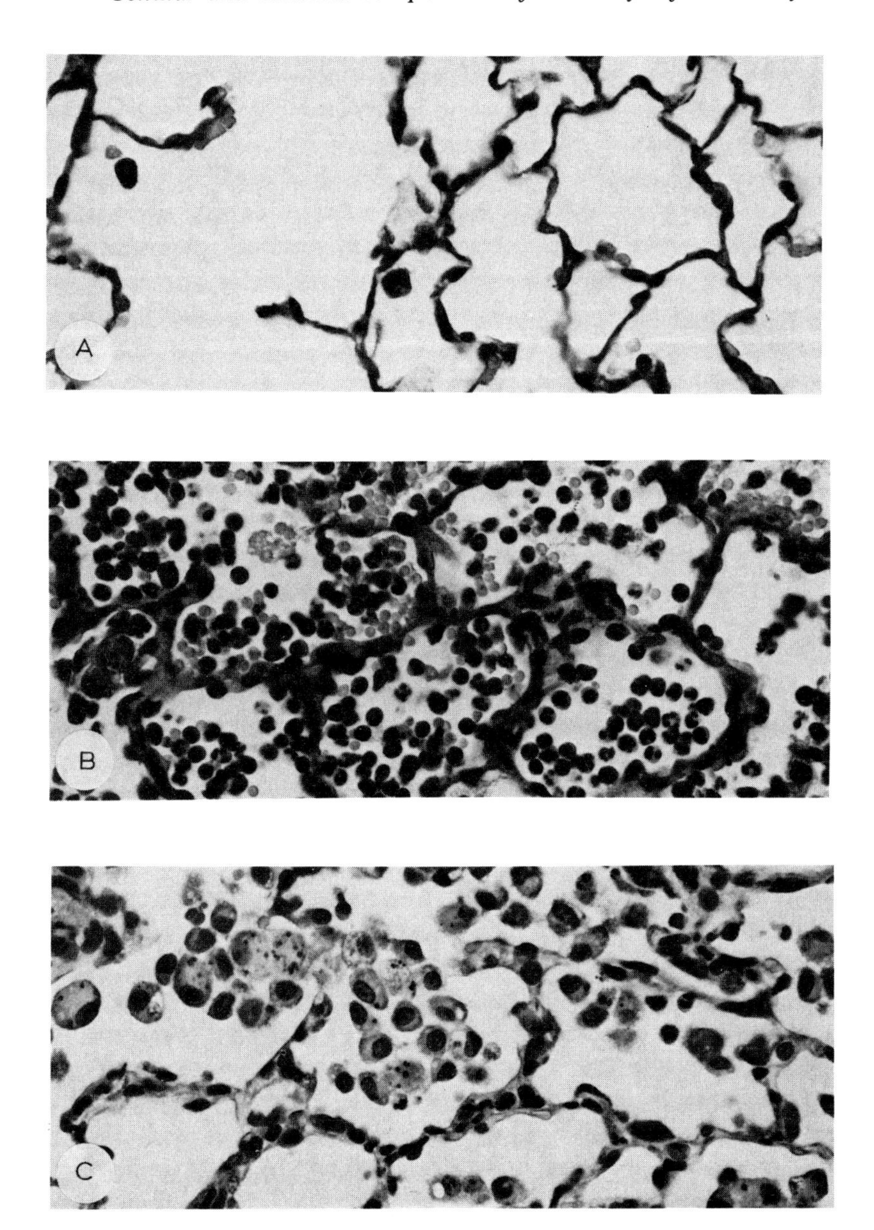

Fig. 1. Morphology of the AP-response in rabbit airways following instillation of C5a des-Arg. (A) Normal lung in which resident macrophages but not neutrophils are present. (B) Six hours after administration of C5a des-Arg neutrophils and protein-rich fluid accumulate in airways. (C) Between 24 and 48 h after C5a des-Arg instillation, mononuclear cells replace the infiltrating neutrophils. Finally, the alveoli clear and the alveolar walls return to their normal thickness. (After Larsen and Henson, 1983.)

1.1. Inflammatory exudates

Vascular leakage can arise directly as a result of injury or indirectly as a result of histamine release (vasoactive mediators). Other vasoactive mediators are serotonin and kinins. These mediators induce brief episodes of vascular leakage which: (1) occur immediately, (2) are of brief duration (15 – 30 min); and (3) affect the venules. Vasoactive mediators cause leaks in layers of endothelial cells, apparently by causing the endothelial cells to contract. Vasoactive mediator-induced leakage is specific for venules, as compared with capillaries and arterioles.

Direct vascular injury is of longer duration and plasma and plasma proteins will seep out, however, red cells will be retained if the endothelium alone is damaged, and the basement membrane remains intact. Peripheral to the area of direct injury, histamine-type mediators will also stimulate venule leakage. Within seconds of the injury, in the area of the dead tissue, circulation stops and fluid is lost. The early constriction or narrowing of all arterioles that sharply reduces the blood flow in the area of injury lasts only a few minutes. Subsequently, the arterioles dilate in and around the injured area leading to delivery of an increased volume of blood to this area. At the center of tissue damage where all small vessels are injured there is direct leakage. At the periphery of the injury the venules leak fluid because they are stimulated by histamine-type mediators released from the areas of cell necrosis and tissue injury. Vasodilation is of importance to development of the acute inflammatory reaction, because the amount of exudate produced is determined largely by the local blood flow.

Following their initial constriction of short duration, the arterioles dilate causing the blood pressure to rise throughout the microcirculatory network. In areas of intact endothelium, the increased blood pressure results in increased transudation (ultrafiltration across normal capillary walls). However, where there are gaps in the endothelium, either directly resulting from injury or in response to injury, increased vascular permeability, plasma leakage occurs and the protein concentration of the interstitial fluid is increased. The presence of large amounts of protein in tissues leads to an increased uptake of water and salts. Normally, when vessel walls are not leaky, the osmotic pressure of the circulating proteins would tend to retain fluid within the blood stream, counterbalancing the hydrostatic pressure of blood which tends to push fluid out of the circulation. The loss of fluid from the circulation results in thickening, sluggishness and coagulation of the blood. This resistance of blood flow in turn results in increased blood pressure and fluid loss. The immediate response to exudation is the plugging of endothelial gaps by platelets and a subsiding of arteriolar dilation and an exhaustion of the supply of histamine mediators. When the pressure of the interstitial fluid equals that of the blood, excess fluid which escapes is drained off through the lymphatic vessels.

TABLE 1
Characterization of inflammatory exudates

Type	Characteristics	Origin
Serous	Few cells, like blood serum	Vascular injury
Purulent	High neutrophil concentration	Bacterial infection that attracts neutrophils
Hemorrhagic	High red cell or hemaglobin content	Severe capillary drainage
Fibrinous	Whitish fibrin layer on serous surface	Exudation of fibrinogen, followed by breakdown to fibrin

The types of exudate can vary, according to the stimulus of the inflammation and the above categorization (Table 1) is frequently employed.

1.2. Blood leukocytosis

Within minutes of the initiation of the inflammatory process, as exudation of fluid from the vessels to tissue occurs, there is a concentration of red blood cells in the capillaries. As these red cells are packed, the rate of blood flow is diminished and white cells leave the center of the blood stream and move to the outer boundary and begin to adhere to the endothelial cells of the vessel wall. This process is called margination. If the injury is very mild, the process goes no further. Once adhered, leukocytes can begin the process of migration to the tissues.

The process of adherence to the vessel appears to be independent of histamine-induced vascular permeability which occurs up to 30 min or so after injury. However, independently, in the presence of bacteria or when significant tissue damage occurs, there is massive leukocytic infiltration. It is not understood how the surface of the endothelium is altered so that negatively charged leukocytes can adhere to normally negatively charged endothelial cells. The alteration may result from a chemical alteration of the endothelial surface or may involve the formation of ionic bridges. Divalent ions such as calcium have been implicated in this process, since treatment of animals with the chelating agent EDTA inhibited sticking (Thompson et al., 1967) as does local administration of anesthetic drugs (Giddon and Lindhe, 1972).

The integrity of the vascular bed is essential for leukocytic delivery to tissues and in disease states such as diabetes, distribution of neutrophils to sites of tissue injury is impaired. Within minutes of injury, monocytes and neutrophils adhere to the vascular endothelium. They are localized primarily in the post-capillary venules and there they insert footlike projections (pseudopodia) into the tight junction between two endothelial cells and force their way through the extracellular space reaching the extravascular space after penetration through the basement membrane. There are no visible openings (Marchesi and Flory, 1960), but the passage is rapid taking 2 to 8 – 12 min.

Although neutrophils and monocytes may move through gaps existing due to the action of histamine-like mediators, they do not need pre-existing gaps. Furthermore, diapedesis does not cause vascular leakage despite the temporary fissures caused by the passage of cells through the basement membrane, perhaps via secretion of an enzyme that alters the membrane to allow passage. Although no patent holes remain, residual damage to the basement membrane can be demonstrated. For example, Cochrane and Aikin (Cochrane and Aikin, 1966; Cochrane, 1968) have used immune-complex mediated injury to study neutrophil migration from the vasculature. They found that extracellularly secreted neutrophil enzymes destroyed portions of the basement membrane. They also presented evidence that collagenase, elastase, cathepsin D and E, and neutral proteases are leukocyte enzymes that are capable of degrading proteins found in basement membrane.

The neutrophil is the predominant cell to enter the area of injury in the first stages of the acute inflammatory response. A significant neutrophil response is usually observed after 6 h, whereas macrophages begin to arrive 12 h following injury and will continue to accumulate for 48 – 72 h. The half-life of neutrophils is short ($t_{1/2} = 6$ h) and as they are incapable of cell division, they must be supplied as needed from the blood stream.

Although they begin to emigrate at the same time, macrophages arrive at the area of injury later than neutrophils because they move more slowly. As well as being longer lived than neutrophils, macrophages are capable of division. While the slow speed and long life of macrophages may be responsible in part for the delay in monocyte infiltration, it is highly probable that the biphasic leukocyte response results from the sequential action of specific chemotactic mediators. One such specific agent is the monocyte chemotactic factor released from neutrophils.

Neutrophils are produced in the bone marrow deriving from a single pluripotent granulocytic stem cell which also gives rise to the eosinophil, the basophil and the monocyte (Cline and Golde, 1979). The development of mature neutrophils takes almost 2 weeks (reviewed by Davis and Gallin, 1981). As neutrophil maturation is completed, their negative surface charge decreases with a resultant increase in membrane deformability. The bone marrow serves as the site of storage as well as for formation of neutrophils, with more than 90% of the body's total of these cells in this location. Upon appropriate stimulation, the neutrophil leaves the bone marrow and enters the circulation where it is equilibrated with the marginated pool.

Several factors have been implicated in the regulatory mechanism for stimulation of maturation and release of cells from the bone marrow. These factors include: diffusable granulopoietic substance, colony stimulating factor, endotoxin and etiocholanolone which probably promote release from the marrow via interleukin 1 (IL-1) and both neutrophil releasing factor (Boggs et al., 1966) and leukocyte endogenous pyrogen (Kampschmidt et al., 1972) which may be forms of IL-1. Furthermore, fragments from the third component of complement (C3) stimulate leukocyte release (Davis and Gallin, 1980).

The number of naturally occurring chemotactic factors is very large. In addition to those factors cited above that stimulate maturation and release of cells from the bone marrow, those chemotactic factors that initiate leukocyte aggregation and subsequent adherence in capillary beds include microorganisms, and cellular and plasma products produced in response to noxious stimuli. Plasma contains three major mediator producing systems, complement, coagulation and kinin release which are orchestrated to interact and to generate phlogistic peptides and cell-derived mediators which may be either preformed and stored in granules, e.g. histamine in mast cells, or may be newly synthesized by the cell (reviewed by Larsen and Hanson, 1983). C5 and its fragments are extremely important for neutrophil influx. C5 convertases and other proteases generate a 74 amino acid terminal fragment termed C5a. Human C5a des-Arg is the fragment derived from C5a after removal of the C-terminal arginine by carboxypeptidase digestion. Both C5a and C5a des-Arg are potent stimulators of neutrophil influx.

The kinin-generating pathway results in formation of plasminogen activator which together with kallikrein is chemotactic for the neutrophil as are fibrinopeptide B and fibrin-degradation products.

Other chemotactic factors are produced by macrophages (polypeptide and lipid) and by bacteria as products of normal prokaryotic protein synthesis (*N*-formylated methionyl peptides). The formyl-methionine peptides have been used extensively as well defined probes of neutrophil function such as oxidative metabolism, stimulus-secretion coupling and chemotaxis (Weissman et al., 1982).

Neutrophils themselves appear to provide an amplification to chemotaxis both by complement activation and by production of leukogressin (Hayashi, 1975). Neutrophil specific granules can activate the alternate complement pathway and azurophil granule products released late in the phagocytic process can inactivate C5a and other chemoattractants and thus may serve to modulate the inflammatory process (Davis and Gallin, 1981).

Macrophages are distinguished from neutrophils in that they undergo a process of differentiation in response to environmental change leading to so-called 'macrophage activation'. The conversion of a monocyte to a macrophage is outlined in Fig. 2 (after Rosenstreich, 1981). In the absence of inflammation, a blood monocyte having left the bone marrow and entered the circulation will eventually enter the tissues to remain quiescent as a resident macrophage. However, during inflammation both monocytes and resident macrophages undergo a series of changes

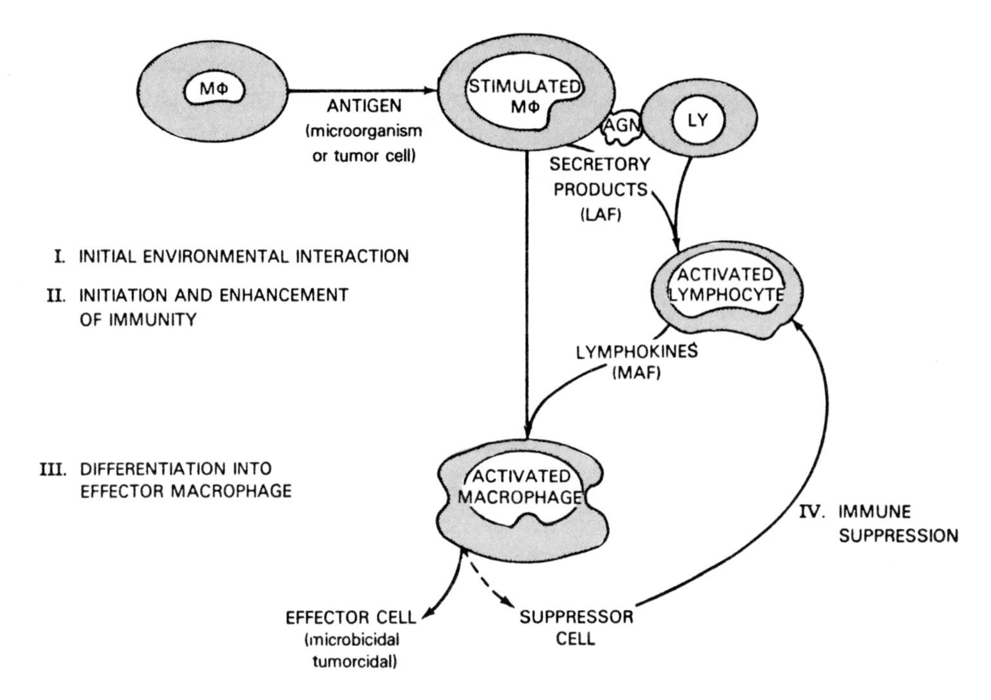

Fig. 2. Sequence of macrophage differentiation. The exact relationship between activated and suppressor macrophages is not clear and is indicated by a broken line. (After Rosenstreich, 1981.)

TABLE 2
Macrophage differentiation (after Cohn, 1978; Rosenstreich, 1981)

	Stimulated macrophage	Activated macrophage
Blood monocyte	Spreading	H_2O_2 secretion
Resident macrophage	Phagocytosis Oxidative metabolism Secretory enzymes	Microbicidal Tumorcidal

resulting in stimulated macrophages. Cohn (1978) has outlined the properties of stimulated macrophages (Table 2). There are several agents (chemoattractants) that cause monocytes to leave the blood stream and enter tissue spaces. Because chemoattractants are generated during inflammation, they provide for a specific localization of monocytes. Many of the chemoattractants for neutrophils including bacterial products, C5a and a lymphocyte-derived chemotactic factor (LDCF) are also chemotactic for monocytes (Rosenstreich, 1981). As mentioned above, synthetic peptides of the formyl methionine-leucine-phenylalanine family are potent chemoattractants as are other small molecules from bacteria. C5a can be generated by several means, by activation of the complement cascade either by the classical or alternative pathway or directly from C5 in limited proteolysis by C5 convertases, plasmin, trypsin, kallikrein and bacterial proteases. LDCF is distinct from C5a and is produced by mitogen or antigen-stimulated lymphocytes. LDCF has been isolated from inflammatory sites and is thought to play a chemotactic role during the later stages of inflammatory reactions (Rosenstreich, 1981).

While there are normally 4000 – 11,000 leukocytes per cubic millimeter in circulating blood, during acute inflammation, the count can reach 100,000. Normally, this increase is due to neutrophils, but in some situations of infection or allergy, a high eosinophil count can be found. Factors implicated in initiation of leukocytosis include: epinephrine-stimulated release from the marginated pool; release from the bone marrow as a result of neutrophil releasing factors; stimulation of neutrophil proliferation by colony stimulating factor. Leukocytosis can be considered to provide an abundant source of neutrophils for microbicidal and other activities at the center of the site of injury. At the periphery of damaged tissues, macrophages are responsible for numerous effector functions including removal of debris and fibroblast proliferation.

2. The tissue function of neutrophils: respiratory burst and degranulation

Once neutrophils have arrived at the site of tissue injury, a series of events is initiated which leads to destruction of invading bacteria. The plasma membrane of the neutrophil contains receptors for the opsonins, most important among these are the Fc portion of immunoglobulins and the complement component C3b. Also, there are nonimmune mechanisms by which particles adhere to the neutrophil cell surface which, may involve hydrophobic interactions between the cell and particle surface. This has been demonstrated experimentally with polystyrene latex beads and with lectins such as concanavalin A which bind to sugar moieties on the cell surface (Hoffstein et al., 1982).

Within the neutrophil there are systems of two types which are responsible for antimicrobial action. One is oxygen dependent and is activated during the metabolic burst associated with the phagocytic process. The other is oxygen independent and involves sequestration of the phagocytic vacuole and degranulation. Recently enormous advances have been made in understanding the biochemical basis of the tissue function of neutrophils. These studies have been favored by the fact that neutrophils remain rather metabolically dormant until opsonization occurs and also by the recognition of numerous clinical disorders of chemotaxis and cell killing. Such 'experiments of nature' have been important to understanding the biology of normal neutrophil function in inflammation.

2.1. The respiratory burst

Phagocytosis in neutrophils is accompanied by a marked increase in the uptake of oxygen (Sbarra and Karnovsky, 1959). This respiratory burst may be the result of membrane stimulation due to opsonization but, as shown by Klempner et al. (1978), neither phagocytosis nor surface attachment is essential (cf. Chapter 8, Section 1).

A major effector function mediated by IgM and IgG is facilitation of the phagocytic process, the first step of which is the binding of particles and immune complexes containing IgG. IgM activates complement by the classical pathway. After accumulation of only a few molecules of IgM or IgG on the surface of a bacterium or particle has triggered extensive deposition of C3 molecules and the membrane attack proteins C5, C6, C7, C8, and C9, neutrophils (and macrophages) attach to those C3 molecules via C3 receptors and phagocytosis then follows.

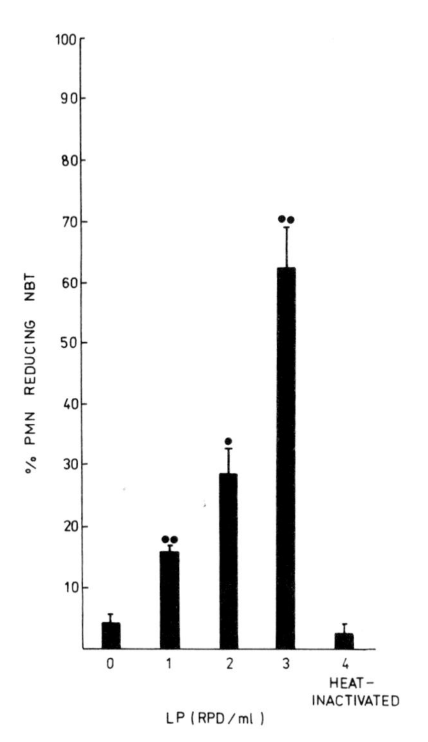

Fig. 3. Dose response of IL-1/EP-(leukocyte pyrogen, LP)-induced human neutrophil NBT dye reduction in vitro. (After Klempner et al., 1979.)

IgM is much more effective than IgG in bringing about bacteriolysis and bacteriocidal effects. The larger size of IgM renders it less able than IgG to gain access to extravascular sites. Also the high concentration of IgG in both the circulation and in the cellular compartment probably makes IgG a more important effector of opsonization in vivo. Furthermore, IgG sensitized organisms can be opsonized in the absence of complement (Potter, 1981).

The primary oxidase enzyme(s) responsible for the respiratory burst is to be found on the outer surface of the neutrophil plasma membrane (Briggs et al., 1975; Goldstein et al., 1977; Dewald et al., 1979). Both NADH oxidase (Karnovsky et al., 1966) and NADPH oxidase (DeChatelet et al., 1975) are believed to contribute to the increase in oxygen consumption by neutrophils that is termed the 'respiratory burst', although other oxidases may contribute (Davis and Gallin, 1981). The initiation of increased oxygen uptake occurs during the membrane stimulation part of opsonization prior to ingestion and degranulation (Rossi et al., 1972).

The fact that IL-1 can bring about the metabolic burst when brought in contact with neutrophils was first reported by Klempner et al. (1978) (Chapter 8, Section 1). As shown in Fig. 3, proportionality exists between the amount of IL-1 added to human neutrophils and reduction of nitro blue tetrazolium (NBT) obtained. Similar results were obtained when measurements of hexose monophosphate shunt were carried out. NBT dye reduction in rabbits, compared with fever at various times after dosage with IL-1, is shown in Fig. 4. Maximum NBT reduction occurred one hour later than maximum fever, suggesting non-involvement of the respiratory burst in fever, but rather that IL-1 may augment the microbicidal activity of neutrophils that begins to be exhibited 6 h after initiation of an inflammatory episode.

Furthermore, the studies of Weiss et al. (1983) indicate that human neutrophils

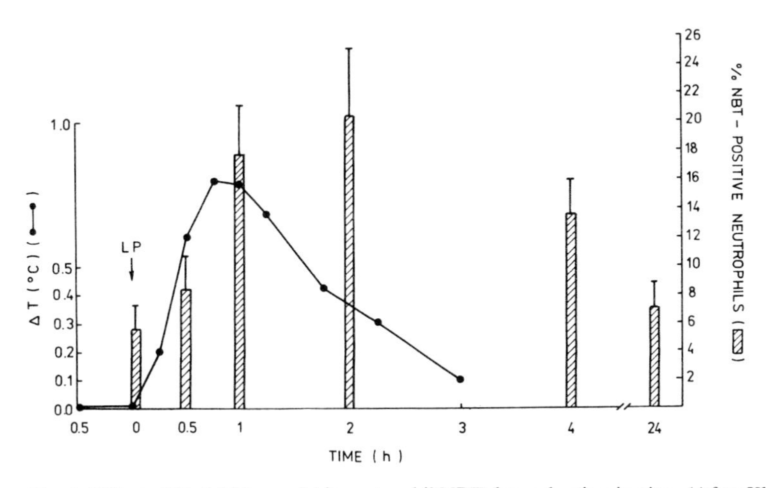

Fig. 4. Effect of IL-1/EP on rabbit neutrophil NBT dye reduction in vivo. (After Klempner et al., 1979.)

stimulated to generate oxygen metabolites and release lysosomal enzymes by addition of serum-opsonized zymozan particles or phorbolmyristic acetate (PMA), generate a long-lasting class of oxidants ($t_{1/2}$ = 18 h) that is similar or identical to *N*-chloroamines. These amines exhibit some selective reactivity to thioether containing molecules such as *N*-formylmethionyl-leucylphenylalanine or lambda 1-protease inhibitor, which regulate the activity of leukocyte elastase.

The cytoplasm of neutrophils contains large amounts of glycogen. During the respiratory burst, the amount of glucose metabolized via the hexose monophosphate pathway (HMP) increases to 10% from a resting level of 1%. If the respiratory burst occurs by way of NADPH oxidase which catalyzes the oxidation of NADPH, then both H_2O_2 and $NADP^+$ would be formed, the latter being required for the HMP. If, however, the key enzyme involved is the more likely NADH oxidase which catalyzes the oxidation of NADH to form H_2O_2 and NAD^+, then the $NADP^+$ required for the HMP might be generated by transhydrogenation, from NADPH or the oxidation of NADPH by lactic dehydrogenase or by oxidized glutathione. This is an area of active investigation (cf. Badwey et al., 1979; Babior, 1984).

Changes in membrane potential precede the respiratory burst. Evidence that alteration of neutrophil membrane potential may be the initiating signal has been produced by Korchak and Weissman (1978). Products of the respiratory burst include superoxide anions ($O_2^{\bar{}}\cdot$), hydrogen peroxide (H_2O_2), singlet oxygen ($^1O_2$) and hydroxyl radicals ($OH^{\bar{}}\cdot$) (Badwey et al., 1979; Davis and Gallin, 1981). This accumulation of superoxide producing phagocytic cells at the site of injury may result in damage to other granulocytes and surrounding tissue. The production of superoxide as part of the microbicidal activity of activated phagocytes can be expected to prolong and exacerbate the inflammation process (Fridovich, 1978). Interaction of superoxide with hydrogen peroxide leading to formation of hydroxyl radicals and other reactive species, not easily removable by enzymatic scavengers, then leads to damage of cell components.

Superoxide dismutases (SOD) catalyze the conversion of superoxide radicals to hydrogen peroxide plus oxygen. SOD are extraordinary efficient catalysts, but normally their concentration in extracellular fluids is low.

Almost all aerobic organisms have been found to possess one or more SOD, however, it remains somewhat controversial as to whether superoxide is the agent responsible for oxygen toxicity or whether SOD activity is actually the primary physiological role of the three distinct SOD metalloproteins which were isolated from eukaryotic cells and are inducible in prokaryotic cells by hyperoxia and by increased levels of superoxide. These proteins were isolated before the SOD activity of either of these cells was recognized (Baum, 1984).

Passive administration of SOD has been reported to have anti-inflammatory effects and is sometimes used therapeutically. Conversion of hydrogen peroxide to water by catalase plays an important role in microbicidal function by way of Klebanoff's H_2O_2-myeloperoxidase-halide system (Klebanoff and Clark, 1978). Patients who have chronic granulomatous disease lack NADH oxidase and thus exhibit impaired H_2O_2 production. Because light is emitted by reactive oxygen species when electrons in excited states fall to more stable orbits, chemiluminescence assays are frequently employed for the respiratory burst.

Two main systems within the neutrophil are responsible for antimicrobial action, one oxygen dependent and an alternative which is oxygen independent (Davis and Gallin, 1981). The oxygen dependent systems either involve myeloperoxidase halide and H_2O_2 or are myeloperoxidase independent and involve superoxide anion, hydroxyl radical, singlet oxygen and hydrogen peroxide.

Aerobic systems of killing have a limited range of activity and may proceed by lysosomal digestion or by creating an unfavorable environment for the bacterium. H_2O_2 plays a central role in the microbicidal activity of neutrophils, however, the cell must also protect itself from the effects of excessive free H_2O_2. This protection can occur via the oxidation of reduced glutathione, but most H_2O_2 is broken down in the cytoplasm by catalase (Davis and Gallin, 1981).

Neutrophils are the first inflammatory cells to enter the area of injury. As discussed earlier, the complement, coagulation and kinin systems modulate the egress of neutrophils from the vasculature. While neutrophils are important in phagocytosis and wound healing, healing can occur and the normal sequence of inflammatory events is not altered when an experimental animal is depleted by neutrophils and then wounded. Thus, neutrophils do not appear to be necessary for completion of an inflammatory episode.

2.2. Degranulation

The most prominent feature of inflammation is the cellular infiltration of injured tissues, effected first by neutrophils and later by mononuclear phagocytes. Tissue injury can arise during the inflammatory response due to production of oxygen radicals by these cells as discussed above, and by release of enzymes from the cytoplasmic granules, the primary lysosomes of infiltrating cells.

Degranulation occurs after attachment and internalization of a particle during phagocytosis. The phagosome, as it is being formed by the fusion of pseudopods extended from the cell around the particle, is converged upon by the cell's cytoplasmic granules. The granules fuse with the phagosome and discharge their contents into the space around the particle. Often granule contents are released from neutrophils during phagocytosis probably because leakage occurs from incompletely closed phagosomes.

There are two main types of neutrophil granules: azurophil or primary granules which appear first in cellular development and specific or secondary granules which appear after the development of primary granules and which are smaller and less dense than the azurophils. The azurophils contain acid hydrolases, neutral proteases, cationic proteins, myeloperoxidase, and lysozyme; the specific granules contain alkaline phosphatase, lysozyme and lactoferrin, but no lysosomal hydrolases or myeloperoxidase.

During phagocytosis, lysosomal enzymes and proteases are secreted (Weissman et al., 1972); a similar secretion occurs when cells attach to surfaces or particles too large to ingest (Henson, 1971). Many lysosomal enzymes exhibit optimal activity at acid pH, and thus may be expected to play a significant role only when the pH at the site of injury is sufficiently acidic.

Among the neutral proteases are collagenase, elastase and cathepsin G. It has been suggested (Larsen and Henson, 1983) that these enzymes allow neutrophils and monocytes to emigrate through the basement membrane to the site of injury. The specific or secondary granules which are found only in neutrophils fuse with the phagocytic vacuole before the azurophil granules (Bainton, 1973). This fusion results in the release of enzymes involved in killing including myeloperoxidase, lysozyme and lysosomal hydrolases (Davis and Gallin, 1981).

3. *Local and general stimulation of mononuclear phagocytes*

As already mentioned in the early stages of the acute inflammatory response, the predominant cell type infiltrating tissues is the neutrophil. However, after a day or so, as the intensity of the response subsides, the predominant cell alternates to the mononuclear phagocyte, derived from the blood monocyte, and called a macrophage when found in tissues. Thus, in the acute inflammatory episode, there is a distinct biphasic response with neutrophils always appearing at an inflammatory focus before monocytes. The biphasic response is thought to be regulated by mediators which probably vary according to the nature of the initiating inflammatory agent. The characteristics of mononuclear phagocytes and the maturation process starting with recruitment of promonocytes from bone marrow to activation in tissues has been reviewed by Rosenstreich (1981).

All mononuclear phagocytes originate in the bone marrow and are derived from the same stem cell precursor that gives rise to cells of the granulocyte series (Virolainen, 1968). The stem cell differentiates into monoblasts and then into promonocytes possessing three properties characteristic of macrophages: glass adhesiveness, phagocytic ability, and immunoglobulin receptors (Cline and Sumner, 1972). The promonocytes divide and further differentiate into monocytes. At this stage, they leave the marrow and circulate in the blood (Fig. 5, Rosenstreich, 1981).

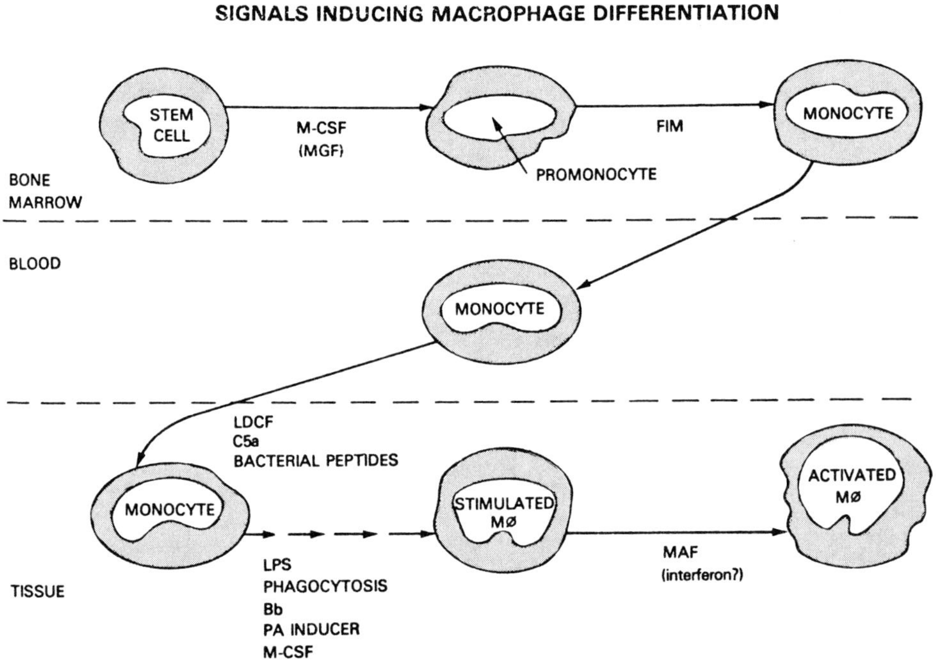

Fig. 5. Outline of macrophage differentiation from bone marrow stem cells. The multiple arrows indicate a multistep process. (After Rosenstreich, 1981.)

The peripheral blood monocyte is approximately 15 μm in diameter with a single kidney-shaped nucleus and an abundant cytoplasm. The cytoplasm is characteristic of a metabolically active cell that constantly produces and secretes proteins. This cytoplasm enlarges as monocytes enter the tissues and differentiate into macrophages, thus, accommodating extra lysosomal granules and endoplasmic reticulum.

Peripheral blood monocytes are characterized by their ability to phagocytose particles and to adhere to glass or plastic surfaces. All mononuclear phagocytes are marked by and characterized by the presence of a cytoplasmic esterase. An intracellular peroxidase is absent in promonocytes but develops and is localized within the rough endoplasmic reticulum of peripheral blood monocytes. Peroxidase activity decreases with further differentiation of monocytes into macrophages, and thus peroxidase activity can be used as a marker for cells that are newly arrived at an inflammatory site. Various surface antigens, especially the Ia antigens, are used to distinguish peripheral blood monocytes from lymphocytes; while others permit identification of activated macrophages.

Rosenstreich (1981) has reviewed the macrophage with respect to its function in immunity and inflammation. The peripheral tissues contain macrophages with different names which vary in functional activity (Table 3). All tissue mononuclear phagocytes are derived from circulatory monocytes and undergo differentiation when they are localized to various anatomic sites. Furthermore, even within one site, macrophages vary in morphology (Table 3).

During inflammation the process by which macrophages differentiate, and are passaged from bone marrow to blood or tissue, is greatly accelerated. There are numerous agents in the area of injury which can cause macrophages to become stimulated and activated. These include bacterial products such as lipopolysaccharide and peptides, complement components C5a and factor Bb and a LDCF (Fig. 2). (Cf. p. 9 for the role of these substances as chemoattractants.)

Bacterial products probably play a prominent role initially in directing migration of monocytes to tissues before the full inflammatory reaction has developed and components and lymphokines become involved.

As the inflammatory response develops, C5a, generated by the cleavage of C5 during activation of either the alternative or classical complement pathways, is thought to be responsible for the rapid recruitment of monocytes into the site of injury. The studies of Snyderman and co-workers (1975) suggest that the frequently used experimental model of elicitation of macrophages by intraperitoneal injection of thioglycollate or endotoxin is mediated by the action of C5a.

LDCF is produced by mitogen or antigen-stimulated lymphocytes, and plays some part during the latter stages of an inflammatory reaction (Rosenstreich, 1981).

TABLE 3
Tissue distribution of macrophages and related cells

Location	Name
Bone marrow	Macrophages
Connective tissue	Histiocytes
Liver	Kupffer cells
Lung	Alveolar macrophages
Lymph node	Free and fixed macrophages
Serous cavity	Pleural and peritoneal macrophages
Spleen	Free and fixed macrophages
Bone tissue	Osteoclasts
Brain	Astrocytes
Kidney	Mesangial cells
Lymphoid tissue	Dendritic cells
Nervous system	Microglial cells
Skin	Langerhans cells

3.1. Differentiation of monocytes to macrophages and production of stimulated and activated macrophages

During inflammation, either newly arrived monocytes or resident macrophages at the site of tissue damage undergo changes as a result of which they become larger and more metabolically active. They are then known as stimulated macrophages.

Late in an inflammatory response or in situations of chronic inflammation, signals generated by activated lymphocytes result in macrophage activation (Mackaness, 1970; Rosenstreich, 1981). Activation of macrophages has been described by Cohn (1978) as the 'conversion of a resting macrophage to an angry, bloodthirsty killer of microbes and tumour cells'. It appears that a T-cell product, called macrophage activating factor (MAF), plays an important role in the terminal differentiation of macrophages to the activated state (North, 1973; Karnovsky and Lazdins, 1978; Leonard et al., 1978).

Although macrophages always obtain their energy from respiration and glycolysis according to their different tissue situations, they develop different metabolic products (e.g. alveolar macrophages consume more oxygen than do peritoneal macrophages). When resident macrophages have been stimulated, they exhibit the increased oxidative metabolism and hexose monophosphate shunt characteristic of neutrophils, including increased ability to secrete superoxide anion. However, there is not a proportional increase in the amounts of hydrogen peroxide secreted by stimulated macrophages. Secretion of hydrogen peroxide is a property exhibited only by activated macrophages.

Once monocytes have been localized to a tissue and have progressed to the stimulated stage, they proceed to spread. This change is brought about, at least in part, by metabolic intermediates of the complement or coagulation systems. Complement dependent spreading appears to be the result of a cleavage product of the B component of the properdin system, factor Bb (Gotze et al., 1979). These macrophage changes are initiated by fluid phase cleavage of factor B to Bb and since macrophages secrete factor B this stimulation can be self-perpetuating. Cleavage of factor B can be brought about as a result of complement consumption, by macrophage proteinases and by plasmin. Another factor implicated in macrophage spreading is serum macrophage stimulatory protein (Leonard and Skeel, 1978).

In certain circumstances, experimentally generated inflammatory macrophages constitutively secrete products such as lysosomal hydrolases, plasminogen activator and colony stimulating factor, but production of most such mediators may require additional signals. When these are available, the following additional products are formed, collagenase, elastase, IL-1 (EP/LAF/SAA inducer) interferon and glucocorticoid antagonizing factor (GAF). Stimulation of production and secretion of these products occurs with phagocytosis and can be brought about in vivo by endotoxin (lipopolysaccharide), the major component of the cell walls of gram negative bacteria. Bacterial products play an important role in vivo (e.g. certain strains of *Staphylococcus aureus* associated with toxic shock syndrome elaborate material that induce human blood monocytes to secrete IL-1 (Ikejima et al., 1984)).

Also, upon stimulation, macrophages produce derivatives of arachidonic acid, either by way of the cyclooxygenase or lipoxygenase pathways. Several different types of prostaglandins arise as well as hydroxyeicosatetraenoic acids and slow-reacting substance of anaphylaxis.

As discussed above, the activated macrophage is specifically defined as a cell with the ability to kill tumor cells or intracellular organisms (Karnovsky and Lazdins, 1978). Activated macrophages are also marked by the ability to secrete hydrogen peroxide but the acquisition of this property usually involves activated lymphocytes, e.g. production of MAF by activated T cells. Probably because of their requirement for activated lymphocytes, activated macrophages are generated late in an inflammatory response and are always present in situations of chronic inflammation. Experimentally, both in vivo and in vitro activated macrophages are responsible for increased production of IL-1 (Sipe et al., 1982).

Damaged collagen and collagen-digestion products at an injury site may also play a role in macrophage localization. Recovery from localized injury depends to a great extent upon macrophage recruitment, stimulation and finally activation. This occurs as bacteria, damaged cells and connective tissue matrix are phagocytozed and is followed by a process of increased adhesiveness, phagocytosis and production of hydrolytic enzymes which digest and transport debris from the cell site. The production of plasmin activator by macrophages results in increased formation of plasmin which removes fibrin deposited in the area of tissue damage and which acts on clotting and complement components.

Both stimulated and activated macrophages produce collagenase (Wahl, 1981) which is then able to remove the damaged collagen usually present at a local site. Production of collagenase is regulated by prostaglandins and modulated by cyclic nucleotides (cf. p. 116.

3.2. *Production of systemic mediators by stimulated and activated macrophages*

In addition to IL-1, macrophages synthesize many other proteins that act locally and systemically in the inflammatory process. These include substances as diverse as hydrolytic enzymes and lysozyme.

Also included are several complement components that are produced by macrophages and in the liver, e.g. C3, however, much the greater proportion of this protein is synthesized in the liver (cf. p. 130). The most important of these monokines that regulate the functions of many types of cell are IL-1, colony stimulating factor and the factors that modulate fibroblast activating factor.

4. Conclusion

Resolution of the local inflammatory reaction is dependent on the sequential participation of several inflammatory cell types which infiltrate the wounded area and proceed to eliminate debris. This process is subjected to precise regulation with the result that the activity of hydrolytic enzymes is limited to damaged tissues and cells. Macrophages influence both vascular proliferation and fibroblast proliferation.

Macrophage products alter hepatic protein synthesis and regulation of body temperature, connective tissue metabolism and muscle proteolysis. Thus macrophages play a primary role in the local response to injury and moreover, during the latter part of the local response the soluble mediators which they produce modulate many target cells thus assisting the repair of wounded tissue.

Microvascular proliferation involves division of the endothelial cells lining small blood vessels in order that new vessels may develop in granulation tissue. Because there is close proximity between inflammatory cell infiltrates and areas of endothelial cell replication, it has been postulated that macrophages can contribute to the control of vascular proliferation (Wahl, 1981). Granulation tissue arises toward the end of the inflammatory response when the area of the wound is replaced by rapid fibroblast proliferation.

References, p. 62.

Gordon/Koj (eds.) The acute-phase response to injury and infection.

CHAPTER 2

Interleukin 1 as a key factor in the acute-phase response

J.D. SIPE

1. Historical perspective

The macrophage product IL-1 was first recognized during in vitro studies of factors modulating thymocyte proliferation (Gery et al., 1971). The biological activity, LAF, was observed to be a product of lymphocytic cells that was mitogenic for mouse thymocytes and that synergistically enhanced the mitogenic effects of Con A and phytohemaglutin on both mouse thymocytes and splenic T cells.

Further studies revealed that the several monokine activities studied variously as mitogenic protein, T cell-replacing factor, B cell-activating factor, and B cell-differentiation factor were actually different manifestations of semipurified concentrates then believed to contain a single active constituent, LAF. The term interleukin 1 was proposed to designate the 12,000 – 16,000 M_r polypeptide produced by macrophages that can exhibit all of these activities (Aarden et al., 1979). However, the wheel has now turned almost full circle and it is recognized, not only that IL-1 consists of a group of very similar molecules but that certain of these have distinct biological activities (cf. p. 303). When IL-1 was defined, it was stated that the term interleukin (between leukocytes) was not meant to imply that these factors can act only on lymphocytes. It is now apparent that both in its origin and function, IL-1 plays a role beyond the immune system.

Subsequent studies (reviewed by Gery and Lepe-Zuniga, 1984 and by Dinarello, 1984a,b) have shown that IL-1 in addition to modulating DNA synthesis by T and B lymphocytes, the oxidative metabolism and degranulation of neutrophils, protein synthesis by hepatocytes, the thermoregulatory activity of temperature-sensitive neurons, also regulates specific aspects of muscle proteolysis and the metabolism of mesenchymal cells. These further activities attributed to IL-1 include stimulation of proliferation and collagenase secretion by fibroblasts, stimulation of synoviocyte and chondrocyte proliferation and prostaglandin and protease release and also the enhancement of bone resorption (Gowen et al., 1983).

IL-1 was originally and is still generally thought to be a product of the mononuclear phagocyte line (Aarden et al., 1979). It has been shown to be secreted by keratinocytes (Luger et al., 1981), by mesangial cells (Lovett et al., 1983), by glial cells (Fontana et al., 1982), and by dendritic cells (Rollinghoff et al., 1982). Surprisingly, a protein with some of the properties of mammalian IL-1 has recently been found to be contained in prokaryotic cells as endotoxin-associated protein (EAP) (Johns et al., 1984; Sipe et al., 1984) (cf. p. 29).

In addition to exhibiting variation in both cell of origin and target cell, IL-1 also exhibits species cross reactivity. This property has led to the widespread use of the mouse thymocyte costimulator assay and definition as IL-1 of anything active in this assay regardless of species of origin. Examples of target cell non-specificity are that human IL-1 stimulates SAA synthesis in the mouse (Sztein et al., 1982), alters protein synthesis by rat hepatocytes (Koj et al., 1984), and that rabbit IL-1 stimulates SAA synthesis in mouse (Sztein et al., 1981).

Even earlier than the in vitro identification of IL-1 as LAF, it had been studied as granulocyte and endogenous pyrogen (EP) from the mid-1940's, as reviewed by Dinarello (1984a,b). These studies were based in part upon the early studies of Homburger (1945) which demonstrated that leukocytes derived from an abscess secreted soluble factors that stimulated fibrinogen synthesis. The early studies of what is now recognized to be, at least in part, IL-1, are described in detail in the Introduction.

Concerned with regulation of the AP-response, Kampschmidt's laboratory (Kampschmidt et al., 1983) studied the in vivo effects of an IL-1-like substance, leukocyte endogenous mediator (LEM) prepared from glycogen elicited peritoneal exudate cells in rabbits, on body temperature in rats and rabbits, and on concentrations of plasma proteins and metal ions.

Bornstein (1982) has studied the in vivo effects of an IL-1-like preparation named leukocyte pyrogen and shown it to lead not only to fever, but also to granulocytosis, and changes in serum metal and protein concentration. He and his collaborators have identified an interaction between the central nervous system and hepatic acute-phase protein synthesis occurring after ICV administration of semipurified IL-1.

The recognition of the role of IL-1 in SAA and other acute-phase plasma protein changes led to a realization of the need for a unifying concept in which host defense and restoration results form a carefully synchronized sequence involving the interaction of IL-1 and virtually all body tissues.

In addition to its modulation of the immune system, IL-1 interacts with cells that are involved with wound healing and response to infection, such as fibroblasts, hepatocytes and hypothalamic cells. Thus, IL-1 may play a central role in mediating processes which are related to wound healing and inflammation. This concept is supported by, in addition to the nature of the target cells of IL-1, the well known fact that IL-1 production is generally initiated and enhanced by agents which are injurious to cells. Furthermore, it appears that in cases of marked cell injury a large proportion of IL-1 remains internalized by the producing cells and that both production and secretion is enhanced when a combination of inflammatory factors is present. This chapter will be concerned with the various aspects of the major role which IL-1 plays in orchestrating the body's AP-response.

2. Cell sources of IL-1

It is evident as discussed above and reviewed by Dinarello (1984b) and Gery and Lepe-Zuniga (1984) that in addition to being produced by cells of the mononuclear phagocyte system, IL-1 can be produced by numerous other phagocytic cells which may share some properties with macrophages but cannot be so classified. In some instances, the cytokine is produced spontaneously as well as after stimulation. For example, epidermal cell-derived thymocyte activating factor (ETAF) production by epidermal cells occurs constitutively during the beginning of the log-growth phase and ceases when growth plateaus. Production recommences, however, when cells are disturbed as with PMA (Luger et al., 1981). In another system, catabolin is produced from pig monocytes stimulated with the mitogen Con A, rather than by the usually employed inflammatory agents such as lipopolysaccharides (LPS) (Saklatvala et al., 1983). This chapter will summarize the several species of cells and cell lines that have been used as sources of IL-1/LEM, LAF, EP, etc. for diverse studies on the AP-response.

2.1. Human

Until 1978, human EP/IL-1 was always prepared from blood monocytes (Dinarello et al., 1984). More recently, a number of leukemia cell lines, mainly of monocyte origin, have been preferred. One of these, THP1, derived from the blood of a one-year-old male with acute-monocytic leukemia (Krakauer and Oppenheim, 1983) was shown to be monocytic in character by its possession of Fc and C3b receptors. Stimulation of these cells with endotoxin or with silica led to only modest increases of IL-1 in the supernatant as measured by the thymocyte costimulator assay. However, that these direct measurements were serious underestimates was shown when such supernatants were applied to a column of Ultragel AcA54 and greatly increased activity was found in the eluates emerging at volumes corresponding to M_r value of 12,000 – 18,000. In a very recent study, cells of non-monocytic lineage have been shown to be capable of producing IL-1. Most interestingly, a promyelocyte cell line HL 60 and a 'histiocytic' line U937 which did not spontaneously produce IL-1, acquired this property when induced to differentiate along the granulocytic and/or the monocytic pathways (Wagasugi et al., 1984). This work represents an important advance towards a much fuller understanding of the cell types which can form IL-1. It is important to note not only that cells other than monocytes and macrophages such as human dendritic cells (Voorhis et al., 1982) are efficient producers, but also that whether or not IL-1 is formed depends on the precise maturational stage of the cells in question. A summary of human cell types and cell lines which have been identified as IL-1 producers and the various activities involved are contained in Table 1.

2.2. Rabbit

The rabbit has been centrally employed in studies of the pathogenesis of fever. The

TABLE 1
Sources of human IL-1

Cell or cell line	Activity	Reference
Peripheral blood monocytes (PBM)	EP	Dinarello et al., 1977; Bodel and Atkins, 1967
PBM	LAF	Lachman and Metzgar, 1980
PBM	BAF[a]	Wood and Cameron, 1978
PBM	HSF	Ritchie and Fuller, 1983
PBM	MCF[b]	Krane et al., 1982
Placental mononuclear phagocytes	LAF	Flynn et al., 1982
Hodgkins tumor cells	EP	Bodel, 1974c
U937 – histiocytic lymphoma	EP	Bodel et al., 1980
U937 – histiocytic lymphoma	LAF	Amento et al., 1982
U937 – histiocytic lymphoma	MCF[b]	Amento et al., 1982
THP-1 monocyte leukemia	LAF	Mizel and Anderson, 1983
COLO-16 skin carcinoma	HSF	Koj et al., 1984
CM-S bone marrow derived	LAF	Butter et al., 1983
Normal epidermal cells	HSF	Sauder, 1984
Alveolar macrophages	LAF	Koretsky et al., 1983
Gingival exudate cells	LAF	Charon et al., 1982
Renal carcinoma cells	LAF, EP	Rawlins et al., 1970

[a] BAF = B-cell activating factor.
[b] MCF = Mononuclear cell factor.

monokine EP has been shown to be produced and released by activated peritoneal exudate cells, peripheral monocytes and alveolar macrophages (cf. Chapter 3, Section 2.1) and to be present in the blood of febrile rabbits (reviewed by Dinarello, 1984b). Although it was originally thought that neutrophils were the source of EP, subsequent quantitation studies indicated that EP is produced by the much smaller number of macrophages in the culture. However, a report by Nakamura et al. (1982) has also described an IL-1 activity associated with neutrophils (cf. also Section 2.5 for a report of IL-1 formation by human large granular lymphocytes). Thus, it remains to be seen whether IL-1 is synthesized by or is taken up by neutrophils (cf. Section 2.5). Some of the rabbit cell types known to produce IL-1 are outlined in Table 2.

TABLE 2
Sources of rabbit IL-1

Cell	Activity	Reference
Peritoneal exudate cells	EP, LEM	Bennett and Beeson, 1953; Bornstein et al., 1963
Mononuclear cells	EP	Atkins et al., 1967
Alveolar macrophages	LAF	Ulrich, 1977
Corneal epithelial	CETAF[a]	Grabner et al., 1982
Kupffer cells	LEM	Haeseler et al., 1977

[a] Corneal epidermal cell-derived thymocyte activating factor.

2.3. *Mouse*

As was mentioned earlier, the mouse played a prominent role in the in vitro studies which identified and defined IL-1 as LAF activity (Gery et al., 1971) and was also used for in vivo studies of pyrogenicity in the prewarmed state (Bodel and Miller, 1976). Although primarily investigated in rat and rabbit, IL-1 has been studied for its LEM activity in mice (Kampschmidt et al., 1980; Flynn, 1983). Murine IL-1 has been purified to apparent homogeneity from the supernatant of P388D$_1$ cells (Mizel and Mizel, 1981) (cf. Chapter 3, Sections 3.2 and 3.4); however, in general, the lack of sufficient material has hampered structural studies of mouse IL-1.

LPS treated mouse peritoneal exudate cells have been used as a source of IL-1 in many experiments (Vogel and Sipe, 1982). Primary cultures of mouse astrocytes stimulated with LPS release an IL-1 activity measured as LAF (Fontana et al., 1982). Sztein and coworkers (1982) have shown that normal mouse keratinocyte lines spontaneously produce ETAF which exhibits LAF and IL-1 activities. Mouse cells or cell lines in which IL-1 has been studied are outlined in Table 3.

TABLE 3
Sources of mouse IL-1

Cell or cell line	Activity	Reference
Peritoneal macrophages	LAF	Gery et al., 1971
$P388D_1$	LAF	Mizel and Mizel, 1981
	MCF	Mizel et al., 1981
	SAA inducer	Sztein et al., 1982
Bone marrow macrophages	LAF	Lee et al., 1981
Epidermal cells	ETAF	Luger et al., 1981
	SAA inducer	Sztein et al., 1982
Astrocytes	LAF	Fontana et al., 1982
J744.1	LAF	Bodel, 1978
Monocytes	LEM	Kampschmidt, 1981

2.4. Rat and other species as sources of IL-1

Rat hepatocytes have frequently been used for in vitro studies of regulation of acute-phase protein synthesis by IL-1 and hepatocyte stimulating factor (HSF) (Ritchie and Fuller, 1983; Koj et al., 1984). Although studies with rat hepatocytes have frequently used human mononuclear cells or cell lines as a source of IL-1 and HSF, LEM activity has been purified from rat Kupffer cells (Kampschmidt et al., 1973) and rat mesangial cells have been shown to elaborate a cytokine that exhibits LAF activity (Lovett et al., 1983). As outlined in Table 4, blood monocytes or buffy

TABLE 4
IL-1 from rat and other species

Cell or cell line	Activity	Reference
Rat mesangial cells	LAF	Lovett et al., 1983
Rat Kupffer cells	LEM	Kampschmidt et al., 1983
Pig mononuclear cells	LEM	Kampschmidt, 1981
Pig mononuclear cells	Catabolin, MCF, LAF	Saklatvala et al., 1984
Pig synovial cells	MCF	Pilsworth and Saklatvala, 1983
Lizards	EP	Kluger et al., 1975
Ox mononuclear cells	LEM	Kampschmidt, 1981
Cat mononuclear cells	LEM	Kampschmidt, 1981
Dog mononuclear cells	LEM	Kampschmidt, 1981
Goat mononuclear cells	LEM	Kampschmidt, 1981
Monkey mononuclear cells	LEM	Kampschmidt, 1981
Salmonella Minnesota	SAA inducer	Sipe et al., 1984

coated cells from ox, cat, dog, goat and monkey, have all been used to produce EP/LEM/IL-1 (Kampschmidt, 1981).

Another IL-1-like biological activity is known as catabolin, this is due to a protein produced by pig synovial tissue explants that induces proteoglycan resorption in bovine nasal cartilage (Saklatvala and Dingle, 1980). Catabolin obtained from pig monocytes has been purified and shown to bring about cartilage and bone resorption, thymocyte proliferation, and prostaglandin and collagenase production by synovial fibroblasts (Saklatvala et al., 1984) (cf. Chapter 3, Section 5.2).

Preparations of endotoxin consisting mainly of LPS, the major component of gram negative bacterial cell walls, may contain varying amounts of an outer membrane protein called EAP. EAP was found to be clearly distinguishable from LPS with respect to initiation of acute-phase SAA biosynthesis in LPS-nonresponder C3H/HeJ mice (Sipe et al., 1984). This resistant strain, after EAP administration, exhibits a degree of SAA elevation comparable to that of endotoxin-sensitive mice given LPS, although they do not mount a significant SAA response to LPS when it does not contain EAP. Unlike LPS, but as expected for an IL-1-like molecule, EAP did not elicit SAA inducer production by cultured peritoneal macrophages from endotoxin-sensitive mice. In agreement with the concept that EAP is a prokaryotic form of IL-1, it has been found to exhibit weak IL-1-like activity in the LAF assay (Sipe, Johns, unpublished observations) (cf. Chapter 7, Section 5).

2.5. *Elaboration of IL-1 from polymorphonuclear leukocytes*

As mentioned above in Section 2.2, because so many of the early studies of IL-1 production used acute rabbit peritoneal exudates induced by injecting shellfish glycogen into the peritoneal cavity and because 95% of the cells in this peritoneal exudate were neutrophils, most attention was focused on the neutrophil as the source of EP. However, studies from Murphy's laboratory indicated that the contaminating mononuclear phagocytes were solely responsible for the EP activity which was formed. Indeed when completely free from macrophages, rabbit neutrophils did not produce EP or LAF (Hanson et al., 1980; Windle et al., 1983). However, more recently, elaboration of a lymphocyte IL-1/LAF activity by neutrophils has been described (Yoshinaga et al., 1980; Nakamura et al., 1982). In view of the limited protein synthetic capabilities of neutrophils and the observed differences between IL-1 and the neutrophil-associated LAF activity, it remains to be determined whether the LAF activity which was ascribed to neutrophils may have been formed in macrophages and then ingested or bound by neutrophils. Recently, however, a small class of human large granular lymphocytes have been identified as IL-1 producers (Scala et al., 1984).

2.6. Possible presence of IL-1 in normal and post-exercise plasmas

IL-1/SAA inducer was shown to be maximally detectable in blood 90 min after injection of mice with LPS (Sipe et al., 1979), but was not detected in blood from unstimulated mice. However, by analogy with other hormones at least trace concentrations of IL-1 might be expected to be present in blood of physiologically normal individuals. The issue is complicated, as mentioned on p. 293, by the presence of inhibitors of the thymocyte costimulator assay in plasma, which until very recently made detection of circulating IL-1 impossible by this technique.

Using an even more sensitive method based on inhibition of leukocyte migration inhibitory factor, Bendtzen et al. (1984) have examined plasma from normal individuals and from those with fever. Slight and very variable responses were obtained from normal bloods suggesting the presence of traces of IL-1 in some individuals.

Recent work on blood obtained after strenuous exercise has thrown light on one of the types of stress which can lead to synthesis of IL-1 and most interestingly has added evidence for the functional heterogeneity of IL-1. Upon fractionating blood from post-exercise individuals by gel filtration, Cannon and Dinarello (1984) observed stimulation of murine thymocytes in the costimulator assay. They succeeded in separating an inhibitory factor having M_r >50 kDa from stimulatory materials eluting at 11 – 13, 2 – 4 and approximately 1 kDa. They also found that the 2 – 4 kDa differed from the 11 – 13 kDa material in that it was stable to heating at 70° C for 30 min and noted that the 2 – 4 kDa material may be similar to that reported by Gordon and Parker (1980). This latter pyrogen obtained from normal human monocytes by treatment with zymosan was assayed by means of fever in mice (cf. Chapter 3, Section 4.5). Induction of fever in rats has recently been used to demonstrate the presence of IL-1 in human post-exercise plasma (Cannon and Kluger, 1983). Temperature increases, mean 0.4° C, were obtained by injecting 1 ml of such plasma per rat and shown to be due to a factor of 13 – 15 kDa. The post-exercise plasma also led to decreased Fe and Zn concentrations in the rat plasmas. Heating of the post-exercise plasma to 65° C for 90 min led to inactivation of the fever response, but not that for Fe and Zn. Thus, IL-1 heterogeneity was again demonstrated.

These findings demonstrate clearly the presence of IL-1 in post-exercise plasma, but leave uncertain its relationship to IL-1 formed after injury and during inflammation, and thus that of special interest in relation to the AP-response.

The question as to whether trace concentrations of IL-1 exist in the blood of non-stimulated, physiologically normal individuals is also unresolved. A relevant finding is that isolated human monocyte/macrophages maintained in culture require only very gentle stimulation such as surface attachment to form small amounts of IL-1 (Treves et al., 1983).

3. *Induction of IL-1 synthesis and secretion*

3.1. *Nature of agents which can stimulate IL-1 production*

IL-1 producing cells of many types (cf. page 18), are distributed in nearly all tissues and as a result IL-1 is thought to play a direct and vital role in host survival. As mentioned above, it is not known whether a low level of IL-1 is produced by resting mononuclear phagocytes. On the other hand, numerous studies have demonstrated that when stimulated these cells can produce a large amount of IL-1 in vitro. Dinarello (1984a) has extensively reviewed the agents that induce IL-1 production; they fall into several broad categories: *Viruses, Microorganisms* and *Microbial Products* including endotoxins and exotoxins, peptidoglycans, polysaccharides, muramyl dipeptide; *agents* that promote, or arise as a result of *cell necrosis* including etiocholanolone, silica, urate crystals, bleomycin and bile salts; *plant lectins* such as phytohemagglutinin and concanavalin A, and *lymphokines* such as colony stimulating factor and MAF and the *interferon inducer* poly I: poly C. In addition, ultraviolet radiation was found to enhance the release of ETAF/IL-1 by keratinocytes (Gahring et al., 1984). In many situations, the induction of IL-1 may involve generation of C5a from cell-bound complement. Macrophages have been shown to have receptors for C5a which induces IL-1 formation (cf. Chapter 11, Section 7) (Chenoweth et al., 1982; Goodman et al., 1982). The surface active tumor promoter PMA has frequently been used to stimulate IL-1 production by mononuclear phagocytes and macrophage-derived cell lines (Mizel et al., 1978) (cf. Chapter 3, Section 3.1). Gery and Lepe-Zuniga (1984) have analyzed the mechanism involved in initiating IL-1 production and the role of cell damage in this response. They have studied IL-1 production and release by mouse macrophage and human monocytes and have found dissociation between extracellular (EC) and intracellular (IC) IL-1 activity according to the nature of the stimulating agents employed and to a certain extent according to the nature of the cell producing IL-1.

3.2. *Intracellular versus secreted IL-1*

Unanue and Kiely (1977) initially observed a dissociation of IC and EC IL-1 activity in cultures of unstimulated mouse macrophages. They found little EC activity throughout 24 h of culture, but observed a burst of IC IL-1 production during the first 4 h of incubation which declined to neglible levels by 24 h of culture. Recently, further investigations of unstimulated macrophages have shown that part of the total IL-1 occurs as a component of the macrophage membrane. This has been demonstrated using peritoneal macrophages which had been lightly treated with formaldehyde. Such cells and also preparations of their plasma membranes were found to retain their ability to stimulate T-cell growth.

Further experiments were carried out to determine the earliest time after removal from the peritoneum that IL-1 could be demonstrated as a membrane component. Freshly obtained cells showed no membrane IL-1 and it was only after surface attachment or endocytosis of a stimulator that this form of IL-1 made its appearance. The effect did not occur after pretreatment with antisera specific for IL-1. These results are of special interest because they raise the question as to how far and in what circumstances IL-1 may act by cell-to-cell contact (Kurt-Jones et al., 1984).

Gery and Lepe-Zuniga (1984) were able to classify agents according to their pattern of IL-1 production. Resident mouse macrophages were an example of cells for which several agents were found to elicit roughly equivalent amounts of IC and EC IL-1 (Fig. 1). On the other hand, with the same cells they found that LPS elicited predominantly IC IL-1 and that silica particles elicited primarily EC IL-1 (Fig. 1).

In the case of human monocyte cultures, silica particles and zymosan were both found to stimulate roughly equivalent amounts of IC and EC IL-1. Unlike mouse macrophages, unstimulated human monocytes produced some IC IL-1 (Fig. 2) and a significant portion of LPS-stimulated IL-1 was secreted (Fig. 2).

Silica particles are known to bring about significant cellular damage (Allison et al., 1966), however, their pronounced effect on IL-1 release cannot be due only to increased leakage of preformed IL-1 because there is an increase in both IC and EC IL-1 (Figs. 1 and 2).

As mentioned in the preceding sections, activated macrophages, upon stimulation, secrete greatly increased amounts of IL-1. Depending upon the cell type, the IC IL-1 is heterogeneous in respect to molecular weight. Mizel and Rosenstreich (1979) reported constituents of kDa 13, 26, 39, 50 – 70 and > 70. Whether the larger IL-1 species are precursors for the 12 – 16 kDa IL-1 or are aggregates or polymers of the smaller species remains to be determined. There is no indication that IL-1 is

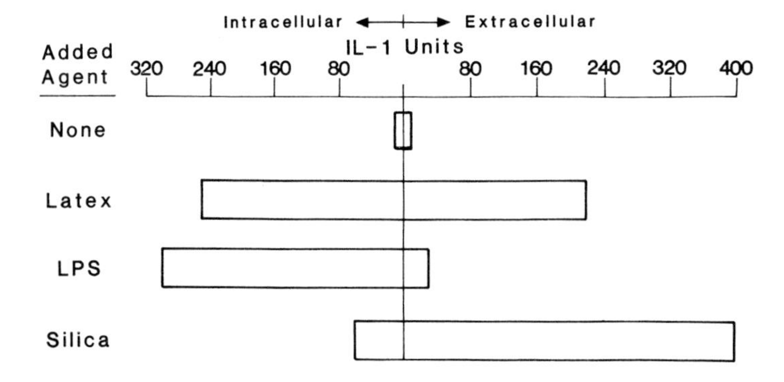

Fig. 1. Patterns of IL-1 production and release by mouse macrophages cultured with various agents. (After Gery and Lepe-Zuniga, 1984.)

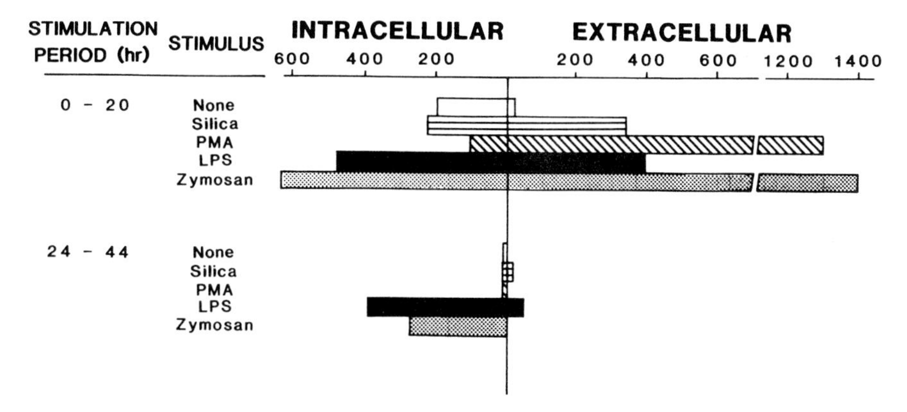

Fig. 2. Patterns of IL-1 production and release by human macrophages cultured with various agents. (After Gery and Lepe-Zuniga, 1984.)

associated with lysosomal or other granules of the macrophage, rather its production seems to be analogous to that of other cell stimulatory proteins such as colony stimulating factor, interferon, tumor necrosis factor and glucocorticoid antagonizing factor.

Thus, the various inflammatory agents outlined by Dinarello (1984a, b) can act differently upon production and release of IL-1. There can be synergism between different types of agents, e.g. the increase in IL-1 production and release by activated macrophages such as those from BCG infected animals exposed to LPS in vitro or in vivo (cf. Vogel et al., 1980, Sipe et al., 1982). This phenomenon may play an important role in chronic inflammation in view of the observation of Wood and Cameron (1978) that monocytes of different human donors vary substantially in their response to LPS.

3.3. Kinetics of IL-1 production and release

Mononuclear phagocytes begin to produce IL-1 within 60 min after stimulation and the process is believed to proceed via changes involving membrane perturbation. Evidence is available that IL-1 is an inducible protein requiring both RNA and protein synthesis since actinomycin, cycloheximide, and puromycin all block IL-1 production (Dinarello and Wolff, 1982; Oppenheim et al., 1982) and there is little evidence for preformed or stored IL-1 in unstimulated mononuclear phagocytes.

IL-1 induction requires approximately 60 min contact with the inducing agent, and substantial amounts of IL-1 can be detected in the supernatant medium after 2 h (Bettens et al., 1982). Most IL-1 inducing agents require a specific concentration and a specific minimum period of exposure (Dinarello and Wolff, 1982). There is a decrease in the rate of IL-1 production after 18 h which can be reversed by the addition of fresh medium. While no feedback inhibitor of IL-1 production has been demonstrated (Bodel, 1974b) it seems certain that inhibitors of IL-1/LAF activity (Shou et al., 1980; Dinarello et al., 1981; Liao and Rosenstreich, 1983; Cannon and Dinarello, 1984) are produced.

Dinarello and Wolff (1982) have reviewed the evidence that IL-1 is similar to other inducible proteins, in that its synthesis is repressed in the resting monocyte. If so, a large and varied number of inducers are able to derepress the genome responsible for synthesis of IL-1 mRNA. However, some, but not all, macrophage tumor cell lines have been observed to form IL-1 spontaneously (Bodel, 1978).

The conclusion that IL-1 production involves synthesis of mRNA from a DNA genome is based on experiments with human mononuclear cells (Dinarello and Wolff, 1982). Addition of actinomycin D at the time of stimulation was shown to suppress completely IL-1 production. However, 2 h later the actinomycin D had no effect. Recently, it has also been shown that mRNA from stimulated but not from unstimulated mononuclear phagocytes can be translated by *Xenopus oocytes* with resultant IL-1 activity in the incubation medium (Dayer et al., 1983; Windle et al., 1984).

The mechanism by which multiple and varied agents bring about IL-1 production remains to be defined and it is hard to imagine that a single such mechanism exists, although cell injury and membrane perturbation seem to be common factors (Dinarello, 1984a; Gery and Lepe-Zuniga, 1984).

4. *Summary and conclusions*

IL-1 is strictly defined as a 12 – 16 kDa monokine that possesses hormone-like activities and affects the metabolism of T lymphocytes and augments their responses to other stimuli (Aarden et al., 1979). In this chapter, we have reviewed the diverse in vivo and in vitro biological activities of IL-1 and the species and cells in which the various manifestations of IL-1 activity have been examined. On the basis of these very numerous studies, the role of IL-1 as a key mediator of the AP-response has been established. The consequences for the host are multifarious. The many different biological activities of IL-1 allow for a fine tuning of the macrophage-hepatocyte, hypothalamus, fibroblast interaction all of which are important aspects of the body's defense reaction. As might be expected, the cells which produce IL-1 are strategically placed in primary defense locations. The fact that IL-1 can be found in non-mammalian species such as birds, reptiles, amphibians and bony fish, argues that it is an ancient molecule that has been conserved because it plays a vital role in host-defense mechanisms (Kluger et al., 1975). As research on IL-1 and its various effects on the host response to infections, inflammation and trauma expands over the coming years, it should become possible to separate the beneficial from aspects of IL-1 activity which may be harmful and to intervene therapeutically when indicated.

References, p. 62.

Gordon/Koj (eds.) The acute-phase response to injury and infection.

CHAPTER 3

Purification and biochemical properties of interleukin 1

A.H. GORDON

1. Introduction

For more than 20 years work has continued, aimed at isolation of pure IL-1. It is convenient to divide this period into two stages at approximately 1973.

By 1974, using traditional separation methods, highly purified EP/IL-1 had been prepared from rabbit peritoneal cells (Murphy et al., 1974). After separation of this material on acrylamide gels and staining with Coomassie Blue, little or no colour could be detected in sections of gel from which pyrogen capable of producing fever in rabbits could be extracted. Thirty-50 μg of such pyrogen, injected into a rabbit, was sufficient to produce a brief fever easily distinguishable from fevers caused by bacterial endotoxins. During this period EP/IL-1 was also obtained in semi-purified form from human leukocytes (Dinarello et al., 1974).

EP/IL-1 at this level of purity provided material for initial characterization and stability studies. These led to the view that EP/IL-1 is a small protein molecular of limited stability, very easily adsorbable on glass surfaces, and thus hard to purify. The second period opened when immunologists identified a factor present in leukocyte culture fluids able to stimulate [^{3}H]thymidine uptake by mouse thymocytes (Gery et al., 1972). Even though subsequent work has shown that assays based on this function are not completely specific for IL-1, the extreme sensitivity of this procedure has made it a very valuable method for assay of IL-1, especially during attempts at its purification. Of equal importance to the availability of this new sensitive assay was the identification of new sources of IL-1. The first, and still very important alternative source, is the P388D_1 mouse cell line of (see Chapter 2.2). When suitably stimulated the claimed rate of production of IL-1 by these cells is 1 μg/24h/4 $\times$ 10^6 cells (Mizel and Mizel, 1981). However, despite the use of sophisticated separation methods (to be described below), the IL-1 constituents thus produced have not been sequenced. Heterogeneity of IL-1 with respect to size and charge has been observed for the species from human, mouse and rabbit. Upon gel filtration, the predominant molecule from all species determined according to the various assays is eluted in the 12,000 – 17,000 M_r range. As discussed in Chapter 2, higher molecular weight forms of IL-1 are observed in the intracellular fraction of macrophage cultures. These higher molecular weight forms of 26,000 – 70,000 M_r exhibit EP, LEM and LAF activity and share antigenic determinants with the 15,000 M_r species (as reviewed by Dinarello, 1984b).

Although, in general, only the lower molecular weight species of IL-1 is secreted, higher molecular weight species are also observed in the incubation medium. In some cases this may be due to IL-1 binding to serum proteins such as albumin, although a 35,000 – 40,000 M_r species has been shown to be released in the absence of serum (Dinarello, 1984b). Recent attempts at purification of IL-1 have been greatly aided by the production of antisera and the use of immunoadsorption. The first such antisera against IL-1 was obtained in 1977 by Dinarello et al. (cf. p. 303).

2. *Rabbit EP/IL-1*

2.1. *Cellular sources and conditions of formation of rabbit EP/IL-1 (cf. Chapter 2, Section 2.2)*

Incubation of stimulated rabbit peritoneal cells leads to the formation of at least three forms of EP/IL-1 which differ slightly in respect to p*I* (Cebula et al., 1979), these range from p*I* 5.0 to 7.3. Much more of the p*I* 7.3 EP/IL-1 was obtained if the peritoneal cells were removed at 16 h instead of at 4 days after injection of sterile mineral oil. Similar results were obtained by Kampschmidt et al. (1983) who used shellfish glycogen as the initial stimulant.Alveolar macrophages were found to produce mainly EP/IL-1 of p*I* 4.5 – 5.0. The acute peritoneal cells were suspended in saline at 1 $\times$ 10^8 cells/ml and incubated with shaking for 2 h at 37°C. The cells removed after 4 days were separated from the oil, filtered through gauze and washed once with MEM. Next they were activated by incubating 5 $\times$ 10^6 cells/ml in MEM containing 5% rabbit serum and 0.2 μg/ml of *Salmonella abortus* endotoxin. After 2 h activation at 37°C they were washed with, and suspended in, serum-free MEM. The supernatant was separated after a further 16 h incubation.

2.2. Methods for purification of EP/IL-1

Attempts at isolation of EP/IL-1 constituent proteins from supernatants obtained as above involved gel filtration on Sephadex G-50 and Thiol Sepharose (Murphy et al., 1981a), electrophoresis in acrylamide gels and isoelectric focusing in glycerol gradients and in beds of Sephadex (Cebula et al., 1979).

A method depending on preparative electrophoresis in acrylamide gels has been described (Pacak and Siegert, 1982). High yields of EP/IL-1 from rabbit peritoneal cell supernatants were claimed based on an induction of fever in rabbits. However, as work from other laboratories clearly indicates that the use of multiple stage purification, including, in the case of human pyrogen immunoadsorption are needed to yield IL-1 in a nearly homogenous state, there is need for confirmation of the report by Pacak and Siegert (1982).

As already mentioned, by means of repeated focusing and gel filtration, EP/IL-1 constituents with different p*I* values have been obtained by Murphy et al. (1981a). Using endotoxin stimulated peritoneal cell supernatant their first step was focusing in a bed of G-75 Superfine Sephadex. This was sufficient to separate the p*I* 7.3 constituent from the group of constituents with much lower p*I* values. In order to remove endotoxin, both fractions were filtered on Sephadex G-50, as shown in Fig. 1 and a clean separation was achieved. A further gel filtration was then carried out followed by focusing in a glycerol gradient. Finally, for the p*I* 4.5 – 5.0 material, the focusing was repeated in a very shallow ampholine gradient (Fig. 2, Murphy et al., 1981b). As also shown, the degree of incorporation of [^{3}H]amino acids was found to coincide with pyrogenicity in rabbits. Approximately 150 μg of the p*I* 7.3 EP/IL-1 was thus obtained which was sufficient to raise antisera in three goats.

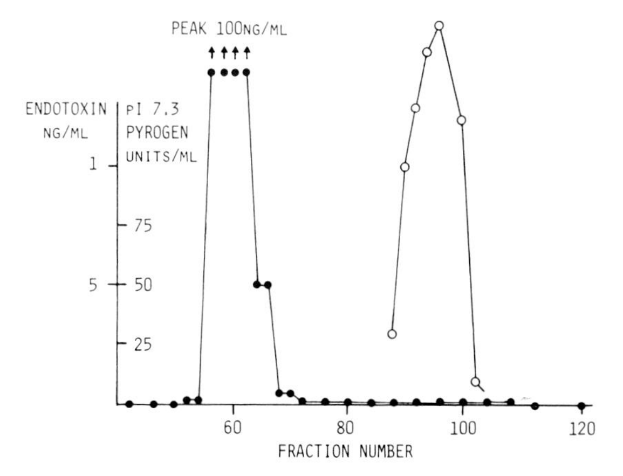

Fig. 1. Gel filtration of p*I* 7.3 pyrogen on Sephadex G-50. ●, Endotoxin; ○, endogenous pyrogen (Murphy et al., 1981a).

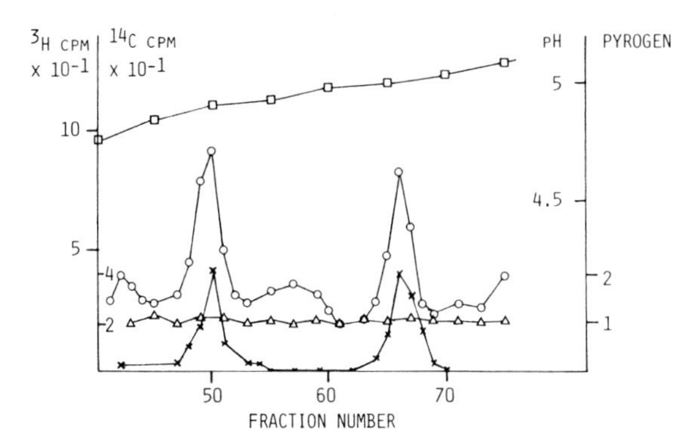

Fig. 2. Isoelectric focusing of fast-moving p*I* 4.5 – 5.0 pyrogens on a very shallow pH gradient. X, Pyrogen; ○, ^{3}H cpm/100 μl; △, ^{14}C cpm/100 μl; □, pH (Murphy et al., 1981b). ^{3}H incorporated from ^{3}H amino acids, ^{14}C from ^{14}C glucose and ^{14}C glucosamine.

2.3. Physico-chemical properties of rabbit EP/IL-1

The IL-1 constituents with p*I* 7.3 and those with p*I*s in the range between 4.5 and 5 have been shown to differ in a number of other ways. Thus, the p*I* 7.3 constituent has slightly smaller molecules with approximate M_r of 13 kDa and contains at least one free SH group per molecule which is responsible for the reversible inactivation by O_2 and pH 9. As expected, the presence of this group allows binding to thiol Sepharose from which it can be eluted in active form by mercaptoethanol. Further evidence for the difference between the p*I* 7.3 constituent and those with lower p*I*s is that only the former is inactivated by iodoacetamide. Another difference also reported by Murphy et al. (1981a) is in respect to stability. Thus, the p*I* 7.3 constituent loses activity if frozen slowly. All constituents retain activity if rapidly frozen and kept at the temperature of liquid nitrogen. None of the antisera raised in any one of the three goats against the p*I* 7.3 constituent was found to affect the pyrogenicity in rabbits of the constituents with lower p*I* values (Murphy et al., 1981a).

An attempt to ascertain whether heterogeneity of the several constituents of rabbit EP/IL-1 might be due to differences in glycosylation, led to a negative result (Murphy et al., 1981b). In fact, neither [^{14}C]glucose nor [^{14}C]glucosamine were found to be incorporated into EP/IL-1 by peritoneal exudate cells.

Further evidence for the absence, or at least for the irrelevance of glycosylation for the biological activity of the rabbit EP/IL-1 constituents, was obtained by treatment with periodate and neuraminidase. In neither cases was there any effect on pyrogenicity or activity in the thymidine costimulator assay. This evidence together with that obtained with the antisera mentioned above, strongly suggests that the difference in the p*I* of the constituents separable by focusing must lie in their amino acid sequences. However, as treatment with phenylglyoxal led to inactivation of all constituents, presence or absence of arginine cannot be the explanation. For a report of a difference between the activities of individual constituents in respect to the stimulation of synthesis of an AP-protein, see p. 303. Undoubtedly future work aimed at identification of the active centre of the EP/IL-1 molecule will be aided by these observations, reflecting slight differences between individual constituents and the corresponding changes in biological activity.

3. Murine EP/IL-1

3.1. Sources of murine EP/IL-1

The existence in the serum of mice of a factor which later proved to be IL-1 was first demonstrated by Sipe et al. (1979). The mice were of the C3H/HeN strain and had been injected with endotoxin. The presence of IL-1 in the plasma of these mice was demonstrated by injection of their plasma into C3H/HeJ mice, a strain known to be unresponsive to endotoxin and thus unable themselves to form IL-1. However, because of their small size, mice could not provide a practical source of work on isolation of IL-1. Thus, prior to the availability of supernatants from $P388D_1$ cells grown in culture, purification of murine EP/IL-1 was not attempted. The advent of these cells, derived originally from a methylcholanthrene-induced lymphoid neoplasm in a D13A/2 mouse, which had been carried in vitro for 18 years and

TABLE 1
$P388D_1$ mouse cell line

Macrophage-like characteristics	
I	Markers
	1. Nonspecific esterase
	2. Fc receptor
	3. C3 receptor
	4. Lymphocyte receptor
II	Functional
	1. Phagocytosis of latex beads
	2. Mediates antibody-dependent cell-mediated cytotoxicity
	3. Enhances T-cell activation by mitogens
	4. Produces LAF
	5. Produces EP
	6. Produces collagenase, plasminogen activator, and elastase

which retained most of the characteristics of macrophages, provided a practical source of IL-1 (Table 1) (Mizel, 1979). Not only could these cells be grown on a large scale, but even more important, after stimulation were found to produce IL-1 at rates far above that shown by monocytes from normal mice. As shown in Table 2 (Mizel and Mizel, 1981) afer 5 h 'Superinduction' in medium containing 1% foetal calf serum (FCS), PMA, cyclohexamide, sodium butyrate and actinomycin D, the cells were washed twice with RPMI 1640 and then incubated for a further 24 h in RPMI 1640 containing 1% FCS and sodium butyrate.

A subculture of $P388D_1$ cells and also another murine macrophage-like cell line RAW 264.1 have been successfully grown in serum-free medium (Estes et al., 1984). The latter cells spontaneously produce macrophage-derived competence factor (MDCF) which can cause 'competence' in the sense of causing quiescent fibroblasts to leave the G_0 stage. Only when stimulated by endotoxin do these cells start to produce IL-1. For certain properties of MDCF and a method for its assay see p. 301. Certain other cells or cell lines which have been investigated as sources of IL-1 are listed in Table 1, Chapter 2.

TABLE 2
Superinduction of interleukin 1[a]

	Time in culture[a]	Units of IL-1
None		< 1
10 μg/ml PMA	0 – 5	< 1
10 μg/ml PMA + 10 μg/ml cyclohexamide	0 – 5	903
+ 10 μg/ml actinomycin D	4 – 5	
10 μg/ml PMA + 10 μg/ml cyclohexamide + 2 mM sodium butyrate	0 – 5	1393
+ 10 μg/ml actinomycin D	4 – 5	
+ 2 mM sodium butyrate	5 – 29	

[a] 2×10^6 $P388D_1$ cells/ml were incubated in RPM1 1640 + 1% FCS in Costar TC_{24} tissue-culture plates (1 ml/well). After 1 h incubation to permit the cells to adhere to the plates, the various drugs were added.

3.2. Methods of isolation of EP/IL-1

Purification of IL-1 from such cell supernatants to apparent homogeneity was carried out by ammonium sulphate precipitation, chromatography on phenyl Sepharose, gel filtration on Ultragel ACA 54, preparative electrofocusing and gel electrophoresis. This final stage led to three components which were believed to dif-

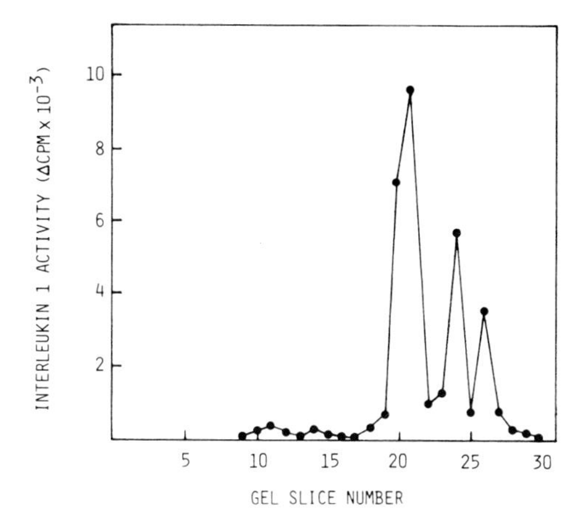

Fig. 3. Resolution of three charge species of purified IL-1. The purified IL-1 was electrophoresed at 2 mA for 3 h. The charge species have been termed α, β and γ (Mizel and Mizel, 1981).

fer slightly in molecular size and which were named IL-1 α, β and γ (Mizel and Mizel, 1981, Fig. 3).

The yield of purified IL-1 obtainable by this procedure was only 1 – 2% of that present in the cell supernatant. For this reason, as soon as an antiserum to murine IL-1 became available, a simplified and much higher yielding method was devised (Mizel et al., 1983). First the cell supernatant was concentrated 50 times and applied to a column of Sepharose AcA 54. The IgG obtained from the antiserum by ammonium sulphate precipitation and ion-exchange chromatography on DEAE cellulose was coupled to Sepharose 4-B. The appropriate fraction from the AcA 54 column was then run through the immunoabsorbent which was held in a small column. All the IL-1 was bound to the column and was not released by washing with 20 mM sodium phosphate at pH 8. However, by means of 3M KSCN approximately 5% of the IL-1 was eluted.

3.3. Definition of IL-1 unit

As defined by Mizel et al. (1983) the unit of IL-1 activity in the thymocyte costimulator assay is the amount required to give half maximal stimulation. This has been found to occur at a concentration of 0.1 ng/ml or 1×10^{-11} M.

3.4. Physico-chemical properties of murine EP/IL-1

Focusing of semipurified murine IL-1 in a pH 4.7 ampholine gradient and estimation by the thymocyte costimulator assay revealed only one band with p*I* between

4.9 and 5.1 gel filtration and gel electrophoresis have indicated that this material has M_r of approximately 30 kDa and consists of at least six components. This degree of heterogeneity has been found for IL-1 obtained using the immunoabsorbent method of Mizel et al. (1983). The final separation was carried out on a Tris-glycinate polyacrylamide gel. The positions in the gel of the IL-1 were revealed both by silver staining (Morrisey, 1981), and by their activities in the costimulator assay. While the separation which occurred is believed to be due to differences in molecular size or presence or absence of amide groups, however, accurate estimates of M_r values are not yet available. Amino acid analysis of IL-1 was reported by Mizel and Mizel (1981), but as this material is now known to have been a mixture of α_1- and α_2-constituents, it is sufficient to mention that serine, glutamic acid and alanine were the three amino acids found to be present in greatest amounts.

4. *Human EP/IL-1*

4.1. *Sources of human EP/IL-1*

These are described in Chapter 2, Section 2.1.

4.2. *Methods for production and purification of EP/IL-1*

Until recently, the methods used in attempts at isolation of human EP/IL-1 were similar to those employed for rabbit EP/IL-1. However, when blood monocytes were the source of IL-1 it was necessary to use a Hypac or Percoll gradient stage to separate the cells from most of the erythrocytes. Subsequently the leukocytes already enriched in monocytes, were kept in plastic or glass containers to allow adherence and thus a further enrichment of the monocyte population. After separation from almost all other cells the monocytes, suspended in 5% human serum, were allowed to phagocytose killed *Staph. albus* or were stimulated with endotoxin. After 1 h the cells were washed and suspended in Hank's balanced salt solution and kept at 37°C for 24 h. The supernatant from these cells served as the starting material for purification of IL-1 (Dinarello, 1984b). As occurred during similar studies of rabbit EP/IL-1, the most effective single purification stage was achieved by means of an immunoabsorbent. The use of an immunoabsorbent for purification of EP/IL-1 was introduced by Rosenwasser et al. (1979) soon after their success in obtaining antisera to EP/IL-1 in rabbits.

As early as 1980, separation of EP/IL-1 by HPLC was reported by Dinarello. Because of the very great resolving power of this type of chromatography, it has now become of general importance in studies aimed at isolation of human EP/IL-1 in pure form. Supernatants obtained from the U937 cell line mentioned above (Knudsen et al., 1984) and from human monocytes (Kock and Luger, 1984) have both been successfully used. The latter cells were stimulated with silica and the supernatant chromatographed first on a Biogel HPHT hydroxyapatite column. Subsequently one further HPLC stage, either on Bio-Sil IEX 540 DEAE anion exchanger or using a Bio-Sil TSK 125 size exclusion column, yielded IL-1 which assayed at 2.6×10^8 units per milligram (thymocyte costimulator assay units, cf. Luger et al., 1983). The main advantages of this procedure are that the use of an antibody to IL-1 is not required; rapidity, and most important, that excellent recoveries of activity were obtained at each stage. Other investigators have used HPLC to fractionate mouse IL-1 (Hansson et al., 1980) and IL-1 from a human B lymphoblastoid cell line (George et al., 1984).

4.3. Physico-chemical and biological properties of human EP/IL-1

Focusing in granulated acrylamide gel as the first stage, resolved the EP/IL-1 into components with p*I* values of 5.5, 5.8, 6.2 and 6.8. The molecular sizes, also ascertained by HPLC, of all four components were within the range of 15 – 20 kDa and did not appear to differ from one another. Biological activity of this material was demonstrated in several systems, including fever in rabbits (Knudsen et al., 1984).

The stability of EP/IL-1 depends on the presence or absence of contaminants. At the highest stage of purity so far achieved it retains activity at −70°C for at least one year. Unlike rabbit IL-1, no evidence for the existence of SH groups has been found and antioxidants do not assist stability. Activity is destroyed by heating to 70°C for 30 min (Oppenheim et al., 1982). Treatment with proteases such as pronase or protease K lead to complete inactivation, whereas sensitivity to trypsin seems to be species and assay variable (Dinarello, 1984b).

Active site studies employing specific reagents have indicated the importance of arginine residues, sensitive to phenylglyoxal, and the γ-carboxyl group of glutamic acid, sensitive to *N*-ethyl-5-phenyl-isoaxazolium-3′-sulphonate (Dinarello et al., 1982).

4.4. Units of measurement of EP/IL-1

For fever in rabbits, amounts of EP/IL-1 have been expressed as 'rabbit pyrogen doses' (RPD), the amount of EP/IL-1 necessary to produce a fever of 1°C in one rabbit. The thymocyte costimulator assay unit, the amount required to give half maximal [^{3}H]thymidine incorporation (see p. 292) is approximately one hundredth part of a RPD.

4.5. Active fragments of human EP/IL-1

IL-1 from human cells is thus far the only species of IL-1 from which fragments with any one of the IL-1 biological activities has been obtained. Evidence for the existence of fragments of IL-1 possessing three of the monokine's characteristic activities has been described (Dinarello et al., 1984). The fragment or fragments of human IL-1 were obtained in three different ways. Firstly, from the culture medium of human leukocytes which had been treated with zymosan, Pyrogen R (Gordon and Parker, 1980), secondly, by trypsin treatment of highly purified IL-1, and thirdly, by spontaneous breakdown of ^{125}I-labelled, highly purified IL-1 (Dinarello et al., 1984). The fragments produced in these ways were found to eluate from Sephadex G-15 and from Biogel P6 at volumes greater than the column volume (V_0) and are thus presumed to have a kilodalton value of not more than 4. The small peptide was shown to be pyrogenic by its ability to induce fever in mice, however, surprisingly when injected into rabbits, even by the intracerebral route, fever did not occur (Gordon and Parker, 1980). In addition, this peptide was shown to be active both in the thymocyte costimulator assay and also in respect to stimulation of muscle proteolysis (Dinarello et al., 1984) (cf. p. 108). A peptide with this property affecting muscle proteolysis has been isolated from patients with fever and named proteolysis-inducing factor (PIF) (Baracos et al., 1983; Clowes et al., 1983). It is similar to the active fragment described above in that it is active in the thymocyte costimulator assay and also is pyrogenic in mice. However, a difference may exist in respect to glycosylation as PIF, but not IL-1, has been found to bind to Con A sepharose and to be sensitive to neuraminidase. Investigation of the structure of these peptides should provide insight into the nature of the active site(s) of EP/IL-1 and also as to whether a single molecular species or polypeptide region is responsible for IL-1's multiple biological activities. As injection of a ^{125}I-labelled human IL-1 into rabbits leads to the urine containing a labelled fragment of similar molecular size to PIF, a starting material for such investigation is available (Dinarello et al., 1984).

5. IL-1-like materials from cells other than monocyte-macrophages

It has recently become clear that although cells of the monocyte macrophage lineage are the main producers of EP/IL-1, certain other cell types also have this potentiality (see Chapter 2, Table 1 – 4). As already mentioned, the development stage at which cells find themselves, at least in culture, is also a critical factor. As in the case mentioned below, certain IL-1-like molecules are formed by cells which differ from typical monocyte macrophages, i.e. catabolin which is formed by synovial cells as well as by monocytes.

5.1. Epidermal cell-derived thymocyte-activating factor (ETAF)

Mediators with little or no differences from murine and human IL-1 have been isolated from both normal epidermal cells (keratinocytes) from a malignant keratinocyte cell line (Luger et al., 1983) and from a human squamous cell carcinoma cell line (Luger et al., 1983). Although these cells share certain functions such as endocytosis with macrophages, they are not part of the immune system.

Purification of ETAF has not been carried as far as has similar work with IL-1 from murine, rabbit or human cells but far enough to indicate clearly that it has similar properties. For example, murine ETAF, like murine IL-1, from $P388D_1$ cells shows a single band on focusing with p*I* value of 5.2. Human ETAF, like human IL-1, can be separated into three constituents with p*I* values of 7.2, 5.8 and 5.0. All biological activities of ETAF so far examined, closely resemble those of IL-1 including activity in the thymocyte costimulator assay, pyrogenicity and induction of SAA production in mice (Sztein et al., 1982). ETAF also stimulated fibroblast proliferation which if local production of ETAF is involved, may be of significance in healing wounds.

5.2. Catabolin

As a result of a long study of breakdown of cartilagenous tissue in organ culture (Fell and Jubb, 1977), a factor with great similarity to IL-1 has been prepared in a highly purified state by Saklatvala et al. (1983) (cf. p. 116). This apparently homogenous protein, known as catabolin, can be prepared both from synovium and from monocytes which have been lectin stimulated. Pig blood was chosen by Saklatvala et al. (1983) as the source for isolation of catabolin. Most of the red cells were removed by sedimentation in 0.75% gelatin, 0.025 mM EDTA followed by lysis of those which remained in the supernatant by pulsing with 0.83% NH_4Cl. After washing, the leukocytes were resuspended in Dulbecco's modified Eagles medium, containing concanavalin A at 0.05 mg/ml and then cultured at 37°C in 5% CO_2 air for 48 h. Catabolin was isolated from the cell supernatant thus obtained by the means of the following 5 stages. Gel chromatography on Ultragel AcA, chromatofusing, chromatography on hydroxyapatite, anion-exchange chromatography on a Mono Q HR 5/5 column (Pharmacia) and finally reversed phase HPLC on a Zorbax ODS column. The physico-chemical properties of pig catabolin, as indicated by its behaviour during the above procedures, were molecular size 21 kDa and p*I* 4.9.

6. Summary

Present evidence suggests that multiple biologic effects are mediated by the IL-1 family of molecules. Definition of homogeneity or heterogeneity of these activities will require amino acid sequence analysis. The demonstration of biologically active fragments of IL-1 presents an additional complication. It may be expected that future studies of genetic control of IL-1 expression will define IL-1 as a family of closely related molecules produced by various cell types to express biologic activities important for host defense.

Addendum

Amino acid sequences of murine and human IL-1

Recently the sequences of murine IL-1 derived from P388D$_1$ cells and human IL-1 from peripheral macrophages have been determined. Using established methods, Lomedico et al. (1984) have cloned and sequenced murine IL-1 and obtained both a precursor peptide of 270 residues and a biologically active 156 residue peptide (underlined). The sequences of the two peptides are as follows:

MET Ala Lys Val Pro Asp Leu Phe Glu Asp Leu Lys Asn Cys Tyr Ser Glu Asn Glu

Asp Tyr Ser Ser Ala Ile Asp His Leu Ser Leu Asn Gln Lys Ser Phe Tyr Asp Ala

Ser Tyr Gly Ser Leu His Glu Thr Cys Thr Asp Gln Phe Val Ser Leu Arg Thr Ser

Glu Thr Ser Lys MET Ser Asn Phe Thr Phe Lys Glu Ser Arg Val Thr Val Ser Ala

Thr Ser Ser Asn Gly Lys Ile Leu Lys Lys Arg Arg Leu Ser Phe Ser Glu Thr Phe Thr

Glu Asp Asp Leu Gln Ser Ile Thr His Asp Leu Glu Glu Thr Ile Gln Pro Arg Ser

Ala Pro Tyr Thr Tyr Gln Ser Asp Leu Arg Tyr Lys Leu MET Lys Leu Val Arg Gln

Lys Phe Val MET Asn Asp Ser Leu Asn Gln Thr Ile Tyr Gln Asp Val Asp Lys His Tyr

Leu Ser Thr Thr Trp Leu Asn Asp Leu Gln Gln Glu Val Lys Phe Asp MET Tyr Ala

Tyr Ser Ser Gly Gly Asp Asp Ser Lys Tyr Pro Val Thr Leu Lys Ile Ser Asp Ser

Gln Leu Phe Val Ser Ala Gln Gly Glu Asp Gln Pro Val Leu Leu Lys Glu Leu Pro

Glu Thr Pro Lys Leu Ile Thr Gly Ser Glu Thr Asp Leu Ile Phe Phe Trp Lys Ser

Ile Asn Ser Lys Asn Tyr Phe Thr Ser Ala Ala Tyr Pro Glu Leu Phe Ile Ala Thr

Lys Glu Gln Ser Arg Val His Leu Ala Arg Gly Leu Pro Ser MET Thr Asp Phe Gln

Ile Ser

The 156 residue peptide has been expressed in E. coli and has yielded IL-1 of specific activity in the thymocyte proliferation assay equal to that previously isolated from P388D_1 cells. The calculated molecular weight of this form of murine IL-1 is 17,992.

Its probable relationship to the 270 residue precursor peptide is indicated because the latter has been detected in the culture fluid of stimulated macrophages. The precursor peptide contains two cysteine residues and a tetrabasic region Lys-Lys-Arg-Arg which may permit its proteolysis to yield the 156 residue IL-1 molecule. There is no sequence homology with any other lymphokine.

The precursor peptide of human IL-1 (Auron et al., 1984)* consisting of 269 residues shows surprisingly little homology with the murine IL-1 precursor.

* Auron, P.E., Webb, C.A., Rosenwasser, L.J., Mucci, S.F., Rich, A., Wolff, S.M. and Dinarello, C.A. (1984) Proc. Natl. Acad. Sci. USA *81,* 7907 – 7911.

References, p. 62.

Gordon/Koj (eds.) The acute-phase response to injury and infection.

CHAPTER 4

Interleukin 1 target cells and induced metabolic changes

J.D. SIPE

The IL-1 cytokines serve as intracellular messengers which help in the regulation of the host's multiple immunological and inflammatory responses. These cytokines are immunologically nonspecific and hormone-like in nature. As covered in Chapter 2, IL-1 cytokines can arise from a wide variety of cell types including macrophages, glial cells, glomerular mesangial cells, and epithelial cells of the skin and cornea. As discussed in Chapter 3, there is biochemical heterogeneity of IL-1 originating from a single species and cell source and post-synthetic proteolysis can give rise to biologically active fragments of IL-1. Thus, the multiplicity of IL-1 effects may be specified both by the target cell and the source and structure of the molecule.

This chapter will deal with the effects which occur when the multiple forms of IL-1, strictly defined as a macrophage-derived 12,000 – 15,000 M_r polypeptide, interact with the various target cells and metabolic changes which follow.

Whether synthesis of IL-1 always occurs or whether the AP-response can sometimes develop due to other cytokines, will also be discussed. Finally, some evidence concerning the mechanism of action of IL-1 particularly with respect to the great variation in target cells and in metabolic effects involved will be given. The synchronization of events following a single acute inflammatory episode will be considered as a series of stimulatory and inhibitory effects that may disappear during chronic or recurrent acute inflammation when a second acute inflammatory episode is superimposed upon the first before it has reached its natural conclusion.

1. Hormone-like nature of IL-1

1.1. Consequences of IL-1 action on thymocytes and on B and T lymphocytes

The IL-1 cytokines act not only on lymphocytes, but also on cells such as fibroblasts, osteoclasts, endothelial cells and skeletal muscle. The mitogenic effect of IL-1 on T cells appears to be mediated by interleukin 2 (IL-2) and IL-1 is known to play a key role in the activation of T-helper cells to release IL-2 (Smith et al., 1980). IL-2 appears to be the ultimate T-cell mitogen (Stadler and Oppenheim, 1982) for both antigenically and polyclonally activated T cells (cf. Chapter 7, Section 3.1). The thymocyte-lymphocyte activating factor assay cannot distinguish between IL-1 and IL-2. However, it appears that the combination of mitogenic lectins and possibly other factors with IL-1 is obligatory for IL-2 synthesis by T cells (Mier and Gallo, 1982) (cf. Chapter 7, Section 4). After IL-1 stimulates production of IL-2 by a T-helper cell subset, IL-1 promotes T-cell growth and also the differentiation and amplification of cytotoxic T cells. IL-1 is considered to be a differentiation signal that together with mitogen or antigen leads to the activation of resting T cells such that receptors for IL-2 are expressed. This process is summarized in Fig.1 after Watson et al. (1982). In the presence of IL-1, IL-2 precursor cells are converted to a state where mitogen or antigen triggers the specific release of IL-2. A distinct subset of pre-effector T cells when stimulated by IL-1 and/or mitogen, exhibit IL-2 receptors. DNA synthesis as measured by the macromolecular uptake of [^{3}H]thymidine and cell division result. The progeny cells also possess IL-1 receptors and can be maintained in continuous proliferation by addition of exogenous IL-2. Future studies will determine how IL-2 mediates DNA synthesis and cell replication, whether by way of direct transport to the nucleus or by way of a secondary messenger such as the arachidonic acid metabolites that are involved in other aspects of IL-1 mediated metabolic changes.

While it appears that in general, the main target cell for IL-1 is the T lymphocyte where changes in differentiation are effected, IL-1 can influence the immune response in a number of different ways. In addition to its effect on T cells, IL-1 has been shown to act as an adjuvant in vivo (Staruch and Wood, 1983) and to assist B cells in the in vitro plaque forming cell response (PFC) as discussed in detail in Chapter 7, Section 3.7. The 15,000 – 17,000 M_r human monokine that directly increases antibody production by splenocytes of nude mice was originally called B-cell activating factor (BAF) as it copurified with LAF, it is now considered to be IL-1 (Wood, 1979b). IL-1 is one of several soluble factors required for antibody production. Thus, when IL-1 activates the immune system, it is triggering potential shut-off mechanisms for the AP-response. For example, in the case of a bacterial infection, IL-1 assists B cells to make more antibodies. This soon removes the invading

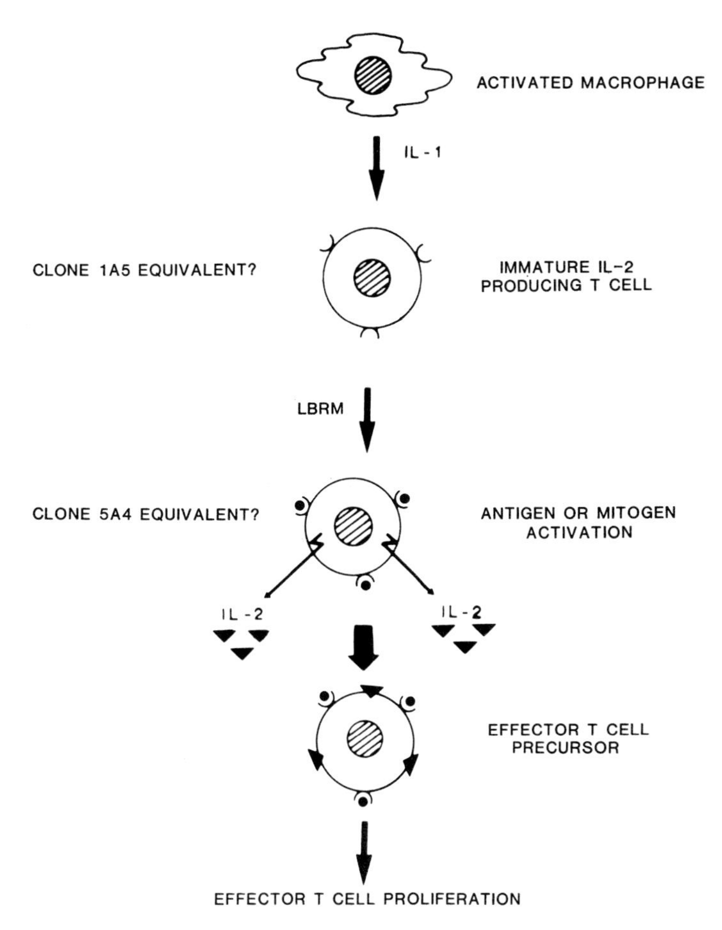

Fig. 1. Scheme for interaction of IL-1 and IL-2 in T-cell development. Clone 1A5 is derived from the original cloning of LBRM 33 and requires both IL-1 and mitogen for IL-2 production. Clone 5A4 also derived from LBRM 33 can be activated to release IL-2 and colony stimulating factor. (Reviewed by Watson et al., 1982.)

organisms and the major source of LPS. Thus, less IL-1 will be formed and the AP-response will die away.

1.2. Consequences of IL-1 action on the thermoregulatory center

Fever accompanies diverse disease processes including infections and inflammatory conditions of traumatic and immunologic origin. The febrile response is a carefully orchestrated sequence of events. Mononuclear phagocytes migrate, become activated and release IL-1. The hypothalamic thermoregulatory center increases its temperature set point, and skeletal muscle generates heat which is conserved by the peripheral vascular system.

IL-1 appears to raise the hypothalamic set point through increased production of prostaglandins. Fever and IL-1/EP are directly correlated with increased levels of prostaglandin E (PGE) in the brain, the cerebral spinal fluid (Bernheim et al., 1980; Coceani et al., 1983) and in the hypothalamic thermoregulatory center. It has been postulated (Dinarello and Wolff, 1982) that the thermoregulatory center located in the preoptic area of the anterior hypothalamus (PO/AH) regulates body temperature at 37°C by balancing heat production and peripheral heat loss. This is accomplished by the heat production from the 'shivering' of muscle contractions and heat conservation from peripheral vasoconstriction which results in the temperature of the blood supplying the hypothalamus matching the higher hypothalamic temperature setting.

Antipyretics inhibit the cyclooxygenase pathway of arachidonic metabolism at the prostaglandin synthetase level (Vane, 1971). Thus, humans with fever caused by prostaglandin E methyl esters do not respond to aspirin (Brenner et al., 1975), and there are studies in animal models which suggest that the site of action of antipyretics is on prostaglandin in the thermoregulatory center of the hypothalamus (reviewed by Dinarello and Wolff, 1982).

It appears that prostaglandins regulate neural pathways in the hypothalamus that are regulated by monoamine neurotransmitters. Blockade of monoamine receptors inhibits PGE-induced fever (Laburn et al., 1975; Lin et al., 1979). Fever is also reduced by protein-synthesis inhibitors (Cranston et al., 1978; Ruwe and Myers, 1979) and this may be due to inhibition of prostaglandin synthetase. The possibility that IL-1 acts directly upon the hypothalamus as well as by way of prostaglandin synthesis has been discussed in Chapter 2, Section 1. It seems possible that biologically active peptides derived from proteolysis of IL-1 might be particularly able to act in this capacity.

Temperature increase is the most general effect of IL-1 which influences all other aspects of its action both in vivo and in vitro. For example, mice are prewarmed in order to measure IL-1/EP activity and IL-1/LAF activity of mouse thymocytes is greater at 39°C than it is at 37°C.

1.3. Consequences of IL-1 action on muscle

The mechanism by which IL-1 stimulates PGE_2 production and protein breakdown is compared with that of fever production (Fig. 2, after Goldberg et al., 1984).

In addition to PGE_2, calcium ions stimulate muscle proteolysis, perhaps at the level of lysosomal activation. The in vitro rate of proteolysis in muscle is dependent upon extracellular Ca^{2+} concentration and IL-1 can stimulate protein breakdown only if calcium is present (cf. Goldberg et al., 1984). As also indicated in Fig. 2, IL-1 stimulated muscle proteolysis is inhibited by Ep-475, an inhibitor of lysosomal proteases. It is noteworthy that at 39°C, both PGE_2 production and protein breakdown show greater sensitivity to IL-1 than at 37°C. Thus, the effects of IL-1 on muscle and on the increase in body temperature are additive in promoting muscle protein loss. The nature of muscle responses to IL-1 is discussed in detail in Chapter 9, Section 1.

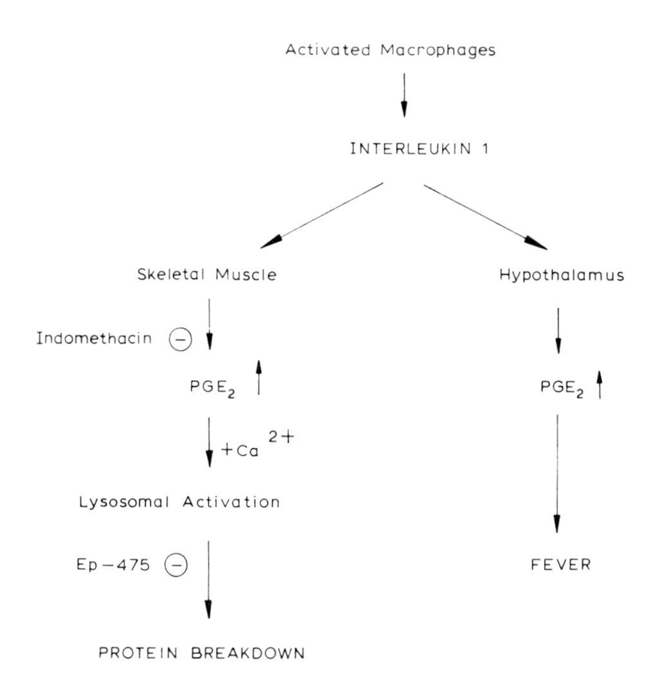

Fig. 2. Effects of IL-1 on body temperature and muscle protein catabolism. (After Goldberg et al., 1984.)

1.4. Probable importance of IL-1 in metabolism of connective tissue

Rheumatoid arthritis is characterized by a gradual onset of inflammatory changes throughout most connective tissues of the body. It is characterized by erosion of bone and articular cartilage. Destruction of articular cartilage originates as changes in synovial tissue and is accompanied by pannus formation. Pannus contains loosely layered fibroblasts, macrophages, lymphocytes and plasma cells, collagen fibers and blood vessels. Invasion and destruction of articular cartilage proceeds by way of the cartilage matrix as this is invaded by macrophages from the pannus. IL-1 is currently being investigated as a key mediator in both the induction of synovitis and in the perpetuation of chronic inflammatory arthritis. Effusions of inflamed joints contain IL-1 (Wood et al., 1983). IL-1 may derive from monocytes/macrophages either in the joint fluid or in the pannus or the IL-1 activity may be due to an IL-1-like cytokine produced by synovial tissue. Several studies have shown that the IL-1 stimulates in vitro the type of cellular responses which presumably play a role in connective tissue destruction. Thus, IL-1 stimulates secretion of collagenase and prostaglandins from synoviocytes (Dayer et al., 1983; Mizel et al., 1981) (cf. Chapter 10) and the release of calcium from bone (Gowen et al., 1983). Catabolin was isolated on the basis of its ability to release glycosaminoglycans from cartilage (Saklatvala and Dingle, 1980; Saklatvala et al., 1984; cf. Chapter 10, Section 3) and has subsequently been shown to exhibit IL-1-like immunostimulatory properties. Similarly, IL-1 stimulates fibroblasts to proliferate and to secrete collagenase (Schmidt et al., 1982; Postlethwaite et al., 1982). The increased synthesis of collagenase by synoviocytes is accompanied by a large increase in production of PGE_2 (Mizel et al., 1981), for similar effects on brain and muscle cf. Part II. The observations of IL-1 mediated processes are consistent with the proliferation of the synovial tissue surrounding the joint which occurs during the development of rheumatoid arthritis and which follows cartilage and joint destruction.

Both lymphokines and monokines have been implicated as mediators of synovial cell proliferation (Pannott et al., 1982), however, it is generally agreed that mononuclear cell factors mediate synovial cell proliferation leading to pannus formation. Increased collagen production by these cells would also contribute to the increased amount of synovial tissue. The cells and soluble mediators involved in chronic inflammatory synovitis are discussed in greater detail in Chapter 22, Section 2.

1.5. Possible consequence of an IL-1-like factor on adipose tissue

In addition to IL-1 mediated systemic effects on lymphocytes, brain, muscle and connective tissue, there has been some suggestion that LPS-treated macrophages inhibit lipoprotein-lipase formation by adipocytes and that they suppress fatty acid biosynthesis (Kawakami et al., 1981; Pekala et al., 1983). This subject is discussed in more detail in Chapter 10, Section 4, and it remains to be demonstrated whether these effects are due to IL-1 or some other macrophage-derived mediator.

1.6. Consequences of IL-1 action on hepatocytes

The AP-response is a series of complex systemic, metabolic, and physiologic alterations, of which one is the change, usually an increase, in concentration of a large number of plasma proteins known as the AP-proteins. The AP-protein serum amyloid A (SAA) is inducible, being minimally present in serum, but within 3 – 4 h of an inflammatory stimulus beginning to rise and finally reaching as much as 1000-fold elevation. The mechanism of SAA biosynthesis was studied with two inbred strains of mice differing in sensitivity to LPS and was shown to proceed sequentially with macrophage stimulation and production of the soluble mediator IL-1, followed by rapidly increasing SAA synthesis (reviewed by Sipe, 1985). The precise relationship of SAA induction by IL-1 to the generally altered pattern of hepatic protein synthesis is an area of current investigation, in particular with respect to induction of C-reactive protein (CRP) synthesis in human and rabbit. As yet, there is no direct evidence for IL-1 mediation of CRP synthesis, although Dinarello (1984a) has suggested that IL-1 induction of CRP may require cofactors.

Evidence for alteration of hepatic protein synthesis by IL-1/LEM, HSF and other cytokines is presented in Part III. In general, with the exception of rat α_2-macroglobulin (α_2-M), most changes both in vivo and in vitro in concentrations of AP-proteins are less dramatic alterrations in concentration of constitutive proteins rather than induction of normally undetectable proteins (cf. Chapter 17, Section 1).

1.7. Consequences of IL-1-like cytokine action on IL-1 producing cells

With the rapid proliferation of research on the origin of IL-1 and IL-1-like cytokines, it is becoming apparent that virtually all cells can make IL-1 and virtually all cells can respond to it. For example, macrophages have been shown to respond to exogenous IL-1 with increased chemotaxis and PGE (Oppenheim et al., 1985). In addition, there is some persistent evidence that neutrophils make (Nakamura et al., 1982), as well as respond to, IL-1 (Klempner et al., 1979) (cf. Chapter 1, Section 1). Oppenheim's laboratory has studied IL-1 production and response in B lymphocytes and a subset of large granular lymphocytes. Currently, fibroblasts, mesangial cells, endothelial and epithelial cells are being studied as IL-1 producers and responders.

2. Mechanism of action of IL-1

Evidence has been presented in this chapter that the mediator IL-1 is non-specific with respect to target cells. Earlier, in Chapter 1, Section 2, the non-specificity of the origin of IL-1 was discussed. However, macrophage-derived IL-1 does appear to be specific in that it is always present during inflammation, cell necrosis and tissue injury, however these may have originated.

There is little direct information about the mechanism of IL-1 action. Current thinking (as reviewed by Dinarello, 1984a,b) suggests that it acts upon cell membranes somewhat analogously to a calcium ionophore. IL-1 appears to alter arachidonic acid metabolism, and Dinarello (1984b) has proposed than an important role of IL-1 is to stimulate release of arachidonic acid from membrane phospholipids. This is plausible if IL-1 acts like a calcium ionophore because activation of membrane phospholipase A_2 (Zenser et al., 1980) has been shown to result from a rapid increase in intracellular calcium.

First, evidence supporting IL-1 as a calcium ionophore will be considered. Many of the biologic activities of IL-1 have been reported for the calcium ionophore A 23187, which transports divalent cations across cell membranes, resulting in an increase in the concentration of intracellular calcium. It has been shown that fluctuation in body temperature is proportional to the concentration of calcium ions in the cerebrospinal fluid (Myers, 1982). Patients treated with calcium channel blockers show a decreased mixed lymphocyte reaction (Milton, 1982) and it was found that A 23187 replaced IL-1 in a human T-cell mitogenesis assay (Koretsky et al., 1983).

There are several studies indicating that A 23187, like IL-1, stimulates prostaglandin synthesis by various cells and tissues, including pituitary cells (Betteridge, 1980) renal medullary cells (Zenser et al., 1980) platelets (Pickett et al., 1977) and rat muscle (Rodemann et al., 1982; Rodemann and Goldberg, 1982; Baracos et al., 1983). However, as Ca ionophores do not lead to IL-2 formation by T cells, their action and that of IL-1 must be considered as being independent (Koretsky et al., 1983a, b).

As discussed in detail (Chapter 7, Section 3.6) there is a T cell membrane lipid viscosity change associated with macrophage-bound antigen activation of primed T cells (Puri and Lonai, 1980). A primary effect of IL-1 on membrane viscosity was implicated by experiments in which added cholesterol (presumably having dissolved in the plasma membrane) increased microviscosity and antigen binding just as much as did added IL-1. This is a second line of evidence that IL-1 may alter membrane protein synthesis.

The IL-1 mediated phenomena of fever and proteolysis are accompanied by increased PGE synthesis; however, lymphocyte and neutrophil responses to IL-1 appear to occur independently of the cyclooxygenase pathway. Instead these functions are decreased by inhibitors of the lipooxygenase pathway of arachidonic acid metabolism (Kelly et al., 1979; Goodman and Weigle, 1982; Payan and Goetze, 1981). Leukotrienes have been shown to affect lymphocyte function and there is little evidence to suggest that lymphocytes synthesize prostaglandins (Lederman et al., 1981).

Dinarello (1984b) has proposed that a possible unifying mechanism for the action of IL-1 would be an effect upon the release of arachidonic acid from membrane phospholipids. Thus, if IL-1 acts like a calcium ionophore, it would act on the target cell membrane to increase the concentration of intracellular calcium which would in turn lead to the activation of membrane phospholipase A_2 (Zenser et al., 1980). The release of arachidonic acid from membrane phospholipids is rate-limiting in the metabolism of arachidonate by either the cyclooxygenase or lipooxygenase pathways. Thus, cells with an active cyclooxygenase pathway such as brain, muscle, and fibroblasts would produce prostaglandins in response to IL-1, and the effect of IL-1 upon these target cells would be blocked by indomethacin. No effect of indomethacin on hepatic AP-protein synthesis has been demonstrated (cf. Chapter 18, Section 1). Other cells such as lymphocytes and neutrophils convert arachidonic acid to leukotrienes. Thus, IL-1 mediated effects on lymphoid cells such as thymocyte proliferation, superoxide production and degranulation are blocked by inhibitors of the lipooxygenase pathways (Serhan et al., 1982; Dinarello et al., 1983).

It remains to be determined whether there is a specific IL-1 receptor (cf. Chapter 18, Section 1) on each of the various target cells and/or whether IL-1 may be a modifier of membrane structure as discussed above. This concept would be supported by the fact that IL-1 is more active at elevated temperatures with respect to thymocyte proliferation, pyrogenicity and muscle proteolysis (Dinarello, 1984a).

3. *Is IL-1 essential for the AP-response?*

The macrophage plays a central role in the initial host response to infection, trauma, inflammation and antigen-antibody reactions. One of the early (within 3 h) (Sipe et al., 1979; Sipe and Rosenstreich, 1981) events is the production of IL-1 which sets into motion (up to 24 h) fever, induction of SAA (and perhaps other hepatic proteins) increased peripheral white cell count and increased muscle proteolysis. A reasonable hypothesis may be that the first response is hormonal or catalytic in nature, while the later systemic response is a more stoichiometric relationship. Thus, IL-1, tumor necrosis factor and glucocorticoid antagonizing factor, being themselves macrophage products, would be expected to be present at levels far less than 1 μg/ml, whereas CRP and SAA, which result from IL-1 action on the liver, increase to several hundred μg/ml. The third stage involves a shut-off of the AP-response as restoration of the host is achieved. This is marked by the inhibitory or suppressive effects of AP-proteins (Sipe and Rosenstreich, 1981). If, as seems extremely probable, this description of the host response represents the most important events in host defense, then the importance of IL-1 needs no further emphasis. However, because the host's defense mechanism is so multisided, particular responses may not each require the release of IL-1.

It is evident from experimental studies of repeated inflammatory stimulation (Macintyre et al., 1982; Sipe et al., 1980; Brandwein et al., 1983) that tachyphylaxis of the CRP, SAA inducer and SAA responses occurs upon repeated inflammatory stimulation (cf. Chapter 18, Section 1). Furthermore, the SAA and CRP levels in patients with chronic inflammation accompanying secondary amyloidosis are never as high as those reached during an isolated inflammatory episode (Gertz et al., 1985; Sipe, unpublished observations). Despite the strong evidence that IL-1 is always present during inflammation and after trauma, IL-1 concentrations in the few instances for which data is available do not correspond with plasma concentrations of AP-proteins, e.g. after chronic UV stimulation (Gahring et al., 1984).

These observations must indicate the likelihood that there are as yet undiscovered factors which either augment or inhibit IL-1 action. It also remains to be determined whether this non-proportionality is due to diminished production of IL-1 or diminished responsiveness of liver cells to mediators. The function of macrophage products other than IL-1 such as hepatocyte stimulating factor (Ritchie and Fuller, 1983) require further investigation.

Furthermore, in view of the similar properties of calcium ionophores and IL-1 discussed above, it may be that changes in metal ion concentration and other factors such as neural effects alter the pattern of 'normal' hepatic protein synthesis. Certainly, with respect to SAA biosynthesis, it has not yet been possible to establish an in vitro system that mimics the in vivo situation of 1000-fold elevation (Tatsuta et al., 1983). This appears to be due in part to a high baseline of SAA synthesis, but also may reflect the requirement for other factors (available in vivo) that mediate IL-1 induction of SAA synthesis. Serum stimulates in vitro SAA synthesis (Tatsuta et al., 1983), however, this appears to be due to a general effect on protein synthesis (L.K.Chaney and J.D. Sipe, unpublished observations).

4. Conclusions

IL-1 is a major mediator of the AP-response with both primary and secondary effects. The interaction of IL-1 with its primary target cells is covered in Parts II and III. In this Part the diversity of IL-1 with respect to species of origin, cell of origin, and target cell has been discussed. Also, some information on the effect of temperature on IL-1 action and on the consequences of modification of IL-1 by proteolysis and other chemical action have been outlined.

The concept that has recently emerged is that the heterogeneity of IL-1 allows for a fine tuning of the AP-response. In addition, the existence of biologically active cleavage products of IL-1 brought about by secretory enzymes arising during the inflammatory episode (Dinarello et al., 1984) may permit additional modulation and control of the AP-response.

The effects of IL-1 can generally be considered to be beneficial to the host if the inflammatory episode eventually reaches its normal conclusion. However, the complex series of events involved may go awry due to superpositioning of a second inflammatory episode before the first response has been completed or to other systemic events which may secondarily modify the normal cause of the AP-response.

TABLE 1
Interrelationship between IL-1 target cells

Primary target cell	Product or effect	Secondary target cell
Hepatocyte	CRP	T-cell responses inhibited
	α_1-acid glycoprotein	(cf. Chapter 13, Section 4)
	SAA	T-cell dependent antibody responses spleen cells[1]
		B-cell suppression[2]
Brain	Temperature increase	Thymocyte proliferation[3]
Muscle, synovial cells	PGE_2	IL-1 effect on fibroblasts uninhibited[4]
		IL-1 and antigen stimulation of T cells inhibited (cf. Chapter 10)
Muscle	Amino acids	Hepatic synthesis AP-plasma proteins stimulated (cf. Chapter 9, Section 3)

[1] Benson et al., J. Exp. Med. *142,* 236, 1975; [2] Martin and Rosenthal, Clin. Res. *3,* 348A, 1983; [3] Duff and Durham, Yale J. Biol. Med. *55,* 437, 1982; [4] Korn et al., Int. J. Dermatol. *19*, 487, 1980.

References (Chapters 1 – 4)

Aarden, L.A., Brunner T.K., Cerottine J.C. et al. (1979) J. Immunol. *123,* 2928 – 2929.

Allison, A.C., Harrington, J.S. and Birbeck, M. (1966) J. Exp. Med. *124*, 141 – 154.

Amento, E.P., Kurnick, J.T., Epstein, A. and Krane, S.M. (1982) Proc.Natl. Acad. Sci. USA *79,* 5307 – 5311.

Atkins, E., Bodel, P.T. and Francis, L. (1967) J. Exp. Med. *126*, 357 – 383.

Babior, B.M. (1984) J. Clin. Invest. *73*, 599 – 601.

Bainton, D.R. (1973) J. Cell Biol. *58*, 249 – 264.

Baracos, V., Rodemann, H.P., Dinarello, C.A. and Goldberg, A.L. (1983) N. Engl. J. Med. *308,* 553 – 558.

Baum, R.M. (1984) Chem. Eng. News, April 9, 20 – 26.

Baumann, H., Jahreis, G.P., Sauder, D.N and Koj, A. (1984) J. Biol. Chem. *259*, 7331 – 7342.

Bendtzen, K., Baek, L., Berild, D., Hasselback, H., Dinarello, C.A. and Wolff, S.M. (1984) N. Engl. J. Med. *310*, 596.

Bennett, I.L. and Beeson, P.B. (1953) J. Exp. Med. *98*, 493 – 508.

Bernheim, H. A., Gilbert, T.M. and Stitt, J.T. (1980) J. Physiol. (London) *301*, 69 – 78.

Bettens, F., Kristensen, F., Walker, C. and de Weck, A.L. (1982) Eur. J. Immunol. *12*, 948 – 952.

Betteridge, A. (1980) Biochem. J. *186*, 987 – 992.

Bodel, P. (1974a) Ann. NY Acad. Sci. *230*, 6 – 13.

Bodel, P. (1974b) Infect. Immun. *10*, 451 – 457.

Bodel, P. (1974c) Yale J. Biol. Med. *47*, 101 – 112.

Bodel, P. (1978) J. Exp. Med. *147*, 1503 – 1516.
Bodel, P. and Atkins, E. (1967) N. Engl. J. Med. *276*, 1002 – 1008.
Bodel, P. and Miller, H. (1976) Proc. Soc. Exp. Biol. Med. *15*, 93 – 96.
Bodel, P., Ralph, P., Wenc, K. and Long, J.C. (1980) J. Clin. Invest. *65*, 514 – 518.
Boggs, D.R., Cartwright, G.E. and Wintrobe, M.M. (1966) Am. J. Physiol. *211*, 51 – 60.
Bornstein, D.L. (1982) Ann. N.Y. Acad. Sci. *389*, 323 – 337.
Bornstein, D.L., Bredenberg, C. and Woods, W.B. (1963) J. Exp. Med. *117*, 349 – 364.
Brandwein, S.B., Sipe, J.D., Skinner, M. and Cohen, A.S. (1984) Fed. Proc. *42*, 3406.
Brenner, W.E., Dingfelder, J.R. and Staurovsky, L.G. (1975) Am. J. Obstet. Gynecol. *123*, 17.
Briggs, R.T., Drath, D.B., Karnovsky, M.L. and Karnovsky, M.J. (1975) J. Cell Biol. *67*, 566 – 586.
Cannon, J.G. and Dinarello, C.A. (1984) Fed. Proc. *43*, 462, Abs. 1034.
Cannon, J.G. and Kluger, M.J. (1983) Science *220*, 617 – 619.
Cebula, T.A., Hanson, D.F., Moore, D.M. and Murphy, P.A. (1979) J. Lab. Clin. Med. *94*, 95 – 105.
Chambers, R. and Chambers, E.L. (1961) Exploration into the Nature of the Living Cell, pp. 143 – 174, Harvard University Press, Cambridge.
Charon, J.A., Luger, T.A., Mergenhagen, S.E. and Oppenheim, J.J. (1982) Infect. Immun. *38*, 1190 – 1195.
Chenoweth, D.E., Goodman, M.G. and Weigle, W.O. (1982) J. Exp. Med. *156*, 68 – 78.
Cline, M.J. and Golde, D.W. (1979) Blood *53*, 157 – 165.
Cline, M.J. and Sumner, M.A. (1972) Blood *40*, 62 – 69.
Clowes, G.H.A., George, B.C., Villee, C.A. and Saravis, C.A. (1983) N. Engl. J. Med. *308*, 545 – 552.
Coceani, F., Bishai, I., Dinarello, C.A. and Fitzpatrick, F.A. (1983) Am. J. Physiol. *244*, R785 – 793.
Cochrane, C.G. and Aiken, B.S. (1966) J. Exp. Med. *124*, 733 – 752.
Cochrane, C.G. (1968) Adv. Immunol *9*, 97 – 162.
Cohn, Z.A. (1978) J. Immunol. *121*, 813 – 816.
Cranston, W.I., Dawson, N.J., Hellon, D.F. and Townsend, Y. (1978) J. Physiol. (London) *285*, 35P.
Davis, J.M. and Gallin, J.I. (1981) in Cellular Functions in Immunity and Inflammation (Oppenheim, J.J., Rosenstreich, D.L. and Potter, M., eds.) pp. 77 – 102, Elsevier/North-Holland, New York.
Dayer, J.M., Zavadil-Grols, C., Ucla, C. and Mach, B. (1983) Clin. Res. *448*, A.
DeChatelet, L.R., MacPhail, L.C., Mullikin, D. and McCall, C.E. (1975) J. Clin. Invest. *55*, 714 – 721.
Dewald, B., Baggiolini, M., Curnette, J.T. and Babior, B.M. (1979) J. Clin. Invest. *63*, 21 – 29.
Dinarello, C.A. (1980) in Fever (Lipton, J., ed.) pp. 1 – 9, Raven Press, New York.
Dinarello, C.A. (1984a) Surv. Immunol. Res. *3*, 29 – 33.
Dinarello, C.A. (1984b) Rev. Infect. Dis. *6*, 51 – 95.
Dinarello, C.A. and Wolff, S.M. (1982) Am. J. Med. *72*, 799 – 819.
Dinarello, C.A., Goldin, N.P. and Wolff, S.M. (1974) J. Exp. Med. *139*, 1369 – 1381.
Dinarello, C.A., Renfer, L. and Wolff, S.A. (1977) J. Clin. Invest. *60*, 465 – 472.
Dinarello, C.A., Rosenwasser, L.J. and Wolff, S.M. (1981) J. Immunol. *127*, 2517 – 2519.
Dinarello, C.A., Bendtzen, K. and Wolff, S.M. (1982) Inflammation *6*, 63 – 78.
Dinarello, C.A., Marnoy, S.O. and Rosenwasser, L.J. (1983) J. Immunol. *130*, 890 – 895.
Dinarello, C.A., Clowes, G.H.A., Gordon, A.H., Saravis, C.A. and Wolff, S.M. (1984) J. Immunol. *133*, 1332 – 1338.
Estes, J.E., Pledger, W.H. and Gillespie, G.Y. (1984) J. Leukocyte Biol. *35*, 115 – 129.
Fell, H.B. and Jubb, R.W. (1977) Arthritis Rheum. *20*, 1359 – 1371.
Flynn, A. (1983) J. Am. Coll. Nutr. *2*, 205 – 213.
Flynn, A., Finke, J.M. and Hilfiker, M.L. (1982) Science *218*, 475 – 477.
Fontana, A., Kristensen, F., Dubs, R., Gemsa, D. and Weber, E. (1982) J. Immunol. *129*, 2413 – 2419.
Fridovich, I. (1978) Science *201*, 875 – 880.

Gahring, L., Baltz, M., Pepys, M.B. and Daynes, R. (1984) Proc. Natl. Acad. Sci. *81*, 1198 – 1202.
George, F.W., Warren, M.K. and Martin, M.J. (1984) Fed. Proc. *43*, 1608.
Gertz, M.A., Skinner, M., Sipe, J.D., Cohen, A.S. and Kyle, R.A. (1985) Submitted for publication.
Gery, I. and Lepe-Zuniga, J.L. (1984) Lymphokines *9*, 109 – 125.
Gery, I., Gershon, R.K. and Waksman, B.H. (1971) J. Immunol. *107*, 1778 – 1780.
Gery, I., Gershon, R.K. and Waksman, B.H. (1972) J. Exp. Med. *13*, 128 – 142.
Giddon, D.B. and Lindhe, J. (1972) Am. J. Pathol. *68*, 327 – 338.
Goldberg, A.L., Baracos, V., Rodeman, P., Waxman, L. and Dinarello, C.A. (1984) Fed. Proc. *43*, 1301 – 1306.
Goldstein, I.M., Cerquiera, M., Lind, S. and Kaplan, H.B. (1977) J. Clin. Invest. *59*, 249 – 254.
Goodman, M.G., Chenoweth, D.E. and Weigle, W.O. (1982) J. Exp. Med. *156*, 912 – 917.
Gordon, A.H. and Parker, I.D. (1980) Br. J. Exp. Pathol. *61*, 534 – 539.
Gotze, O., Bianco, C. and Cohn, Z.A. (1979) J. Exp. Med. *149*, 372 – 286.
Gowen, M., Wood, D.D., Ihrie, E.J., McGuire, M.K.B. and Russell, R.G.G. (1983) Nature (London) *306*, 378 – 380.
Grabner, G., Luger, T.A., Smolin, G. and Oppenheim, J.J. (1982) Invest. Ophthalmol. Visual Sci. *23*, 757 – 763.
Haeseler, F., Bodel, P. and Atkins, E. (1977) J. Reticuloendothel. Soc. *22*, 569 – 581.
Hanson, D.F., Murphy, P.A. and Windle, B.E. (1980) J. Exp. Med. *151*, 1360 – 1371.
Hansson, E., Henriksen, O., Alvarey, V.L., Barnes, M., Bachtold, H., Frey, J.R., Hansson, E., Howie, S., Lefkovits, I., Roitsch, C.A., Soderberg, A. and Young, P. (1980) FEBS Lett. *121*, 157 – 160.
Hayashi, H. (1975) Int. Rev. Cytol. *40*, 101 – 151.
Henson, P.M. (1971) J. Immunol. *107*, 1535 – 1546.
Higgs, G.A. (1982) in Phagocytosis Past and Future (Karnovsky, M.L. and Bolis, L., eds.) pp. 105 – 129, Academic Press, New York.
Hoffstein, S.T., Korchak, K.M., Smolen, J.E. and Weissman, G. (1982) in Phagocytosis, Past and Future (Karnovsky, M.L. and Bois, L., eds.) pp. 44 – 65, Academic Press, New York.
Homburger, F. (1945) J. Clin. Invest. *24*, 43 – 45.
Ikejima, T., Dinarello, C.A., Gill, D.M. and Wolff, S.M. (1984) J. Clin. Invest. *73*, 1312 – 1320.
Johns, M., Sipe, J.D., DeMaria, A., Bjornsen, B. and McCabe, W.R. (1984) Abstracts of the Annual Meeting of the Am. Soc. Microbiol.
Kampschmidt, R.F. (1981) in Infection: the Physiologic and Metabolic Responses of the Host (Powanda, M.C. and Canonico, P.G., eds.), pp. 55 – 74, Elsevier/North-Holland Biomedical Press, Amsterdam.
Kampschmidt, R.F., Long, D. and Upchurch, H.F. (1972) Proc. Soc. Exp. Biol. Med. *139*, 124 – 126.
Kampschmidt, R.F., Pulliam, L.A. and Upchurch, H.F. (1973) Proc. Soc. Exp. Biol. Med. *144*, 883 – 887.
Kampschmidt, R.F., Pulliam, L.A. and Upchurch, H.F. (1980) J. Lab. Clin. Med. *95*, 616 – 123.
Kampschmidt, R.F., Upchurch, H.F. and Pulliam, L.A. (1982) Am. N.Y. Acad. Sci. *339*, 338 – 353.
Kampschmidt, R.F., Upchurch, H.F. and Worthington, M.L. (1983) Infect. Immun. *41*, 6 – 10.
Karnovsky, M.L. and Lazdins, J.K. (1978) J. Immunol, *121*, 809 – 813.
Karnovsky, M.L., Shafer, A.W., Cagan, R.H., Graham, R.C., Karnovsky, M.J., Glass, E.A. and Saito, K. (1966) Trans. NY Acad. Sci. *29*, 778 – 787.
Kawakami, N., Pekala, P.H., Lane, M.D. and Cerami, A. (1981) Proc. Natl. Acad. Sci. *79*, 912 – 916.
Kelly, J.P., Johnson, W.C., Parker, L.W. (1979) J. Immunol. *122*, 1563 – 1571.
Klebanoff, S.J. and Clark, R.A. (1978) The Neutrophil Function in Clinical Disorders, Elsevier/North-Holland, New York.
Klempner, M.S., Dinarello, C.A. and Gallin, J.I. (1978) J. Clin. Invest. *61*, 1330 – 1336.
Klempner, M.S., Dinarello, C.A., Henderson, W.R. and Gallin, J.I. (1979) J. Clin. Invest. *64*, 996 – 1002.

Kluger, M.J., Ringler, D.H. and Anvers, M.R. (1975) Science *188*, 166 – 168.
Knudsen, P.F., Dinarello, C.A. and Strom, T.B. (1984) Abstr. 16th International Leucocyte Culture Conference.
Koch, A. and Luger, T.A. (1984) J. Chromatogr. *296*, 293 – 300.
Koj, A., Gauldie, J., Regoeczi, E., Sauder, D.N. and Sweeney, G.D. (1984) Biochem. J. *224*, 505 – 514.
Korchak, H.M. and Weissmann, G. (1978) Proc. Natl. Acad. Sci. USA *75*, 3818 – 3822.
Koretsky, G.A., Daniele, R.P., Greene, W.C. and Nowell, P.C. (1983a) Proc. Natl. Acad. Sci. USA *80*, 3444 – 3447.
Koretsky, G.A., Elias, J.A., Kay, S.L., Rossman, M.D., Nowell, P.C. and Daniele, R.P. (1983b) Clin. Immunol. Immunopathol. *29*, 443 – 450.
Krakauer, T. and Oppenheim, J.J. (1983) Cell. Immunol. *80*, 223 – 229.
Krane, S.M., Goldring, S.R. and Dayer, J.M. (1982) Lymphokines *7*, 75 – 136.
Kurt-Jones, E., Beller, D.I., Mizel, S. and Unanue, E.R. (1984) Proc. Natl. Acad. Sci. USA, in press.
Laburn, H., Woolf, C.J., Willes, G.H. and Roseendorff, C. (1975) Neuropharmacology *1*, 405 – 411.
Lachman, L.B. and Metzgar, R.S. (1980) J. Retic. Soc. *27*, 621 – 629.
Larsen, G.L. and Henson, P.M. (1983) Ann. Rev. Immunol. *1*, 3335 – 3359.
Lederman, M.M., Ellner, J.J. and Rodman, H.M. (1981) Clin. Exp. Immunol. *45*, 191 – 200.
Lee, K.C., Wong, M. and McIntyre, D. (1981) J. Immunol. *126*, 2474 – 2479.
Leonard, E.J. and Skeel, A.H. (1978) Exp. Cell. Res. *114*, 117 – 126.
Leonard, E.J., Ruco, L.P. and Meltzer, M.S. (1978) Cell. Immunol. *41*, 347 – 357.
Liao, Z. and Rosenstreich, D.L. (1983) Clin. Res. *31*, 492A.
Lin, M.T., Pang, I.H., Chen, S.I. and Chein, Y.F. (1979) Pharmacology *18*, 188 – 194.
Lomedico et al. (1984) Nature (London) *312*, 459.
Lovett, D.H., Ryan, J.L. and Sterzel, R.B. (1983) J. Immunol. *130*, 1796 – 1801.
Luger, T.A., Stadler, B.M., Katz, S.I. and Oppenheim, J.J. (1981) J. Immunol. *127*, 1493 – 1498.
Luger, T.A., Charon, J.A., Colot, M., Mickche, M. and Oppenheim, J.J. (1983) J. Immunol. *131*, 816 – 820.
Macintyre, S.S., Schultz, D. and Kushner, I. (1982) Ann. NY Acad Sci *389*, 76 – 87.
Mackaness, G.B. (1970) Sem. Hematol. *7*, 172 – 184.
Marchesi, V. and Flory, Q.J. (1960) Exp. Physiol. *45*, 343 – 348.
Metchnikoff, E. (1905) Immunity in Infectious Disease. Translated by Frances G. Bennie, Cambridge University Press, London.
Mier, J.W. and Gallo, R.C. (1982) Lymphokines *6*, 137 – 163.
Milton, A.S. (ed.) (1982) in Pyretics and Antipyretics, pp. 151 – 1186, Springer-Verlag, Berlin.
Mizel, S.B. (1979) Ann. NY. Acad. Sci. *332*, 539 – 549.
Mizel, S.B. and Anderson, B.J. (1983) in Interleukin, Lymphokines and Cytokines (Oppenheim, J.J. and Cohen, S., eds.) pp. 401 – 407, Academic Press, New York.
Mizel, B. and Mizel, D. (1981) J. Immunol. *126*, 834 – 837.
Mizel, S.B. and Rosenstreich, D.L. (1979) J. Immunol. *122*, 2173 – 2179.
Mizel, S.B., Rosenstreich, D.L. and Oppenheim, J.J. (1978) Cell. Immunol. *40*, 230 – 235.
Mizel, S.B., Dayer, J., Krane, S.M. and Mergenhagen, S.E. (1981) Proc. Natl. Acad. Sci. USA *78*, 2472 – 2477.
Mizel, S.B., Dukovich, M. and Rothstein, J. (1983) J. Immunol. *131*, 1834 – 1837.
Morrisey, J.H. (1981) Anal. Biochem. *117*, 307 – 310.
Movat, H.Z. (1985) The Inflammatory Reaction, Elsevier Science Publishers (Biomedical Division), Amsterdam.
Munck, A., Gayne, P.M. and Holbrook, N.J. (1984) Endocrinol. Rev. *5*, 25 – 44.
Murphy, P.A., Chesney, P.J. and Wood, W.B. (1974) J. Lab. Clin. Med. *83*, 310 – 322.

Murphy, P.A., Cebula, T.A., Levin, J. and Windle, B.E. (1981a) Infect. Immun. *34*, 177 – 183.
Murphy, P.A., Cebula, T.A. and Windle, B.E. (1981b) Infect. Immun. *34*, 184 – 191.
Myers, R.D. (1982) in Pyretics and Antipyretics (Milton, A.S., ed.) pp. 151 – 186, Springer-Verlag, Berlin.
Nakamura, S., Goto, F., Goto, K. and Yochinaga, M. (1982) J. Immunol. *128*, 2614 – 2621.
North, R.J. (1973) Cell. Immunol. *7*, 166 – 176.
Oppenheim, J.J. and Potter, M. (1981) in Cellular Functions in Immunity and Inflammation (Oppenheim, J.J., Rosenstreich, D.L. and Potter, M., eds.) pp. 1 – 28, Elsevier/North-Holland, New York.
Oppenheim, J.J., Stadler, B.M., Siraganian, R.P., Mage, M. and Mathieson, B. (1982) Fed. Proc. *41*, 257 – 262.
Oppenheim, J.J., Matsushima, K., Onozaki, K., Procopio, A. and Scala, G. (1985) 16th Int. Leuc. Cult Conf. Cambridge U.K.
Pacak, F. and Siegert, R. (1982) Eur. J. Biochem. *127*, 375 – 380.
Pannott, D.P., Goldberg, R.L., Kaplan, S.R. and Fuller, G.C. (1982) Eur. J. Clin. Invest. *12*, 407 – 415.
Payan, D.G. and Goetzl, E.J. (1981) J. Clin. Immunol. *1*, 266 – 270.
Pekala, P.H., Rawakami, M., Angus, C.W., Lane, M.D. and Cerami, A. (1983) Proc. Natl. Acad. Sci. USA *80*, 2743 – 2747.
Pickett, W.C., Jess, R.L. and Cohen, P. (1977) Biochim. Biophys. Acta *486*, 209 – 213.
Pilsworth, L.M.C. and Saklatvala, J. (1983) Biochem. J. *216*, 481 – 489.
Postlethwaite, A.E., Lachman, L.B., Mainardi, C.L. and Kang, A.H. (1982) J. Exp. Med. *157*, 801 – 806.
Potter, M. (1981) in Cellular Functions in Immunity and Inflammation (Oppenheim, J.J., Rosenstreich, D.L. and Potter, M., eds.) pp. 229 – 254, Elsevier/North-Holland, New York.
Puri, J. and Lonai, P. (1980) Eur. J. Immunol. *10*, 273 – 281.
Rawlins, M.D., Luff, R.H. and Cranston, W.I. (1970) Lancet *1*, 1371 – 1373.
Ritchie, D.G. and Fuller, G.M. (1983) Ann. NY Acad. Sci. *409*, 490 – 502.
Rodemann, H.R. and Goldberg, A.L. (1982) J. Biol. Chem. *257*, 1632 – 1638.
Rodemann, H.R., Waxman, L. and Goldberg, A.L. (1982) J. Biol. Chem. *257*, 8716 – 8723.
Röllinghoff, M., Pfizenmaier, K. and Wagner, H. (1982) Eur. J. Immunol. *12*, 337 – 342.
Rosenstreich, D.L. (1981) in Cellular Functions in Immunity and Inflammation (Oppenheim, J.J., Rosenstreich, D.L. and Potter, M., eds.) pp. 127 – 159, Elsevier/North-Holland, New York.
Rosenwasser, L.J., Dinarello, C.A. and Rosenthal, A.S. (1979) J. Exp. Med. *150*, 709 – 714.
Rossi, F., Romeo, D. and Patriaca, P. (1972) J. Reticuloendothel. Sci. *12*, 127 – 149.
Ruwe, W.D. and Myers, R.D. (1979) Brain Res. Bull. *4*, 741 – 745.
Saklatvala, J. and Dingle, J.T. (1980) Biochem. Biophys. Res. Commun. *96*, 1225 – 1231.
Saklatvala, J., Curry, V.A. and Sarsfield, S.J. (1983) Biochem. J. *215*, 385 – 392.
Saklatvala, J., Pilsworth, L.M.C., Sarfield, S.J., Gavrilovic, J. and Heath, J.K. (1984) Biochem. J. *224*, 461 – 466.
Sauder, D.N. (1984) Lymphokine Res. *3*, 145 – 151.
Sbarra, A.J. and Karnovsky, M.L. (1959) J. Biol. Chem. *234*, 1355 – 1362.
Scala, G., Allavena, P., Dieu, J.Y., Kasahara, T., Herberman, R.B. and Oppenheim, J.J. (1984) Nature (London) *309*, 56 – 59.
Schmidt, J.A., Mizel, S.B., Cohen, D. and Green, I. (1982) J. Immunol. *128*, 2177 – 2182.
Serhan, C.N., Radin, A. Smolen, J.E., Korchak, H. Samuelson, B. and Weissman, G. (1982) Biochem. Biophys. Res. Commun. *107*, 1006 – 1012.
Shou, L., Schwartz, S.A., Good, R.A., Peng, R. and Chen, C.L. (1980) Proc. Natl. Acad. Sci. USA *77*, 6096 – 6100.

Sipe, J.D. (1985) Lymphokines, Vol. 12, in press.
Sipe, J.D. and Rosenstreich, D.L. (1981) in Cellular Function in Immunity and Inflammation (Oppenheim, J.J., Rosenstreich, D.L. and Potter, M., eds), pp. 411 – 429, Elsevier/North-Holland Inc., New York.
Sipe, J.D., Vogel, S.N., Ryan, J.L., MacAdam, K.P.W.J. and Rosenstreich, D.L. (1979) J. Exp. Med. *150*, 597 – 606.
Sipe, J.D., Vogel, S.N., Rosenstreich, D.L. and McAdam, K.P.W.J. (1980) in Amyloid and Amyloidosis (Glenner, G.G., Costa, P.P. and Freitas, F., eds.) pp. 505 – 512, Excerpta Medica, Amsterdam.
Sipe, J.D., Vogel, S.N., Sztein, M.B., Skinner, M. and Cohen, A.S. (1982) Ann. N.Y. Acad. Sci. *389*, 137 – 150.
Sipe, J.D., Johns, M., DeMaria, A., Cohen, A.S. and McCabe, W.R. (1984) Fed. Proc. *43*, 1542.
Smith, K.A., Lachman, L.B., Oppenheim, J.J. and Fauta, M.F. (1980) J. Exp. Med. *151*, 1551 – 1556.
Snyderman, R., Pike, M.C., McCarley, D. and Lang, L. (1975) Infect. Immun. *11*, 3883 – 4892.
Stadler, B.M. and Oppenheim, J.J. (1982) Lymphokines *6*, 117 – 135.
Staruch, M.J. and Wood, D.D. (1983) J. Immunol. *130*, 2191 – 2194.
Sztein, M.B., Vogel, S.N., Sipe, J.D., Murphy, P.A., Mizel, S.B., Oppenheim, J.J. and Rosenstreich, D.L. (1981) Cell Immunol. *63*, 164 – 176.
Sztein, M.B., Luger, T.A. and Oppenheim, J.J. (1982) J. Immunol. *129*, 87 – 90.
Tatsuta, E., Sipe, J.D., Shirahama, T., Skinner, M. and Cohen, A.S. (1983) J. Biol. Chem. *258*, 5414 – 5418.
Thompson, P.L., Papadimitrion, J.M. and Walters, W.N.I. (1967) J. Pathol. Bacteriol. *94*, 389 – 396.
Treves, A.J., Basak, V., Tal, T. and Fuks, Z. (1983) Eur. J. Immunol. *13*, 647 – 651.
Unanue, E. and Kiely, J.M. (1977) J. Immunol. *119*, 925 – 931.
Vane, J.R. (1971) Nature New Biol. *231*, 232 – 235.
Virolainen, M. (1968) J. Exp. Med. *127*, 943 – 951.
Vogel, S.N. and Sipe, J.D. (1982) Surv. Immunol. Res. *1*, 235 – 241.
Vogel, S.N., Moore, R.N., Sipe, J.D. and Rosenstreich, D.L. (1980) J. Immunol. *124*, 2004 – 2009.
Voorhis, W.C.V., Hain, L.S., Steinman, R.M. and Kaplan, G. (1982) J. Exp. Med. *155*, 1172 – 1185.
Wagasugi, H., Harel, A., Dokhelar, M.C., Fradelizi, D. and Tursz, T. (1984) J. Immunol. *132*, 2939 – 2947.
Wahl, S.M. (1981) in Cellular Functions in Immunity and Inflammation (Oppenheim, J.J., Rosenstreich, D.L. and Potter, M., eds.) pp. 453 – 466, Elsevier/North-Holland, New York.
Watson, J., Frank, M.B., Mochizuki, D. and Gillis, S. (1982) Lymphokines *6*, 95 – 116.
Weissmann, G., Zurier, R.B. and Hoffman, S. (1972) Am. J. Pathol. *68*, 539 – 564.
Weissmann, G., Serhan, C., Korchak, H.M. and Smoten, J.E. (1982) Ann. NY Acad. Sci. *389*, 11 – 24.
Weiss, S.J., Lambert, M.B. and Test, S.T. (1983) Science *222*, 625 – 628.
Windle, B.E., Murphy, P.A. and Cooperman, S. (1983) Infect. Immun. *39*, 1142 – 1146.
Windle, J.J., Shin, H.S. and Morrow, J.F., (1984) J. Immunol. *132*, 1317.
Wood, D.D. (1979a) J. Immunol. *123*, 2395 – 2399.
Wood, D.D. (1979b) J. Immunol. *123*, 2400 – 2407.
Wood, D.D. and Cameron, P.M. (1978) J. Immunol. *121*, 53 – 60.
Wood, D.D., Ihrie, E.J., Dinarello, C.A. and Cohen, P.L. (1983) Arthritis Rheum. *26*, 975 – 983.
Yoshinaga, M., Nishime, K., Nakamura, S. and Goto, F. (1980) J. Immunol. *124*, 94 – 99.
Zenser, T.Y., Herman, C.A. and Davis, B.B. (1980) Am. J. Physiol. *238*, E371 – 375.

ADDENDUM

A.H. GORDON

Further information has recently become available concerning:

(1) The properties of hepatocyte stimulating factor (HSF) and certain cell lines capable of its formation;
(2) the anti-viral effects of a form of IL-1;
(3) isolation of human IL-1 in high yield without the use of an immunoadsorbant.

(1) Relationship of HSF and IL-1

Minor injury or local tissue damage lead to easily measurable increases in concentration of AP-plasma proteins but little or no rise in body temperature. On the other hand, bacterial LPS is a good stimulator of both fever and AP-protein synthesis. These facts are to be expected if differently stimulated macrophages are assumed to secrete different monokines. Considerable evidence for physico-chemical differences between HSF and IL-1 are already available (Baumann et al., 1984). Thus the HSF activity obtained both from normal human monocytes and from certain neoplastic cell lines has usually been found to be associated with protein molecules larger than the 15 kDa characteristic of the IL-1 main component. This has been confirmed by experiments using well purified constituents from human monocytes and from several leukemia cell lines (Woloski and Fuller, 1985). The constituents with HSF activity were characterised and assayed by their ability to stimulate hepatocytes to synthesise fibrinogen, and those with IL-1 activity by means of the thymocyte costimulator assay. The final stage of purification was by HPLC. The patterns of HSF and IL-1 activities obtained from P388D_1, U937, HL60 and for normal monocyte supernatants were all very different. The overlap between the two activities varied and was particularly small for the supernatant from normal monocytes. Surprisingly the main IL-1 peak from this supernatant eluted at a position corresponding to 12 kDa. The supernatants from the leukemia cell lines all yelded IL-1 with kilodaltons close to 15. On the basis of this information and the effects of superinduction with PMA (cf. p. 42) on P388D_1 cells which led to increased IL-1 but not HSF activity, the two stimulators were supposed to be different. However, for several reasons the important question whether the molecules responsible for HSF and IL-1 have reciprocal effects must remain open. One of these is that the various AP-plasma proteins may require different HSF factors for maximum stimulaton. Furthermore, such individual HSF factors might show differences in respect to their capacity to act like IL-1. A further complication derives from the fact that synthesis of AP-proteins may require the presence of a corticosteroid hormone (Koj et al. 1984).

(2) A form of IL-1 with anti-viral effects

The recent purification and partial sequencing of a factor which under special conditions can induce human fibroblasts to produce β-interferon is of special interest because molecules of the active substance show partial homology with human IL-1 (Van Damme et al., 1985). This factor, originally investigated because of its β-interferon-like activity, in purified form induces many of the effects of IL-1, e.g. fever in rabbits at 10 ng/kg, and [^{3}]thymidine incorporation by human thymocytes at 0.1 ng/ml. In consequence of this discovery careful testing of other forms of IL-1 for possible interferon induction, not only in fibroblasts, will be necessary. Information so derived should deepen understanding of the protective functions of IL-1. As described in Parts II and III, these already include modulation of both cell mediated and humoral immunity, the role of certain AP-proteins as protease inhibitors and as components of the complement cascade. Thus, in view of the numerous and very diverse functions of IL-1 in bodily defence mechanisms, it can be no surprise that IL-1 may, in certain circumstances, be able to assist control of viral infection.

(3) Isolation of human IL-1

Using a simple three-stage purification, IL-1 has been isolated from human mononuclear cells in 31% yield (Kronheim et al. 1985). Yields were monitored by means of three separate cell culture assays. The final product led to fever in rabbits at a dose of 0.5 μg/kg.

References

Baumann, H., Jahreis, G.P., Sauder, D.N. and Koj, A. (1984) J. Biol. Chem. *259*, 7331 – 7342.

Koj, A., Gauldie, J., Regoeczi, E., Sauder, D.N. and Sweeney, G.D. (1984) Biochem. J. *224*, 505 – 514.

Kronheim, S.R., March, C.J., Erb, S.K., Conlon, P.J., Mochizuki, D.Y. and Hopp, T.P. (1985) J. Exp. Med. *161*, 490 – 502.

Van Damme, J., De Ley, M., Opdenakken, G., Billian, A. and De Somer, P. (1985) Nature (London) *314*, 266 – 268.

Woloski, B.M.R.N.J. and Fuller, G.M. (1985) Proc. Natl. Acad. Sci. USA *82*, 1443 – 1447.

PART II

Extrahepatic actions of monokines and injury-derived mediators

Gordon/Koj (Eds.) The acute-phase response to injury and infection.

CHAPTER 5

Response of the brain to Interleukin 1

A.H. GORDON

Investigations of the fever response have advanced steadily since Bennett and Beeson (1953) succeeded in distinguishing endogenous pyrogen (EP) from bacterial endotoxin. The relationship of bacterial endotoxin to EP is described on p. 20 and shown in Table 1. The active fever inducing constituents of EP/IL-1 have yet to be completely purified, they apparently consist of a family of closely similar molecules all of which induce fever (cf. Chapter 3).

1. *Effects of IL-1 on brain cells; formation of prostaglandins (PG)*

Investigations of how brain cells respond to IL-1 have been carried out both in vitro and in vivo. In experiments of the former kind, Dinarello and Bernheim (1981) found that highly purified human IL-1, when incubated with rabbit brain slices, led to 3 to 4-fold increases in prostaglandin E (PGE) in the supernatant fluid. The response was the same whether the slices were from the preoptic area of the anterior hypothalamus (PO/AH) or from other parts of the brain. As shown in Fig. 1, increasing concentrations of IL-1 added to rabbit cortical brain slices led to progressively increasing concentrations of PGE in the supernatants. When endotoxin was used in place of IL-1, slightly larger increases of PGE than those that had resulted from IL-1 were obtained.

Cultured astrocytes and C_6 glioma cells when stimulated with lipopolysaccharide (LPS) produce not only PGE but also a factor with properties very close to IL-1

TABLE 1

Exogenous pyrogens	*Cell sources of IL-1*		
Viruses	*Mononuclear phagocytes*		
Bacteria	Peripheral blood monocytes		
Fungi	Lung and peritoneal macrophages		
Bacterial products	Synovial macrophages		
Endotoxin	Kupffer's cells and splenic macrophages		
Muramyl dipeptide	Bone marrow macrophages		
Etiocholanalone	Myelomonocytic leukaemia cells		
Ag-Ab complexes →	Hodgkin's lymphoma cells		
Polynucleotides		New protein	
Antigens (via lymphokines	*Other sources* →	synthesis →	EP/IL-1
from sensitised	Keratinocytes		
lymphocytes	Langerhan's cells		
	Corneal epithelial cells		
	Gingival exudate cells		
	Astrocytes/glial cells		
	Renal mesangial cells		
	Renal cell carcinomas		
	Dendritic cells		
	Cells in culture		
	Murine 388/D_1 (hybrid monocyte cells)		
	Human monomyeloblastic cells U937 and many others		

(Fontana et al., 1982, 1983). As the possibility that production of the IL-1-like factor by contaminating macrophages was excluded, the potential of glial cells themselves to form both of the types of stimulant which, when injected intracerebroventrically (ICV) lead to fever, seem well established.

Because the formation of PGE and an IL-1-like substance is a property of glial cells in general, and because the formation of PGE when highly purified IL-1 is present, is not restricted to brain slices for the PO/AH area, an explanation of the special characteristics of this area is needed. The most obvious suggestion is that the PO/AH neurones are specially sensitive to IL-1 and/or PG. An alternative explanation might be that formation of an IL-1-like substance by the cells of the PO/AH area may cause an augmentation of the fever response due to entry into this area of IL-1 from the vascular system. In either case, the formation of PGE by IL-1 could occur and an explanation for the reduction in fever observed in vivo when indomethacin is given with IL-1 would be available.

However, despite the ability of prostaglandin E_2 (PGE_2) to bring about fever when injected ICV, there are reasons (cf. Section 3) for the view that PGE_2 is not itself the ultimate mediator of the fever response. The possibility remains that other products of the cyclooxygenase on lipoxygenase pathways, such as thromboxanes, may be more significant in this respect (Laburn et al., 1977). Perhaps the most important consequence of the work with rabbit brain slices is simply that IL-1, can lead to metabolic effects involving release of PGs (cf. Section 3).

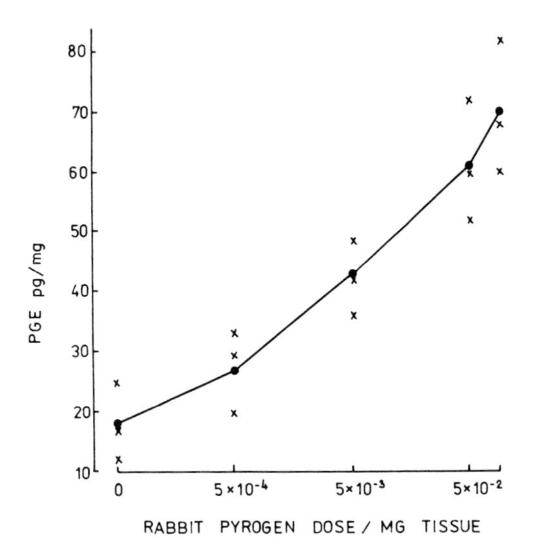

Fig. 1. Dose-response of increasing concentrations of human EP on rabbit cortical brain slices. A single rabbit was used for these studies, and a preparation of 15 kDa EP was assayed prior to these incubations (Dinarello and Bernheim, 1981).

2. *Indentification of the PO/AH area as the primary thermoregulatory centre*

So far, neither in vivo nor in vitro experiments have successfully elucidated the precise changes which IL-1 brings about in those brain cells which are responsible for the control of fever. Microinjection experiments with solutions of EP which may be presumed to have contained IL-1 have served to localise the PO/AH area by demonstrating its uniqueness in respect to the fever response (Cooper et al., 1967).

Localisation and demonstration of the uniqueness in respect to fever of the PO/AH area was also obtained when PGE, instead of IL-1, was injected (Milton and Wendlandt, 1970, 1971b). Following ablation of this area, and injection of EP/IL-1 into the lateral ventricles in rabbits, indication of the existence of certain secondary areas have been obtained (Veale and Cooper, 1975).

3. Evidence for and against PG as the final mediators of fever

Although recent work suggests that the fever response induced by PGE occurs more rapidly than that which follows IL-1, evidence is available, as summarised below, that PGE cannot be the main activator of the thermo-sensitive cells. Before further consideration of the interactions of IL-1 with the thermoregulatory cells, it is important to insist on the difference between effects induced experimentally, e.g. by microinjection techniques and the metabolic events which characterise the various stages of the fever response as it occurs in the absence of experimental intervention.

The concept that when fever develops, PG might themselves activate the thermoregulatory cells, was introduced by Cooper et al. (1967) on the basis of microinjection of, and observation of concentration of, PGE in the cerebrospinal fluid (CSF). This theory found support due to the discovery by Vane (1971), that nonsteroidal anti-inflammatory agents were able to inhibit PGE synthesis by the cyclooxygenase system and the demonstration by Milton and Wendlandt (1968) that intra-peritoneal paracetamol reduced the fever caused by ICV injection of endotoxin. The same result was obtained with fever produced by lipid A (cf. p. 80 and 81) and EP injected ICV or into the PO/AH area. The persuasive concept that PG are the final mediators of the fever response also found support because following the injection of paracetamol, not only was fever reduced but PG in the CSF fell sharply (Feldberg et al., 1972, 1973).

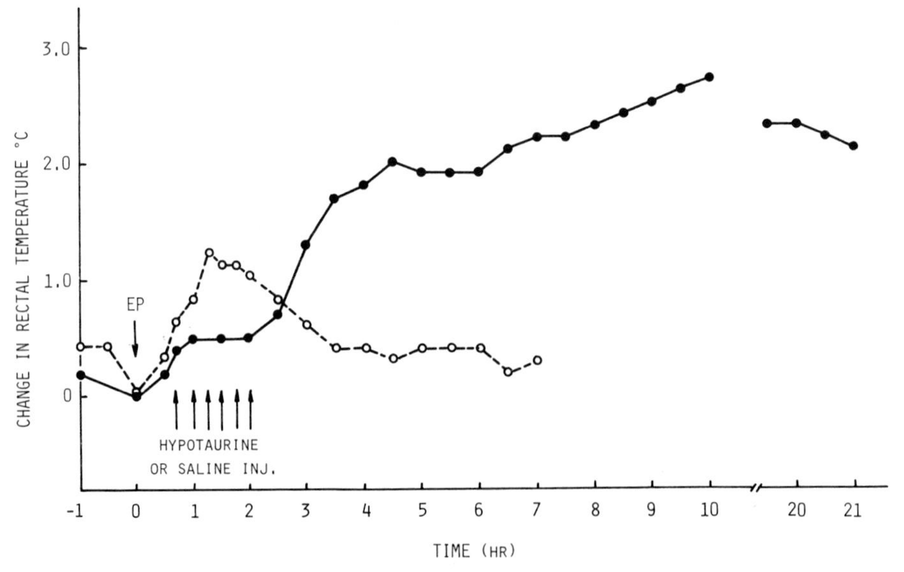

Fig. 2. Effects of repeated ICV, injections of hypotaurine (1 mg) on fever produced by I.V. EP in a single rabbit. ●, EP + hypotaurine; ○, EP + saline (Lipton and Ticknor, 1979).

Although strong evidence against the original view that PGE_1 or PGE_2 might constitute the final mediator of the fever response has now accumulated, it remains true that the presence or absence of PG is more relevant to fever than to any other of the many metabolic changes which are inducible by IL-1. This has been clearly demonstrated by the use of Ibuprofen and other inhibitors of the cyclooxygenase pathway by Sobrada et al. (1983), Tocco et al. (1983) and Poole et al. (1984). For details cf. Section 4 and Fig. 4.

Further information concerning the fevers which are induced by EP and PGE_2 has been derived from experiments in which taurine and certain other inhibitors of transport out of the central nervous system (CNS) have been employed. As shown in Fig. 2, ICV injection of hypotaurine soon after I.V. injection of endotoxin into a rabbit, led to very great prolongation and intensification of fever (Lipton and Ticknor, 1979). This is thought to be due to retention of EP in the CNS due to blocking of the transport system normally responsible for its removal. Using another transport inhibitor, iodipamide, a comparison has been made between I.V. LP and ICV PGE_2. As shown in Fig. 3, the thermal-response index for both substances increased proportionately with the dose of iodipamide (Lipton et al., 1979). This experiment indicates that the effect of PGE_2, like that of EP, can be prolonged by use of an appropriate transport inhibitor. In view of the profound differences between PGE_2, a complex ring structure, and EP now known to be a small protein molecule, it is perhaps surprising that the transport of these very different

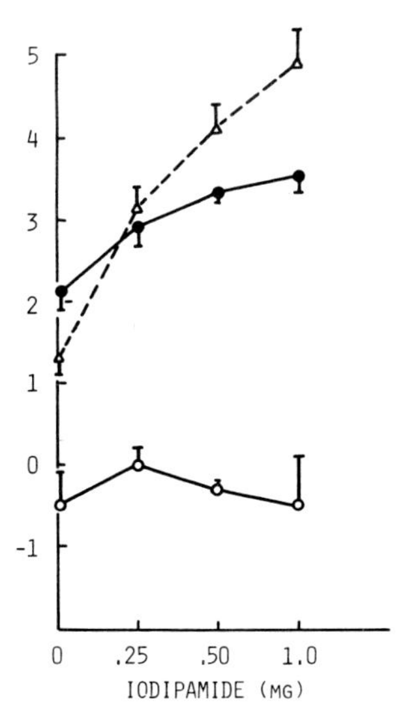

Fig. 3. Effects on the temperature response to I.V. EP, I.V. saline or ICV PGE_2 of 10 min prior ICV injection of iodipamide, or perchloric acid. Scores are mean (± SE) temperature response of five rabbits. ●—●, I.V. EP; ○—○ I.V. saline; △ – – – △, ICV PGE_2 (Lipton et al., 1979).

molecules is similarly affected. A likely explanation may be that neither constitutes the final mediator of the fever response, but that both can lead to the formation of such a substance.

Strong evidence against the theory that PGE_2 is the final mediator of the fever response was obtained by Cranston et al. (1975), who injected EP continuously into rabbits in the presence and absence of a low concentration of sodium salicylate. In both conditions fever was obtained, but only in those animals not receiving salicylate was there an increased concentration of PGE in the CSF. Further evidence, against a role for PG, was that the pretreatment with either one of two PG antagonists SC 19220 and HR 546 prevented fever following PGE_2, but had no effect on fever produced by EP/IL-1 (Cranston, 1976; Hellon et al., 1980). Evidently, EP/IL-1 remained able to induce fever despite blocking of PGE_2 receptors.

4. *Importance of both IL-1 and PG as fever inducers*

Evidence that induction of fever includes a stage involving protein synthesis has been obtained by the use of cycloheximide (Cranston et al., 1978). Given before I.V. injection of EP/IL-1, this protein synthesis inhibitor led to significant attenuation of fever. Fever due to PGE_2 was similarly attenuated. Similar results have been obtained with the inhibitor of protein synthesis, anisomycin (Milton and Sawhney, 1981), which was first used for this purpose by Ruwe and Myers (1979). Because this inhibitor is much less toxic than cycloheximide and because by itself it has no hyperthermic effect, its use has firmly established that if protein synthesis is inhibited, EP/IL-1 will no longer induce fever (Cranston et al., 1980). Some further insight into the metabolic changes which lead up to the development of fever has been obtained by injection of arachidonic acid ICV into rabbits. The resultant fever is unaffected by anisomycin (Townsend et al., 1981). This finding and experiments with mepacrine and parabromophenylacylbromide, drugs known as inhibitors of phospholipase A_2, have led to the concept that fever is affected by anisomycin, both as a result of inhibition of phospholipase A_2 and also by inhibition of the metabolic pathway between EP/IL-1 and the ultimate mediator of the fever response. The evidence for the effect of anisomycin on this latter pathway is that when it was injected ICV together with EP/IL-1 into rabbits, both fever and protein synthesis were inhibited (Cranston et al., 1982). The stronger inhibition under these conditions of protein synthesis compared with fever is explicable as being due to EP/IL-1 acting in two ways, firstly, as a stimulator of PGE_2 synthesis and secondly, as a direct stimulator of fever (Fig. 4). The possibility that inhibition of phospholipase A_2 may prevent not only the formation of arachidonic acid and subsequently the products of both the cyclooxygenase and the lipoxygenase

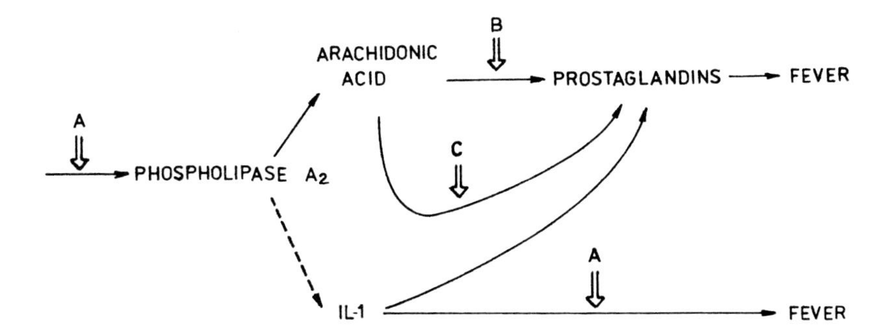

Fig. 4. Pathway leading to formation of EP/IL-1 and direct and indirect pathways by which it may act on the PO/AH (– – –) hypothetical pathway. (A) Pathway inhibited by anisomycin; (B) pathway inhibited by steroidal and non-steroidal anti-inflammatories, but not inhibited by anisomycin; (C) pathway inhibited by steroidal and non-steroidal anti-inflammatories.

pathways but also the formation of EP/IL-1 also deserves consideration. If EP/IL-1 can be shown to act by stimulating the synthesis of proteins which act as the final mediators of the fever response, as is suggested by the effect of anisomycin, there will be an analogy with virus inhibition by interferon. Like EP/IL-1, interferon is formed by macrophages and has been shown to act by inducing the synthesis of anti-viral proteins (Stewart, 1979).

The pathways involving formation of EP/IL-1 and by which it may act on the PO/AH area, are shown in Fig. 4. The probability that EP/IL-1 can act directly on this area raises the question as to how it can pass into this area from the blood stream and also from the 3rd ventricle following injection into this part of the brain. Elegant work on the anatomy of 3rd ventricle and PO/AH brain area has assisted understanding of how this is likely to occur (Brightman et al., 1975).

5. *Passage of pyrogens through the blood-brain barrier*

As shown in Fig. 5, it is the walls of capillaries in most areas of the brain, which themselves constitute the blood-brain barrier, that are impermeable to protein molecules. On the other hand, the capillaries of the *organum vasculosum laminae terminalis* (OVLT) have fenestrated walls, through which molecules such as EP/IL-1 can be expected to pass easily.

A complication which has been partly explained by experiments with horse radish peroxidase (HRP) is the existence of a relatively impermeable barrier between the 3rd ventricle and the PO/AH area. Essentially the problem is that whereas HRP of 40 kDa does not pass rapidly from the 3rd ventricle into the PO/AH (Broadwell et al., 1983) and that a steep concentration gradient of HRP between blood and CSF can be shown to exist for several minutes after its injection, EP/IL-1 when injected into any one of three sites causes rapid fever. These sites are the PO/AH itself, the 3rd ventricle and the vascular compartment. The most likely explanation for the observed failure of HRP to pass out of the 3rd ventricle is that the wall of the 3rd ventricle may act as a graded sieve (Brightman et al., 1975). Later at 90 min after ICV injection some HRP was observable in a region close to the PO/AH. The slower development of fever after injection of EP/IL-1 by this route can be thus explained.

Confirmation of the important part played by the OVLT in passage of EP/IL-1 from the blood into the PO/AH area has been obtained by Blatteis et al. (1983). Following electrolytic ablation of this brain region in guinea pigs, fever which would have been expected after I.P. injection of endotoxin did not occur. The explanation advanced for this effect is that the OVLT which forms the wall of the arterioventral 3rd ventricle and lies in close proximity to the PO/AH, permits entry of EP/IL-1 into this area (Fig. 5). If indeed EP/IL-1 as such, does reach the cells of the PO/AH, then by analogy with hormones and their target cells, a search for EP/IL-1 receptors should be fruitful. If such receptor complexes do exist they may set in mo-

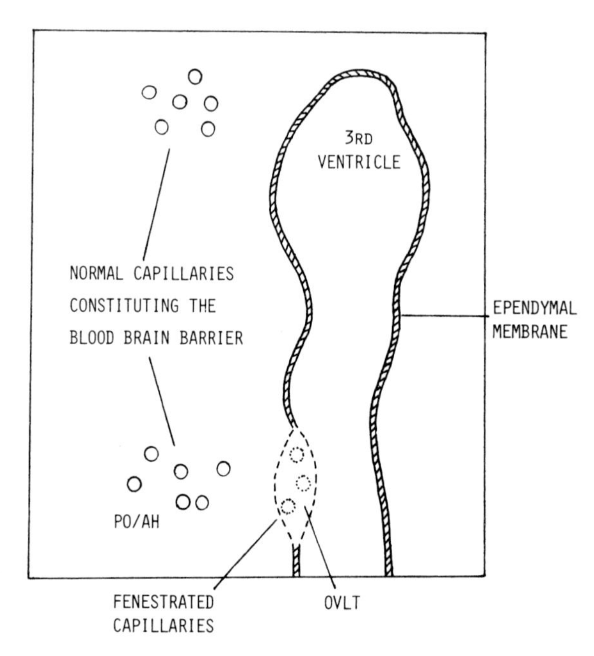

Fig. 5. Diagrammatic representation of blood-brain barrier, constituted by walls of normal capillaries and the fenestrated capillaries of the OVLT through the walls of which EP/IL-1 may reach the thermoregulatory centre in the PO/AH.

tion the metabolic events, including protein synthesis, which finally lead to fever. If these events involve rapid breakdown of EP/IL-1, an explanation may be available for the failure to demonstrate labelled EP/IL-1 in any part of the brain after its I.V. injection (Feldberg and Milton, 1978). An alternative explanation for this failure would be that contact of EP/IL-1 with its receptors is transitory and that it returns to the blood without appreciable concentration in the PO/AH area. Despite failure to locate EP/IL-1 in the PO/AH area, the view that 'LP easily gains access to the hypothalamic thermoregulatory centre' (Dinarello, 1982), is widely held.

6. Other inducers and inhibitors of fever

As already indicated, the metabolic events which lead to the fever response are still largely unknown. Only two general statements can be made with any degree of confidence. These are firstly that EP/IL-1, or alternatively a molecule with similar properties to EP/IL-1 such as epidermal cell-derived thymocyte-activating factor (ETAF) (Sauder et al., 1982), or a metabolic product derived from such a substance is likely to play an important role. The second well established fact concerns the importance of PG or other molecules derived from arachidonic acid. Much of the information concerning the metabolic events involved in the fever response has been obtained using endotoxin injected both ICV and I.V. It is of special interest to examine the available evidence as to how far, and in which circumstances, the formation of EP/IL-1 occurs as a stage between injection of endotoxin and the development of fever. An important early contribution was made by Atkins and Wood (1955) who employed rabbits, tolerant and non-tolerant to endotoxin, in order to distinguish the degree to which fever induced by I.V. injection of endotoxin is the result of endotoxin remaining in the circulation, and how far the fever is due to newly formed EP. They achieved their results by removing plasma from rabbits at various times after endotoxin injection and re-injection of this plasma into tolerant and non-tolerant rabbits followed by measurements of the resulting fever. They found that at times later than 30 min, subsequent to endotoxin injection, the transferred plasma contained only EP. This finding does not wholly explain the reason for the well known diphasic fever response of rabbits which have been injected with endotoxin. However, it is in accord with the concept that the fever, after endotoxin, which lasts much longer than that after EP, is due to continued formation of EP.

The technique of plasma transfer has also been used to investigate the results of injection of *N*-acetylmuramyl-L-alanyl-D-isoglutamine, (muramyl dipeptide). This synthetic adjuvant has been found to induce fever in rabbits. When, however, the animals had been pretreated with indomethacin, fever did not occur but plasma from such an animal, if transferred after dialysis, was shown to lead to fever in the recipient rabbit (Parant et al., 1980). This experiment not only demonstrated the capacity of muramyl dipeptide to bring about formation of EP in vivo, but also indicated that in the absence of PG fever does not occur. In this regard it is important to note that the absence of fever need not necessarily mean that EP has not been formed. Thus, in rabbits restrained in a prone position, Sheagren and Wolff (1966) showed that endotoxin injection did not lead to fever. However, plasma from these rabbits, when injected into unrestrained rabbits, did lead to fever. The absence of fever in the prone rabbits has been explained as being due to restricted conservation of body heat in rabbits restrained in this way.

Indomethacin has also been shown to prevent the fever response due to *S. dysenteriae* (S.dys) in mice (Poole et al., 1984). In these animals, EP/IL-1 gives a fever maximum at 20 min and *S. dysenteriae* at 30 min. This result is of interest because the maximum response to *S. dysenteriae* occurs early, and at a time only 10 min later than the response to EP/IL-1. Whether formation of EP/IL-1 after endotoxin injection, is sufficiently rapid in mice to account for a maximum response at 30 min is not yet certain. Comparison with rabbits and rats, both species in which endotoxin fever occurs later, Splawinski et al. (1977) suggest the existence in mice of a different metabolic pathway. Of special interest is the question whether this other pathway involves EP/IL-1 or alternatively whether endotoxin can itself act directly on the cells of the PO/AH area in such a manner that fever results. If EP/IL-1 is not responsible for PG formation, it would be necessary to postulate that endotoxin its itself directly responsible. While this has been shown to occur with brain slices in vitro (Dinarello and Bernheim, 1981), evidence for similar pathway in vivo is not yet available. The question as to whether EP/IL-1 can cause fever by any pathway not involving PG is discussed in Section 3.

Whether or not a particular PG may be an obligatory stage between IL-1 and fever, there seems little doubt that a product of the cyclooxygenase pathway from arachidonic acid must be involved. Furthermore, since no other protein molecule secreted as a result of endotoxin treatment of isolated cells has been shown to induce fever, the importance of IL-1 as inducer of fever seems to be firmly established.

Some further information regarding the importance of EP/IL-1 as the main cell produced fever inducer, has been obtained from experiments in which endotoxin or EP/IL-1 have been injected ICV or into the area of PO/AH. As reported by Cooper et al. (1967), fever followed at 8 min after injection of EP/IL-1 into the PO/AH area, whereas 24 min elapsed when endotoxin was used. An explanation for the longer period after endotoxin may be that, as at any other injury site, endotoxin causes accumulation of macrophages which, in this case, would be in the capillaries of the PO/AH area. If indeed evidence can be found for macrophage localisation in the neighbourhood of the PO/AH area, an analogous situation will have been demonstrated to that of the Kupffer cells in the liver. As long ago as 1968, Dinarello et al. were able to show that these cells, when appropriately stimulated, are able to form EP/IL-1. As a consequence of activation by the endotoxin escaping from the PO/AH area and passing into the capillaries, both EP/IL-1 and numerous other macrophage-derived mediators would be formed. Conditions would then exist in which these mediators, presumably including PG with relatively short half-lives, could diffuse back into the cells of the PO/AH area.

Evidence for the involvement of EP/IL-1 in both phases of the biphasic fever occurring in rabbits which have been injected with endotoxin has already been mentioned. A biphasic fever, due to endotoxin, also occurs in newborn guinea pigs. Using ICV injection of E.coli endotoxin, Szekely (1978) has investigated the effect of pretreatment with indomethacin. The result clearly differentiated the two fever peaks in respect to requirement for products derived from arachidonic acid. Thus the first fever peak was found to have disappeared completely as a result of the indomethacin treatment, whereas the second peak was retained and appeared at a slightly later time. Thus like the inhibitory effects of indomethacin in respect to fever in mice, rats and rabbits, the first fever peak in the guinea pig has been shown to require PG and presumably therefore the production of PG by IL-1. Because the 2nd fever peak in these animals was found to be unaffected by indomethacin, it would appear likely that an entirely different mechanism is involved. The possibility that endotoxin can itself effect the firing rate of the thermosensitive neurones deserves investigation using isolated cells as described by Hori (1980).

7. Evidence for secondary thermoregulatory areas; neuronal firing rate and fever induction

By whatever mechanism EP/IL-1, or molecules derived from it by proteolysis may achieve passage through the blood-brain barrier, an explanation remains to be found for its ability to initiate fever when injected into the PO/AH area but not into other areas of the brain (except *medulla oblongata*). Is the failure to induce fever in the other areas of the brain attributable to non-production in these areas of a final mediator, perhaps a PG with a very short half-life, or is the most important difference between the PO/AH and other brain areas solely a difference of the neurones?

The existence of important differences between neurones of the PO/AH and those in certain other thermoresponsive areas have been demonstrated. Thus as shown by Barker and Carpenter (1970) thermoresponsive neurones, i.e. neurones which show a changed firing rate on warming or cooling, have been found in the sensorimotor cortex of the cat, an area not known to have a thermoregulatory function. As well as those brain areas apparently lacking thermoregulatory function, there are other areas, into which microinjection, e.g. with PGE, brings about a fall in temperature. This finding, using unanaesthetised rats, is specially noteworthy because a temperature increase of more than 2°C was obtained when PGE_1 was injected into the PO/AH, whereas a fall of just over 2°C followed injection of the same dose into the *medulla oblongata* (Lipton et al., 1973).

Apart from the experiments with brain slices already described, the effects of EP/IL-1, purified to the stage now referred to as IL-1, on different brain areas have yet to be investigated. Of particular interest will be alteration of neuronal firing rates, if indeed such are found to occur, in secondary thermoregulatory areas such as the *medulla oblongata*. In the absence of such data, the changes in firing rates of thermopositive and thermonegative neurones of the PO/AH induced by microinjection of EP/IL-1 are of most interest. As shown in Fig. 6, a single thermopositive neurone in the PO/AH area of a cat brain at 15 min after identification as such, by warming, had returned to its normal firing rate. Microinjection of EP/IL-1 near the neurone then led to a depression of its firing rate. Fourteen out of 16 thermopositive neurones responded in this way, giving a mean fall of firing rate of 51% (Schoener and Wang, 1975). The involvement of PG in such responses was indicated by their being antagonised when sodium acetylsalicylate was injected after the PG. These results suggest that EP/IL-1 itself, or a metabolite derived from or induced by EP/IL-1, lead to rapid changes of firing rate of neurones in the PO/AH area

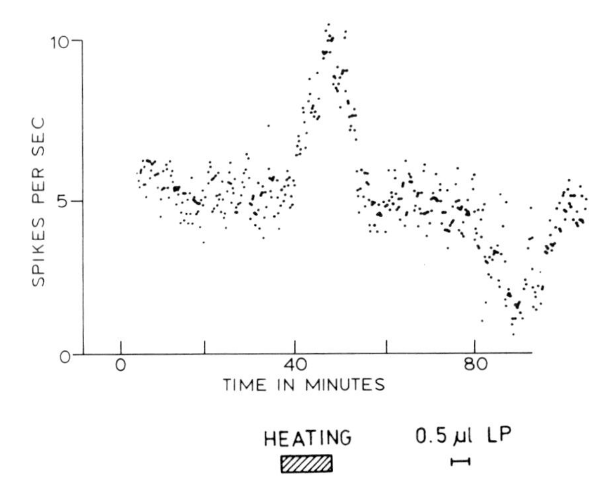

Fig. 6. PO/AH neuron responding with increased discharge to small rise of subcutaneous temperature. Leukocyte pyrogen (LP), 0.5 µl, microinjected near the neuron caused a depression of the firing from control rate. Recovery was complete within 15 min. Time of injection indicated by bar length (Schoener and Wang, 1975).

which in turn are sufficient to bring about the vascular and respiratory changes which lead to fever. It is noteworthy that this work involving single neurones, demonstrates sensitivity at least in the PO/AH area to stimuli of two different kinds i.e. thermal and chemical by the same neurone.

A necessary step towards an understanding of the difference between the cells of the thermoregulatory and those of the non-thermoregulatory brain areas would seem to be similar experiments to those of Schoener and Wang (1975), to ascertain whether the non-thermoregulatory areas may lack sensitivity to one or the other of these types of stimuli. Absence of sensitivity to either type would help to explain the non-function of these areas as thermoregulatory centres. However, it must be emphasized that absence of nervous pathways to effector areas would also constitute an explanation of non-function. The finding mentioned above (Dinarello, 1981), that EP/IL-1 does react with non-thermoregulatory brain areas, is a preliminary indication that absence of certain essential nervous pathways will be found to be most important. Further advance towards an understanding of the functioning of thermoregulatory brain cells can be expected to require comparisons of these cells with those of non-regulatory areas in respect to the types of stimuli mentioned above. In such experiments, measurement of firing rates, and when appropriate, changes in body temperature will also be necessary. Of particular interest, should be results obtained by microinjection of the smaller pyrogen derived from human leucocytes and shown to induce fever in mice but not in rabbits (Gordon and Parker, 1980). Thus injection of the smaller pyrogen and the EP/IL-1, from which it was derived, into rabbits might indicate whether the same brain areas are affected and whether the non-function of the smaller pyrogen is due to failure to be taken up by receptors with slightly different properties from those in mice.

References, p. 132.

Gordon/Koj (eds.) The acute-phase response to injury and infection.

CHAPTER 6

Metabolic changes in other organs following intracerebroventricular injection of endogenous pyrogen/interleukin 1

A.H. GORDON

Although the mechanisms involved have not yet been revealed, there is adequate evidence that ICV injection of crude EP will induce the acute-phase response (AP-response). Using such material, obtained by incubation of 17.5×10^6 rabbit peritoneal cells, injection into rabbits led to increases in C-reactive protein (CRP) and haptoglobins which where respectively 36 and 10.5 times higher when the ICV rather than the I.V. route was used (Turchick and Bornstein, 1980). This work confirmed and extended earlier results in which rabbit EP had been injected into rats (Bailey et al., 1976). Evidence showing the extreme sensitivity of the PO/AH area to EP injected ICV in guinea pigs has been obtained by Blatteis et al. (1983). Using only 1/1000th of the amount required to give a comparable response when given I.P. increases in plasma sialic acid and Cu of 34 and 48%, respectively were obtained. Turchick and Bornstein (1983) also reported small reductions in the numbers of granulocytes and plasma Fe concentrations after ICV injection of EP. These effects were considered as possibly due to leakage of EP from the I.V. space and therefore, unlike the increases in CRP and haptoglobin, provide insufficient evidence for stimulation via a neural pathway. Strong evidence for the involvement of neural pathways controlling the AP-response and increases in plasma copper has been obtained by ablation of the OVLT. As described on p. 78, the fenestrated capillaries in this brain area can permit passage of mediators such as EP/IL-1 from the blood into the PO/AH brain area. It is of considerable interest that following I.P. injection of endotoxin and thus the intravascular release of EP/IL-1 the effect of the ablation was limited to fever, release of AP-proteins and plasma copper. This may be explained as suggested by Blatteis et al. (1983) as being due to EP/IL-1 action on neutrophils leading to release of lactoferrin, and a parallel mechanism for Zn. The above evidence for the existence of neural pathways which can effect the AP-response must not be taken as proof that these pathways constitute the only means by which hepatic synthesis of AP-proteins are controlled. This is because experiments with isolated hepatocytes, (cf. Chapter 17) have demonstrated direct effects of macrophage-derived mediators on these cells. Another finding of Blatteis et al. (1983) that ICV injection of PGE_2 did not lead to increased plasma sialic acid or Cu is also relevant. Although ICV injection of EP into rabbits, rats and guinea pigs has been shown to stimulate synthesis of several AP-proteins and possibly to bring about other changes characteristic of the AP-response, little is yet known about the nervous connections involved except that the liver is well supplied with adrenergic and with parasympathetic fibres, some of which terminate in close proximity to hepatocytes and to Kupffer cells (Forsmann and Ito, 1977). Whether impulses arriving at Kupffer cells may lead to local production of EP/IL-1 which could then rapidly diffuse into the hepatocytes, or whether there are other chemical mediators derived from nerve endings closer to the hepatocytes, remains to be discovered. In either case, there is a need to investigate the relative importance of nervous pathways from brain to liver and direct humoral stimulation by EP/IL-1 as means of control of the AP-response.

The importance of particular brain areas in relation to responses which are mainly under humoral control is known for both antibody and cell-mediated immunity (Stein et al., 1976; Jankovic and Isakovik, 1973). Thus, by analogy, control involving nervous pathways may contribute to an important degree in the AP-response.

References, p. 132.

Gordon/Koj (eds.) The acute-phase response to injury and infection.

CHAPTER 7

Responses of the immune system to interleukin 1

A.H. GORDON

Evidence is now available that IL-1 affects most, if not all, aspects of the immune system, certainly including both humoral and cellular responses. The effects of IL-1 are indirect and result from its ability in appropriate circumstances to bring about synthesis of IL-2 by mature thymocytes and also by certain T cells after sensitisation by antigens or lectins. Before describing the conditions in which IL-1 is formed and its relationship to antibody synthesis by B cells and cellular immune effects, it is important to emphasise that most results have been obtained from cells maintained in vitro. The mass of information thus obtained, almost all resulting from experiments carried out since the characteristics of IL-1 were first appreciated, has illuminated in considerable detail the nature of the interrelation between macrophages and some of the many types of T cell. Evidently this information can aid towards fuller understanding of the immune response as it occurs both in vitro and in vivo. Indirect effects of IL-1 on the immune response due to effects of AP-proteins are discussed in Chapter 13.

1. Interrelationships of IL-1 and PG

Consideration of the far more complex responses which result when IL-1 is formed in vivo ideally take into account the effects of all other mediators. Certain of these will affect the metabolism of B and T cells and thus alter their responses to IL-1. Of primary importance in this respect are the PG. Detailed in vitro studies have shown that in different conditions, PGE_2 will either stimulate or inhibit particular aspects of B- or T-cell function. As shown by Goodwin et al. (1977), PG present at 10 – 1000 ng/ml is sufficient to cause suppression of human T-cell mitogenesis. This concentration of PGE_2 is similar to that occurring in inflammatory exudates. However, whether inhibition of B- or T-cell function actually occurs, has been shown to depend on more complex interrelationship between PG and other macrophage-derived factors.

Considering only the relationship of IL-1 and PG the following facts seem to be most important. Firstly, IL-1 has been shown to bring about increased production of PG when incubated with cells as different as those from brain (Dinarello and Bernheim, 1981) synovium (Dayer, 1981) and muscle (Baracos et al., 1983). In addition, and of special interest, was the finding that IL-1 stimulates PGE_2 formation from macrophages which, when appropriately stimulated themselves, form IL-1 (Dinarello et al., 1983). Secondly, IL-1 has been shown to stimulate proliferation of T cells which respond differently from other cells in that instead of producing PG they have been shown to be inhibited, e.g. by PGE_1 (Fontana, 1982). As shown in Fig. 1, a system of this kind could in theory display negative feedback. Whether any such interrelationship actually exists in any part of the body must depend on the anatomical relationships of the various cell types and also on the half-lives of the PG involved. If the existence of PG-producer cells could be demonstrated in lymph glands, the possible importance of a system such as that indicated in Fig. 1 would become more likely.

The finding that prostaglandins of the E series inhibit many in vitro measurements of immune function, such as direct cytolysis in murine lymphocytes, haemolytic-plaque formation by murine leukocytes and mitogen-induced stimulation of murine lymphocytes (Goodwin et al., 1977), needs qualification in view of the results obtained by Stobo et al. (1979). Using density gradient centrifugation of human T cells, separation into cells with opposite characteristics in respect of PGE_2 was achieved. The cells with the lower densities, constituting 20 – 30% of the total, in presence of PG, gave increased mitogenic responses to PHA. Cells with higher

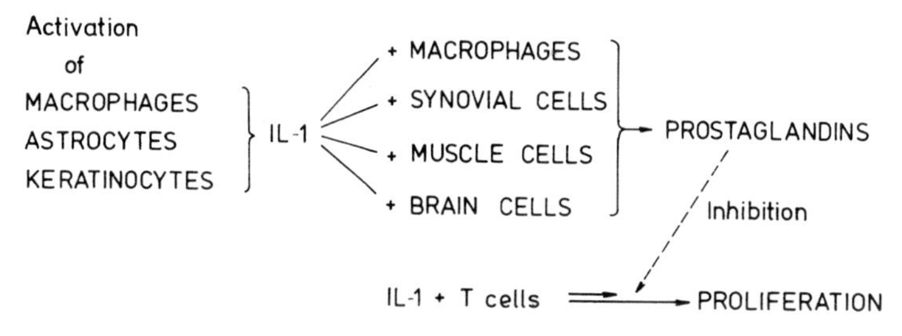

Fig. 1. Potential inhibitory effect of prostaglandins originating from diverse cells on the immune system.

densities, on the other hand, were inhibited by 60%. In this work, although IL-1 was not added, the effect of the PHA (increased incorporation of [^{3}H]thymidine) is likely to have been due in part to the presence of macrophages or IL-1. Thus the existence of subclasses of T cells sensitive to stimulation by IL-1, and with negative responses to PG, seems very probable. Without doubt, further investigation will reveal even more intricate relationships between IL-1 and PG. The above description should however be sufficient to indicate the importance of IL-1 in respect to immunoregulation and also some of the ways by which its functioning can be modulated by PG.

Lymphoid tissue, wherever it occurs, is affected by both IL-1 and IL-2. As already mentioned, the formation of IL-2 in lymphoid tissues does not occur unless IL-1 or an IL-1-like factor is present. As will be described, IL-1 affects thymus, spleen and other lymphoid organs and most important, at the cellular level, has different effects on the T and B cells of the spleen.

2. *Formation of immunocompetent cells in the thymus*

It is convenient to consider first the interaction of IL-1 with the thymus. This interaction is unique because it is here that IL-1 comes into contact with pre-T cells arriving from bone marrow and foetal yolk sac. During this passage through the thymus these cells become immunocompetent. This occurs in the special environment of the thymus which is isolated from the blood by a barrier similar to the blood-brain barrier. During passage through the cortex, the pre-T cells are modified in several important respects, thus they change from being unaffected by polyclonal mitogens such as PHA to sensitivity to such mitogens. When this has occurred, if IL-1 is also present, rapid proliferation will occur. If considered as a whole, the mouse thymus has been shown to consist of approximately 90% of these immature, immunoincompetent cortical cells which are sensitive to cortisone and agglutination by peanut lectin from *Arachis hypoglaea* (Conlon et al., 1982). Thus only 5 – 10% of thymocytes, most of which are concentrated in the medulla, are insensitive to cortisone, are not agglutinatable by peanut lectin and, as just mentioned, are sensitive to polyclonal mitogens. In presence of PHA, these cells are extremely sensitive to IL-1 and make possible the widely used costimulator assay for this substance (Smith 1984). Mature thymocytes are very similar to splenic T cells and can be subdivided into (Lyt-1$^-$23$^+$), cytotoxic suppressor cells and Lyt-1$^+$23$^-$, helper cells according to their surface phenotype (Jandinski et al., 1976).

There is also an important difference between isolated thymocytes and isolated splenocytes in that only the latter cells can be maintained in culture for more than a very short period. For this to be possible the cells must have been exposed to antigen or lectin and in addition IL-2 must be added at intervals. This is despite the fact that these cells can, when appropriately stimulated, manufacture IL-2. Smith et al. (1980) have shown that the comitogenic effect of IL-1 on thymocytes is mediated by its ability to induce IL-2 production by thymocytes. However, because addition of IL-2 to thymocytes does not make possible long-term culture, these cells appear to require a further factor not required by splenic T cells. Evidence concerning the interaction of IL-1 with immunocompetent cells from thymus, spleen and lymph glands, has been obtained using many different in vitro systems of which the following five have yielded the clearest results. The fact that a requirement for IL-1 has been identified in so many in vitro systems suggests, but does not prove, that IL-1 will be found to be of importance in vivo wherever similar cell interactions occur.

3. In vitro systems in which IL-1 plays a part

The following cellular in vitro systems have been identified as being IL-1-requiring. They may be subdivided into those which are antigen specific and those which are not. Many other IL-1-requiring systems have also been described, e.g. Chu et al., 1984.

3.1. Thymocyte costimulator assay (Mizel et al., 1978)

This assay takes advantage of the presence in the murine thymus of 5 – 10% of immunocompetent cells. As mentioned above, these cells are responsive to the presence of polyclonal mitogens such as PHA. When incubated in the presence of such mitogens, in growth media such as that described in Chapter 23, Section 3.1, uptake of [^{3}H]thymidine added after 2 days is increased by 2 – 3 times. This occurs when PHA concentrations of 1 – 2 μg/ml are used (Fig. 2). At higher PHA concentrations much greater [^{3}H]thymidine incorporation takes place. PHA concentration of 1 – 2 μg/ml has been found to provide optimum conditions for assay of IL-1, because in the presence of PHA and IL-1, [^{3}H]thymidine incorporation increases to 10 – 20 times that which occur in the absence of IL-1. In such circumstances, [^{3}H]thymidine incorporation is proportional to the log concentration of IL-1.

In certain circumstances, IL-1 in the absence of any lectin may itself stimulate thymocytes but as yet the relationship between the direct mitogenic and the augmenting effects of IL-1 are unclear. Higher levels of IL-1 result in both direct mitogenicity and synergistically enhanced thymocyte responses to lectins. At lower dilutions, the same preparations of IL-1 can potentiate only the mitogenic response of thymocytes. In contrast, partly purified murine IL-1 has no direct mitogenic effects.

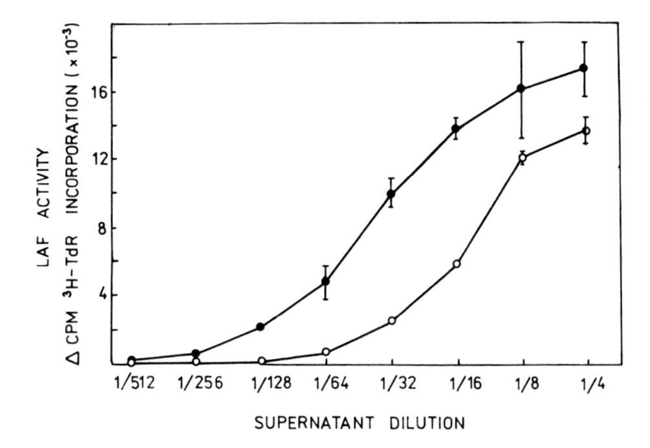

Fig. 2. Concentration dependence of lymphocyte activating factor (LAF)-stimulation of mouse thymocyte proliferation by supernatant from P388D_1/LNL cultures. Thymocytes were incubated for 72 h in medium supplemented with dilutions of a LAF-containing supernatant with (●—●) or without (○—○) 1 μg/ml PHA. The levels of [^{3}H]TdR incorporation in control cultures supplemented with only RPMI 1640 or PHA were 159 ± 10 and 674 ± 93 cpm, respectively (Mizel et al., 1978).

Whether IL-1 can function as a second amplifier signal has been investigated using cyclosporin A which was found to block the potentiation of PHA-induced thymocyte proliferation but not the direct mitogenic effects of IL-1 (Oppenheim and Gery, 1983). Because antigen-antibody complexes affect thymocytes in a manner rather similar to PHA or other polyclonal mitogens, thymocytes from mice that have been exposed to infection may show little or no PHA effect. IL-1 assay using such cells, and without addition of PHA, is sometimes still possible.

It has been mentioned above that IL-1 plus lectin or antigen brings about formation by certain thymocytes of IL-2. Because this occurs, it is important to note that the thymocyte costimulator assay, at least as usually conducted, is not specific for IL-1. The manner of which IL-2, if added as such, influences the uptake of [^{3}H]thymidine is too detailed for discussion. It is sufficient to note that most of the immature cortisone-sensitive thymocytes, which do not respond to PHA when this is added alone, do respond to PHA when IL-2 is also present (Conlon et al., 1982). As already mentioned, mature thymocytes can be divided into, on the one hand, cytotoxic or suppressor cells, Lyt-1^-23^+ and, on the other, helper cells, Lyt-1^+23^-. The former cells respond to IL-2 although they do not produce it and may be presumed to play some part in the thymocyte costimulator assay. On the other hand, the helper cells both respond to, and produce, IL-2 and therefore must also respond in this assay. The importance of IL-2 as a stimulator of [^{3}H]thymidine incorporation is made clear if a Con A treated spleen cell supernatant is added to the cells (cf. Section 3.3). Usually a very high [^{3}H]thymidine incorporation will then occur. The use of such a supernatant known to give high incorporation is of value as a check that conditions of thymocyte viability are fully satisfactory.

Since the thymocyte costimulator assay does not require antigen pretreatment it is, of course, antigen non-specific. Exceptionally, however, if IL-1 is generated in presence of the thymocytes by macrophages bearing antigens, antigen specificity is introduced.

3.2. Antigen presentation by human monocytes to T cells

In the complete absence of monocytes neither antigens nor certain mitogens such as the polyclonal stimulant, *S. aureus* protein A (SpA), can by themselves induce proliferation of T cells. N.B., however, that this failure to stimulate does not include all polyclonal mitogens, e.g. Con A (Resch et al., 1981). In the forgoing circumstances, using SpA as shown by Scala and Oppenheim (1983), the addition of

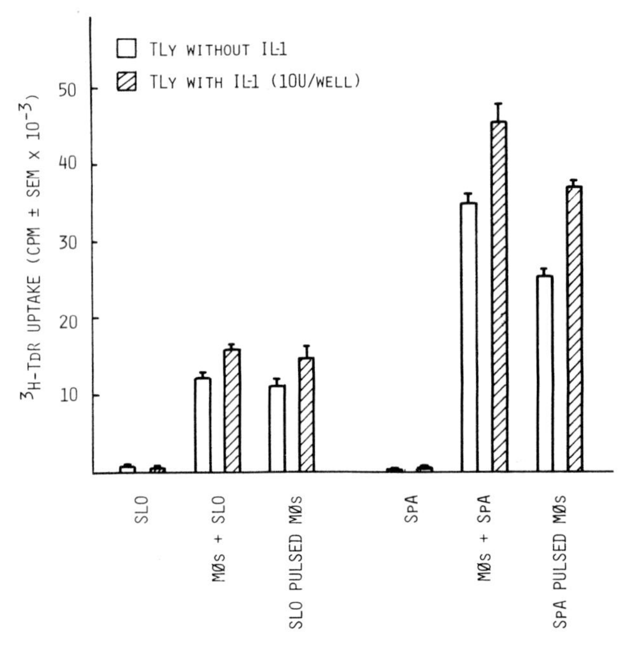

Fig. 3. Monocyte requirements for T-lymphocyte proliferation. 5×10^5 purified T-lymphocytes were cultured with 5×10^4 autologous monocytes in the presence of SLO (1:20 final dilution) or SpA (10 μg/ml). In pulsing experiments, monocytes were pulsed for 2 h with the above concentration of stimulants and washed three times to remove unbound stimulant. MØs = human peripheral monocytes; SLO = soluble streptolysin antigen; SpA = polyclonal stimulant *S. aureus* protein A; TLy = T lymphocytes. (Scala and Oppenheim, 1983.)

IL-1 has no effect on the proliferation response as indicated by uptake of [^{3}H]thymidine (Fig. 3).

For T-cell proliferation to occur it is sufficient for monocytes to have been in contact with the antigen or mitogen. Even after complete removal of unadsorbed stimulant by repeated washing, such cells are able to induce T-cell proliferation. As shown in Fig. 3, the degree of such proliferation is slightly increased if IL-1 is also present. The effect of IL-1 is much more marked if after contact with antigen or mitogen the monocytes are treated with 0.03% paraformaldehyde. In these circumstances, proliferation of T cells occurs only if IL-1 is also present (Fig. 4).

The most likely explanation for these events depends on monocytes having several

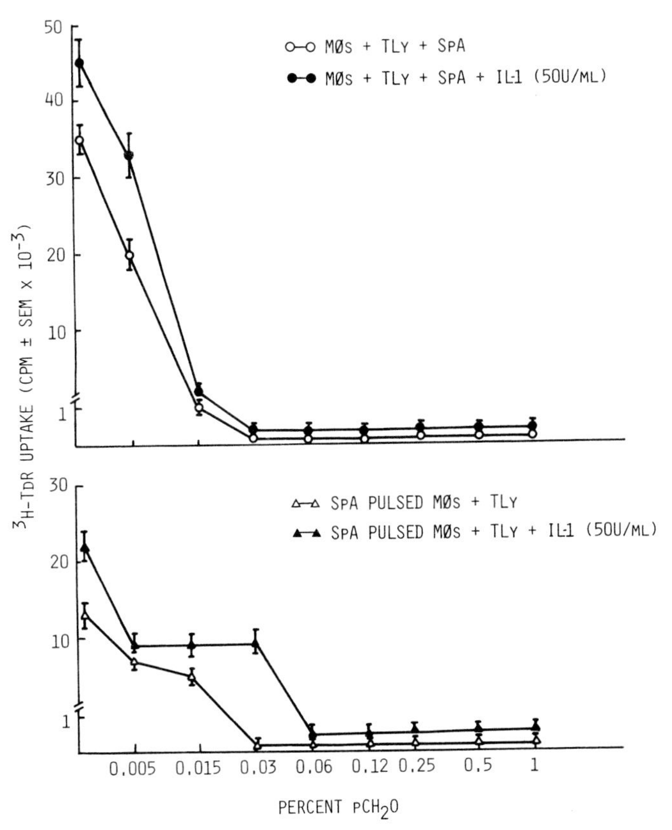

Fig. 4. Effect of paraformaldehyde treatment. Monocytes were treated with graded doses of paraformaldehyde (pCH_2O) before (upper panel) or after 2 h pulsing (lower panel) with (10 μg/ml) SpA. The capacity of these treated monocytes to induce lymphocyte proliferation was then assessed from the [^{3}H]TdR uptake in cpm by the cultured cells MØs = human peripheral monocytes; SpA = polyclonal stimulant *S. aureus* protein A; TLy = T lymphocytes. (Scala and Oppenheim, 1983.)

functions. When soluble antigens are involved, they are internalised and processed by the monocytes and subsequently presented to the T cells. At the same time, some IL-1 is formed. If the cells are treated with 0.03% paraformaldehyde after uptake of the antigen, processing must already have reached a stage at which the effective presentation can occur, but at which IL-1 can no longer be formed. Thus, for proliferation to occur, IL-1 must then be added as such. If the monocytes are treated with 0.03% paraformaldehyde before contact with the antigen, T-cell proliferation cannot be induced even in presence of IL-1. Presumably such cells cannot internalise and process antigen as is required for presentation to the T cell or form IL-1. Indeed, in a separate experiment, Scala and Oppenheim (1983) showed that a concentration of 0.03% paraformaldehyde is sufficient to prevent formation of IL-1. However, for recent evidence of IL-1 activity by formaldehyde-treated macrophages, cf. Chapter 2, Section 3.2.

These experiments are strong evidence that IL-1 constitutes an essential second signal without which antigen or mitogen-driven proliferation of T cells cannot take place.

A similar conclusion derives from the work of Bendtzen and Petersen (1984) who employed antigen-induced production of the lymphokine, leucocyte migration inhibitory factor, as a measure of T-cell activation. They demonstrated that after killing by heating to 56°C for 60 min, human monocytes, in the presence but not in the absence of IL-1, even if pulsed with antigen before killing, were able to induce T-cell production of the lymphokine.

3.3. Splenic T cells cultured with Con A as producers of IL-2

That production of IL-2 by splenic T cells requires the presence of either macrophages or IL-1 has been demonstrated by Smith et al. (1980). Thus when splenic T cells, which had been deprived of macrophages by passage through nylon wool, were cultured in presence of Con A, only minimum amounts of IL-2 were produced. However, when IL-1 was also present, large amounts of IL-2 were formed. Similar amounts of IL-2 were obtained using cortisol-resistant thymocytes but not with spleen cells from nude mice. This system is antigen non-specific.

3.4. Proliferation of primed lymph node lymphocytes induced by antigen

As described above, in the conditions employed in Section 3.7, IL-1 and antigen are both required for activation of B cells and subsequent development of the plaque-forming cell (PFC) response. A similar requirement for IL-1 has been demonstrated by Mizel and Ben-Zvi (1980), using primed lymph node lymphocytes and ovalbumin (OVA) as a soluble antigen. In these experiments, uptake of [^{3}H]thymidine was used as the indicator of B-cell activation. In Table 1 is demonstrated the requirement either for macrophages or for semi-purified IL-1 in presence of OVA and primed T cells. The results given in Table 2 demonstrate the requirement for antigen primed T cells. If these cells are removed using anti-Thy 1 serum, even in presence of OVA, uptake of [^{3}H]thymidine is very low. Following incubation with and without anti-Thy 1.2 serum and complement, 4 × 10^5, viable cells were plated in the presence and absence of OVA ± IL-1. The results in Table 3 indicate that antigen-mediated enhancement of [^{3}H]thymidine incorporation in this system is proportional to IL-1 concentration.

The method for preparation of the lymphatic node cells (LNC) was as follows.

Separation of LNC from popliteal lymph nodes from lymph node adherent cells was achieved by pressing the tissue through 60 gauge wire mesh into RPMI 1640 medium. After removal of cell debris by settling, the LNC remaining in the medium were centrifuged and resuspended in fresh RPMI 1640 containing 10% heat inac-

TABLE 1
Adherent cell dependence of antigen-induced lymph node lymphocyte (LNL) proliferation

Cells	cpm [^{3}H]TdR incorporation	
	− OVA	+ OVA[a]
LNC[b]	5507 ± 1124	12,875 ± 1181
LNL[c]	2170 ± 272	2,660 ± 346
LNL + LAF[d]	1660 ± 442	19,483 ± 566

[a] LNL at 10 – 12 days after the mice had been immunised with OVA.
[b] 4 × 10^5 LNC/well.
[c] 4 × 10^5 LNL/well (nylon wool, plastic non-adherent lymphoid cells).
[d] LAF/IL-1 isolated from murine 388/D$_1$ cells.

TABLE 2
T-cell dependence of OVA plus IL-1-induced LNL[a] proliferation

Additions	Anti-Thy 1.2[b]	[^{3}H]thymidine incorporation cpm
None	−	115 ± 25
OVA	−	160
OVA + IL-1	−	9915 ± 865
OVA + IL-1	+	430 ± 140

[a] LNL at 10 – 12 days after the mice had been immunised with OVA.
[b] Addition leads to removal of T cells from spleen cells.

TABLE 3
Concentration dependence of IL-1-mediated enhancement of antigen-induced proliferation of LNLs

Concentration of IL-1[a] (units/ml)[c]	△ cpm [^{3}H]TdR incorporation[b]
1.8	4,683 ± 916
7.1	7,720 ± 184
15.3	9,442 ± 835
30.5	13,367 ± 752
61	13,909 ± 512

[a] 4 × 10^5 LNL from mice at 10 – 12 days after immunisation with OVA were incubated with 30 μg/ml OVA in the presence and absence of IL-1.

[b] △ cpm = cpm (LNL + OVA + IL-1 – cpm (LNL + OVA). In the absence of OVA and IL-1 LNL incorporated 308 ± 52 cpm and with OVA, 2564 ± 542 cpm. Assay conditions were similar to those described for assay of IL-1 using murine thymocytes (cf. p. 292).

[c] The unit of LAF/IL-1 is the amount needed to give half the maximum [^{3}H]TdR incorporation in the assay conditions employed.

tivated human serum and mercaptoethanol. Adherent cells were then removed in two stages, passage through nylon wool and incubation for 1 h in plastic dishes at 37°C. The non-adherent cells thus obtained, gave the results shown in Tables 1 and 3. In Table 2, some of the cells were also treated with anti-Thy 1.2 serum which removed approximately 93% of the T cells.

Cell preparation of this kind obtained at 10 – 12 days after immunisation with a soluble antigen were plated at 2 × 10^6 cells/ml in Falcon microtest tissue culture plates and incubated with and without antigen and/or IL-1 for 72 h at 37°C in humidified CO_2. Finally, a 4 h pulse of [^{3}H]thymidine was introduced and after 24 h the cells were harvested, washed and estimated as described above.

3.5. Alloantigen specific cytotoxic response to murine parental T cells in presence of stimulator splenic B cells

The cytotoxicity reaction of murine alloantigen specific T cells is IL-1 dependent. The presence of cytotoxic T cells formed in this reaction can be estimated by their ability to release ^{51}Cr from certain tumour cells labelled with chromium. As described by Farrar et al. (1980), purified parental T cells (C57 BL/6) were prepared by two successive passages through nylon wool. Stimulator splenic F_1 B cells derived from DBA/2 and C57 BL/6 mice were used after passage over Sephadex G-10 and incubation with nylon wool. Some red blood cells (RBC) remaining with the spleen cells served as markers. These procedures removed all but trace numbers of macrophages. The T and B cells thus obtained were mixed using a ratio of 2 B cells per T cell and incubated in culture medium supplemented with 10% foetal calf serum (FCS) and mercaptoethanol. After 5 days in a 5% CO_2 95% air incubator, ^{51}Cr labelled tumour cells were added for an additional 4 h incubation. Finally, the cell supernatant was harvested using an automatic cell washer and ^{51}Cr radioactivity was estimated. As shown in Fig. 5, addition of partially purified IL-1, obtained as $P388D_1$ cell culture supernatant, to macrophage depleted, and therefore substantially IL-1 free, cytotoxic T cells led to complete restoration of the response to that characteristic of unfractionated spleen cells.

3.6. Macrophage-bound antigen, activation of primed T cells and concomitant T cell membrane lipid viscosity change

The mitogenicity of lymphocyte activating factor (LAF) IL-1 for T cells was first reported by Gery et al. (1972). Further work using murine spleen cells has revealed that antigen recognition by primed T cells depends on macrophage-derived products of two kinds (e.g. Puri and Lonai, 1980). These are, firstly, genetically determined products of the major histocompatibility complex, soluble products consisting of antigen which has been processed by the monocytes, associated with the genetically determined factors. Secondly, a non-genetically determined factor which may be presumed to be IL-1 is also required. Only when both are present can T cells proceed

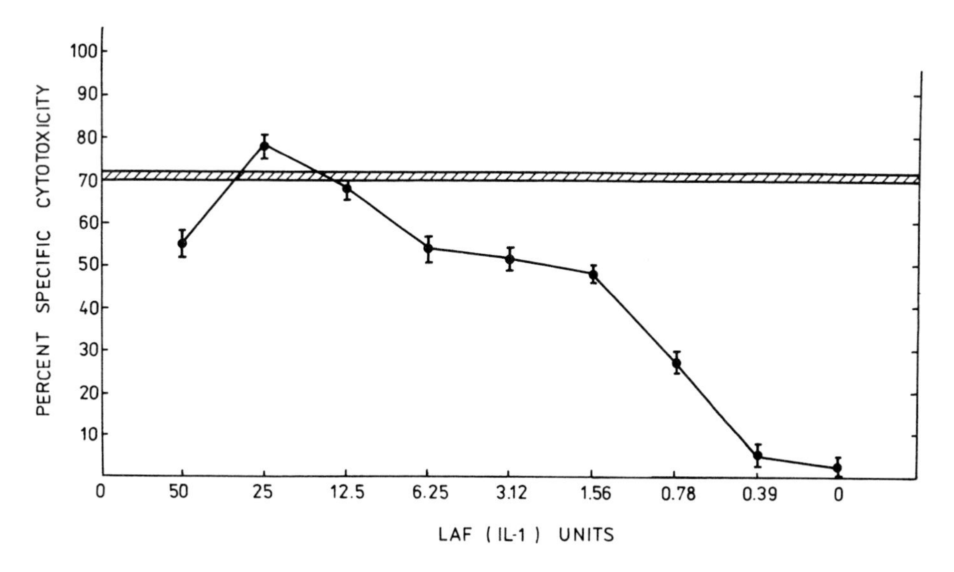

Fig. 5. LAF (IL-1) dose-response of the cytotoxic T-lymphocyte responses. Partially purified LAF (IL-1) was prepared from stimulated $P388D_1$ cultures by successive ammonium sulfate, DEAE cellulose chromatography, and S200 gel filtration. The resulting material was then titrated into macrophage-depleted cytotoxic T-lymphocyte cultures. (Farrar et al., 1980.)

to target cell lysis, cell multiplication, delayed type hypersensitivity or cause B cells to mature to antibody-forming plasma cells. These effects which begin within 2 h are characteristic of T cells and are not given by B cell enriched spleen cells. They involve changes in membrane lipid microviscosity which can be measured by

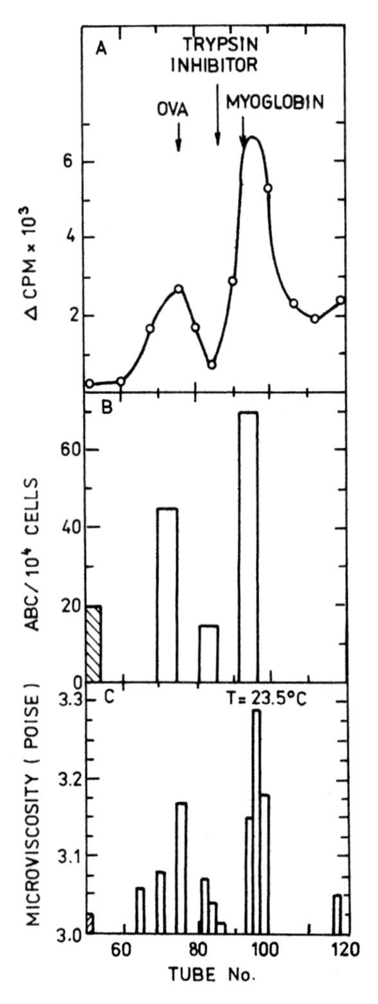

Fig. 6. Gel filtration on Sephadex G-75 of LPS-induced IL-1 containing medium from $P388D_1$ cells. (A) Mitogenicity fractions diluted to ¼ were assayed on thymocyte cultures in the presence of PHA. Incorporation in the absence of IL-1 was 159 cpm. (B) Antigen binding. Tubes corresponding to the main mitogenic peaks, and the intermediate fractions were pooled and tested at 1/60 dilution. [^{125}I]NIP-FGG was used as antigen. (C) Microviscosity. C3H/DiSn T cell-enriched splenocytes were tested after incubation with the indicated fractions at a dilution of 1/60. Shaded columns in (B) and (C) show the untreated control values. NIP-FGG = 4-Hydroxy-3-iodo-5-nitrophenyl acetic acid-14,2-fowl-γ-globulin. (Puri et al., 1980.)

fluorescence depolarisation after addition of 1,6-diphenyl-1,3,5-hexatriene for 1 h (Puri et al., 1980).

That the primary effect of IL-1 is on membrane viscosity was strongly suggested by experiments in which cholesterol was added instead of IL-1. Presumably, as a result of the cholesterol having dissolved in the plasma membrane, both microviscosity and antigen binding were increased just as under the influence of IL-1. Treatment of splenic T-cell cultures with appropriate antisera indicated that the cells which respond to IL-1 are T cells of the Lyt-1^{+}, 23^{-} class (cf. p. 89). In Fig. 6 are shown comparisons of [^{3}H]thymidine incorporation, antigen binding and microviscosity of T cells using individual fractions obtained from crude IL-1 by chromatography over Sephadex G-75.

3.7. Reaction of antigen activated B cells with T-helper cells

As shown by Hoffmann (1980), the differentiation of B cells in the in vitro PFC response to RBC antigens occurs in two distinct phases. In the first, the antigen reactive B cells acquire the ability to interact with helper T cells; for this to take place IL-1 or macrophages must be present. In the second phase the B cells are converted into antibody secreting PFC. This transformation requires the presence either of T-helper cells or mediators such as T cell replacing factor (TRF) that they release. The first phase in which either macrophages or IL-1 must be present lasts for some 40 h. After this IL-1 is no longer required and the response of the B cells depends on T cells or factors derived from them. That more than one T cell-derived factor may be required for transformation of B cells into antibody secreting plasma has been found by Puri et al. (1983). They were able to show that B cells from adult but not from neonatal mice proliferate when treated with B cell growth factor (BCGF). Other factors themselves specific for IgM or IgG secretion, were found to be necessary for B cells from neonatal mice, i.e. B cell differentiation factor (BCDF) as well as BCGF. The effect of both BCGF and BCDF on B cells of such immature mice was restricted to differentiation, and not to stimulation of proliferation. T cells have been shown to function in both phases of this system because not only are they essential for the second phase, but also in the first, in which being antigen-primed, they induce the release of IL-1 by macrophages. According to Hoffmann (1980), substitution of IL-1 for macrophages leads to maximum PFC response. The most probable relationships of the several cell types are shown in Fig. 7. This system is antigen dependent with respect to the B cells. In absence of IL-1, unprimed T cells showed minimal activity, approximately 1/30th of that given by primed T cells.

Further investigation of the manner in which IL-1 assists B cells in the PFC may well reveal that the scheme in Fig. 7 requires further elaboration. Thus, Farrar and Halfiker (1982) refer to three T cell-derived mediators which act on B cells. These

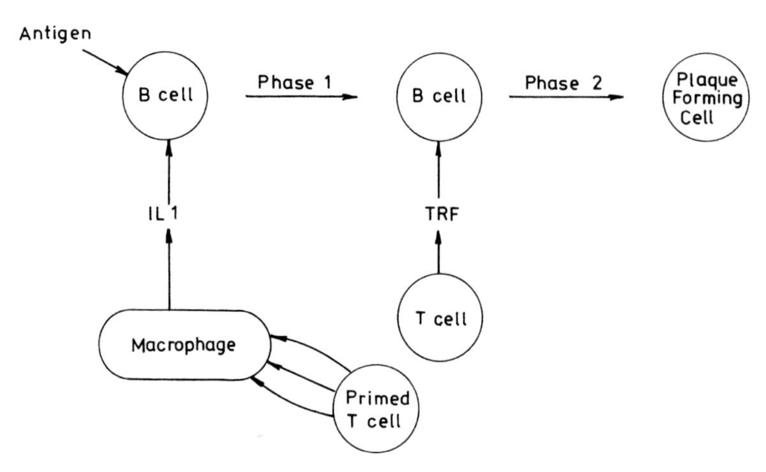

Fig. 7. IL-1 and TRF-controlled phases of B-cell differentiation in the anti-SRBC IgM plaque forming cell response. (Hoffman, 1980.)

are IL-2, TRF and colony stimulating factor, the importance of this latter factor is that it can induce macrophages to form PGs and IL-1. The stimulation of IL-1 production may explain the effect of colony stimulating factor in the PFC response of spleen cell cultures which contain macrophages.

Suppression of the in vitro antibody response such as that demonstrated using serum from acute phase mice raises an interesting question regarding the function of IL-1. This is because the effect of acute phase serum which has been shown by Benson et al. (1975) to suppress the PFC response may be due to inhibition of the effect of IL-1 or to suppression of its formation. Whichever explanation is true, the mechanism involved is of special interest because serum amyloid A (SAA), the plasma protein responsible, is an AP-protein. The existence of an IL-1 inhibitor with very different properties has been identified by Dinarello et al. (1981).

Evidence such as that of Hoffmann (1980) showing that addition of IL-1 can substitute for macrophages in the first phase of B-cell differentiation, does not exclude the possible existence of other factors with similar potentialities. Thus Wetzel et al. (1982) have described a factor prepared from T cells cultured with Ia-positive stimulator cells which functions in the absence of IL-1. Experiments of Lachman and Maizel (1983) in which a B cell stimulating factor from conditioned medium, derived from 72 h lectin-stimulated peripheral blood lymphocytes, was separated from IL-1 and IL-2, are also relevant.

As shown in Table 4, murine B cells, after removal of macrophages by Sephadex G-10 and T cells by antiserum (anti-Thy 1 serum and complement), require IL-1 for reactivity with helper T cells. As also shown in Table 4, before the B cells from

TABLE 4
The frequency of SRBC-reactive B cells responding to T-cell help is increased by IL-1[a]

	IL-1	SRBC reactive B cells per 10^7 cultured cells
–	–	1
–	+	35.00 ± 13.37
TRF	–	34.50 ± 9.95
TRF	+	342.12 ± 95.67
T cells	–	29.14 ± 12.18
T cells	+	358.23 ± 86.35

[a] C57BL/6 spleen cells were passed over Sephadex G-10 columns (for the removal of macrophages) and treated with monoclonal anti-Thy 1.2 and complement (for the removal of T cells). 10^5 SRBC-primed (0.2 ml. 1% SRBC solution i.v., 8 days) spleen cells treated with 40 μg/ml mitomycin C for 30 min, 37°C were added per well as a source of helper T cells. TRF was added in a concentration of 3%, IL-1 5%.

spleens of mice previously immunised with sheep red blood cells (SRBC) can go on to form antibody, a further activation stage is necessary.

4. *Suggested mechanisms involved in the formation of IL-2: the roles of IL-1 and mitogens*

An essential requirement in the above systems, without which cell proliferation cannot occur, is the presence of IL-2. Evidence concerning the role of IL-1 in the formation of IL-2 has already been mentioned. However, according to the cell type in question, factors additional to IL-1 and IL-2 are also necessary. Thus polyclonal mitogens, e.g. the plant lectins, alloantigens and cell supernatant factors such as BCGF and BCDF, some of which are still in need of proper characterisation, act in this way. Thymocytes on the one hand, and splenic and peripheral T cells on the other, differ in respect to the effect of mitogenic lectins. Thus only the mature cell types can be maintained in long-term culture. For the success of such cultures, regular additions of IL-2 are also necessary. Thymocytes are also affected by mitogens such as PHA in presence of which a limited increase in incorporation of [^{3}H]thymidine takes place. Presumably formation of some IL-2 must occur. If this is so, it is insufficient for the cells to be maintained in culture. The important effect of PHA on thymocytes is to sensitise them to IL-1. As mentioned above, when in this condition, addition of IL-1 leads to proliferation and large increases in incorporation of [^{3}H]thymidine. However, IL-2 also leads to increases in incorporation of [^{3}H]thymidine (Fontana et al., 1983), thus proving, as mentioned above, that the thymocyte costimulator assay is not specific for IL-1.

The fact that IL-1 added to thymocytes in presence of PHA and also to T cells leads to proliferation of mature thymocytes and T cells and uptake of [^{3}H]thymidine has been known for several years, but the sequence of changes which bring this about is still unclear. The most probable course of events, at least in T cells, is that IL-1 acting in presence of an antigen or lectin stimulates the expression of surface receptors for IL-2 on certain of these cells and concurrently stimulates the secretion of IL-2 by a different T-cell subset. An unresolved question is whether these two functions are carried out by the same cells or by others with different properties. Evidence for the existence of two types of splenic-helper cells (subgroups of Lyt-1$^+$ 23$^-$ cells) is referred to by Bendtzen (1983).

As mentioned above, although IL-1 is required for formation of IL-2, other factors are also necessary. Using cytofluorimetric analysis, Kristensen (1982) found that PHA + IL-1 is no more effective than is PHA only, in moving murine thymocytes from the G_0 to the G_{1a} cell cycle stage. Thus at least with RNA synthesis as indicator of incipient cell proliferation, IL-2 formation has been shown to require the presence of other factors in addition to IL-1 and PHA.

If, as seems probable, murine thymocytes of the mature type mentioned above are very similar to splenic T cells in their responsiveness to IL-1, a question must arise regarding the choice of organ from which such cells should be prepared. Evidently an important practical consideration is that the thymus, although it consists of 90% of cells which are phylogenetically immature, the remaining 10% are rather similar to mature T cells from spleen and other organs and most important B cells and macrophages are absent or very few in number. Often therefore, thymocytes will be chosen because employment of splenic T cells usually requires preliminary removal of splenic B cells and macrophages.

The reaction of IL-1 with B cells, unlike with thymocytes and T cells, does not require activation with a polyclonal mitogen. Autogenic activation or T cell contact are sufficient. As described by Hoffman (1980), B cells, from the spleen of a mouse previously injected with SRBCs, can be used. Activation of these cells occurs in the presence of IL-1 so that a higher proportion of B cells become capable of responding to T-cell help. However, IL-2 and factors from T-helper cells are likely to be required before rosette formation from SRBC can take place. These results apply only to the effects of particulate antigens, e.g. SRBC. For soluble antigens, more complex relationships, involving two classes of T-helper cells have been demonstrated.

5. *Adjuvant effects of IL-1*

Investigations carried out in vitro have indicated the involvement of IL-1 with both B and T cells and thus with both the humoral and the cell-mediated arms of the immune response. Despite the apparently central role of IL-1, suggested by such experiments, evidence of an effect on any aspect of the immune system as it exists in vivo is of great interest. Such evidence is now available in respect to adjuvanticity of IL-1.

IL-1 can influence the immune response in a number of different ways among which perhaps the most clearly demonstrated is its ability to act as an adjuvant. This effect has been investigated in mice by Staruch and Wood (1983). The IL-1 was injected I.P. 2 h after S.C. injection of BSA. Increases in antibody to BSA of between 4 and 30 times were thus obtained. Freund's complete adjuvant and muramyldipeptide (MDP) when tested concurrently gave much higher responses. Certain of the ways by which IL-1 stimulates antibody synthesis have been indicated above cf. Section 3.7. They can be divided between direct effects on B-cell activation and proliferation and indirect effects on T-helper or T-suppressor cells (Dinarello, 1984).

The full significance of IL-1 as an adjuvant only became apparent when it was demonstrated that MDP is highly active as a stimulator of macrophages to form IL-1 (Chedid et al., 1978). Because MDP has been shown to be the minimal structure that can replace whole mycobacterial cells in Freund's complete adjuvant (Ellouz et al., 1979) and because it leads to IL-1 formation, its role as an adjuvant has become clear. However, only after a further series of discoveries has the special significance of MDP and its derivatives for the AP-response finally emerged. The first of these concerned the butyl ester of MDP which was found to retain adjuvanticity but not the ability to induce fever or bring about release of IL-1 in vitro (Chedid et al., 1982). Stimulation of the AP-response by this substance required 100 times more than for MDP (McAdam et al., 1983). Later it has also been possible, using a desmuramyl derivative of MDP to dissociate adjuvanticity and pyrogenicity from non-specific immunity (Parant et al., 1980) (cf. p. 201). These authors found that DP-L-alanyl-glycerol mycolate protected mice against injections of Klebsiella pneumoniae type 2 but did not show either of the other effects.

The second discovery concerning the AP-response was that IL-1 consists of a group of closely similar molecular species, some of which have different biological effects. Most important in this respect was the finding that whereas rabbit IL-1 of p*I* 7.3 can induce increases in fibrinogen concentration in rabbits, the p*I* 4.6 IL-1 component lacks this function (Kampschmidt et al., 1983).

The above experiments strongly suggest that the different biological responses which IL-1 can produce will be found to be due to different forms of IL-1 and furthermore that different proportions of these forms of IL-1 may be produced according to the nature of the stimulant. It is relevant in this regard to notice that both LPS and MDP are constituents of bacterial cell walls. The further possibility that in certain instances MDP or its derivatives may act in parallel with, or in place of, IL-1 also cannot be ignored. A thorough study of the molecular requirement for adjuvanticity has shown the muramyl residue to be essential, however, as already mentioned, the desmuramyl derivatives were found to have retained non-specific ability to kill infective agents and tumour cells, (Parant et al., 1980). How far production of AP-responses may be involved is not yet known but would appear to be very probable.

Work with MDP has led to the further interesting and potentially very important discovery, that antisera against MDP are able to bind IL-1. This now well established fact makes it hard to avoid the conclusion that a muramyl-containing peptide must form part of the structures of most, or indeed probably all, forms of IL-1. If this can be proved it will be a fact of exceptional interest as it is well known that muramic acid has been found only in prokaryots. The only exception to this is the isolation from human urine and from CSF of sleep-deprived animals of *n*-acetyl-muramyl-1-alanyl-d-isoglutamine-diaminopimelic acid, i.e. MDP with a lipophilic-end group (Krueger et al., 1982). This substance has been shown to induce slow wave EEG patterns in rabbits.

Having carefully excluded the possibility that the constituent muramic acid found could have been due to bacterial contamination of the urine or CSF, its origin in eukaryots is discussed. Whether it derives from bacteria in the gut or has been retained in mammalian tissues after their divergence from prokaryots as a component of certain hormones especially concerned with bodily defence is of special interest. Two further relevant observations deserve mention. The first is that phenol extraction of LPS from *Salmonella minnisota* has yielded proteinacious material that acts like IL-1 in the [^{3}H]thymidine assay but does not stimulate macrophages to produce IL-1 (Sipe et al., 1984) (cf. Chapter 2, Section 2.4). Whether this contains a muramyl residue, MDP or is an IL-1-like molecule of bacterial origin, is not yet known. Doubtless the search for molecules smaller than IL-1 but with some of its properties will continue (cf. Chapter 3, Section 4.5) for evidence that trypsinisation of IL-1 can lead to such molecules.

References, p. 132.

Gordon/Koj (eds.) The acute-phase response to injury and infection.

CHAPTER 8

Responses of cells other than those of the brain and the immune system to interleukin 1

A.H. GORDON

That IL-1 can affect the metabolism of a number of different cell types has already been mentioned briefly. Further details concerning these cells and their various responses to IL-1 follow.

Cells as different as muscle cells, granulocytes and hepatocytes are affected. Not surprisingly the consequences of the interactions of these very different cell types with IL-1 are diverse. The question must then arise as to whether IL-1 merely accelerates the synthesis by each cell type of one or more of the products which these cells normally make, in which case other stimulators might be expected to have similar effects, or whether IL-1 action is highly specific. The specificity of IL-1 can be of two kinds. Firstly, effect or absence of effect, on apparently very similar cells, e.g. the cells of PO/AH and other brain areas. Secondly, production by a particular cell type of a substance of which production does not occur in any other circumstance.

1. Stimulation of blood granulocytes

Release of specific granule contents from human granulocytes treated with human EP/IL-1 has been demonstrated by Klempner et al. (1978). As this has been eluted from a specific immunoabsorbant, and showed only a single band on SDS polyacrylamide gel, it can be assumed to have been highly purified EP/IL-1. The release of the granule contents was accompanied by oxidative metabolic changes usually known as the 'respiratory burst' (cf. Chapter 1, Section 2). This effect was to be expected because many different treatments of granulocytes including endocytosis of particulate materials and stimulation by soluble factors, e.g. C5a, are known to lead to the metabolic events which constitute the metabolic burst (Fantone and Ward, 1982). 'IL-1 represents – naturally occurring small molecular weight protein which can initiate release of specific granule contents without other perturbation of the cell membrane' (Klempner et al., 1978). These facts make possible a partial answer to the question posed above in respect to the specificity of action to IL-1. In respect to granulocytes, IL-1, and other secretogogues lead to release of similar specific granule contents, i.e. IL-1 acts non-specifically.

The effects of IL-1 on the release of lactoferrin from neutrophils has been studied both in vitro and in vivo (Klempner et al., 1979). Because hypoferraemia is typical of patients with high fever, the release of lactoferrin induced by IL-1 is of special interest. Using human granulocytes at 5×10^7 cells/ml and excess IL-1 about 30% of the lactoferrin in the cells was released (Klempner, 1978).

Because of the known avidity of apolactoferrin for serum iron and also the ability of lactoferrin to remove iron from transferrin, and the fact that the Fe-lactoferrin is preferentially cleared from the circulation, an explanation is available for the hypoferraemia characteristic of periods of inflammation and fever (cf. Chapter 15). A consequence of the extra number of granulocytes arriving in the blood as a result of EP/IL-1 (Kampschmidt and Upchurch, 1977), can be assumed to be further augmentation of such hypoferraemia. The effects of hypoferraemia are further discussed in Chapter 21.

References, p. 132.

Gordon/Koj (eds.) The acute-phase response to injury and infection.

CHAPTER 9

Responses of muscle to interleukin 1

A.H. GORDON

1. Role of PGE_2 in degradation of muscle proteins

When strips of rat muscle are incubated in vitro with and without added human IL-1, protein synthesis and degradation and PGE_2 formation all increase (Baracos et al., 1983). Because in both cases the increases in degradation are the greater, net degradation occurs and when IL-1 has been added this is increased (Table 1). A similar result is obtained if PGE_2 is added in place of IL-1. Addition of IL-1 to certain other tissues and cells in culture has been shown to lead to increased formation of PG, e.g. from brain slices (Dinarello and Bernheim, 1981), and from synovial cells (Dayer et al., 1976). The use of inhibitors of protein synthesis in vivo has shown that the fever which follows injection of IL-1 does not occur in absence of protein synthesis (Cranston et al., 1978). These findings suggest that muscle may be unique, not because IL-1 leads to formation of PG and increases in protein synthesis, which may occur in many tissues, but because only in this tissue degradation predominates over synthesis.

2. Proteolysis inducing factor (PIF) and its relationship to IL-1

In view of these results and the fact that macrophages, when stimulated in vitro by pyrogens such as the bacterial endotoxins, produce large amounts of IL-1, it might be assumed that such IL-1 would be identified as the agent responsible for muscle degradation. However, despite many efforts, evidence for the presence of IL-1 in

TABLE 1
Effects of IL-1 on protein synthesis and degradation and PGE_2 formation and effect of PGE_2 on protein synthesis and degradation by rat muscle

		Experiment 1 nmol tyrosine/mg muscle/2 h			Experiment 2 nmol tyrosine/ mg muscle/2h	pg/mg muscle/2h
Experimental muscle	Addition	Protein synthesis	Protein degradation	Net protein balance	Net protein balance	PGE_2 production
	None	0.157 ± 0.019	0.364 ± 0.015	0.207 ± 0.012	0.145 ± 0.010	21 ± 2
Soleus	IL-1[a]	0.184 ± 0.042	0.528 ± 0.065	0.344 ± 0.028	0.289 ± 0.016	75 ± 8
	% Change	NS	45	−66	−99	+257
	Experiment 3					
	None	0.123 ± 0.010	0.613 ± 0.034	0.481 ± 0.042		
Diaphragm	PGE_2[b]	0.158 ± 0.010	0.746 ± 0.041	0.588 ± 0.046		
	% Change	NS	22	−22		

In experiments 1 and 2 the muscle strips were preincubated in Krebs-Ringer bicarbonate buffer at pH 7.4, supplemented with 5 mM glucose for 30 min. The measurements are for the subsequent 120-min period. In experiment 3 there was a longer preincubation period and a shorter (90 min) measurement period.
[a] Human IL-1 at 0.5 rabbit pyrogen dose/ml.
[b] PGE_2 final concentration was 2.8×10^{-6} M.
Results from Baracos et al. (1983) and Rodeman et al. (1982).

the blood during fever has only become available (Cannon and Kluger, 1983; Bendtzen et al., 1984; Cannon and Dinarello, 1984). Possible reasons for the initial failure of such attempts, including the discovery of an inhibitor of the thymocyte costimulator assay (Dinarello et al., 1982), have been discussed by Dinarello (1984). Fortunately further experiments with rat muscle strips have revealed the existence, in blood from patients with sepsis or trauma, of a peptide of approximately 4 kDa i.e. approximately one third of that of IL-1 (Clowes et al., 1983). The properties of this peptide, known as proteolysis inducing factor (PIF), have already been carefully compared with those of IL-1. Because both are active in the thymocyte costimulator assay and are pyrogenic in mice as well as having PIF activity, it is likely that they are structurally similar at least in part (Dinarello et al., 1984) (cf. Chapter 3, Section 4.5 for a difference).

Whether PIF derives from IL-1 by proteolysis, or whether both pyrogens derive from an even larger precursor molecule, is not yet known (cf. also Chapter 17). The existence of proteolytic fragments of molecules secreted by macrophages at injury sites is to be expected because of the presence there of proteases from the numerous neutrophils also present. Evidence for the view that PIF is a proteolytic fragment of IL-1 has been obtained by Dinarèllo et al. (1984), who have shown that trypsinisation of IL-1 leads to formation of a factor with properties very similar to those of PIF and also that the addition of protease inhibitors reduced the amount of this factor deriving from IL-1. PIF can be separated from IL-1 by a simple chromatographic procedure using Biogel P6 (cf. Chapter 3, Section 4.5). The use of antisera to IL-1 and to preparations active as PIF has revealed a further difference. This is that the PIF activity isolated from plasma is blocked by an antibody to IL-1, whereas an antibody raised against PIF is only partially effective against IL-1. Thus, while IL-1 pyrogenicity is blocked, its effect in the thymocyte costimulator assay is not affected (Dinarello et al., 1984).

3. Consequences of the release of amino acids from muscle

Amino acids derived from muscle are important both as sources of energy and because as amino acids they are the units from which all proteins are built. When released from muscle as part of the AP-response, amino acids serve both of these purposes. In such conditions, e.g. in patients suffering from infection or inflammatory disease, amino acids are specially important as sources of energy. This is because in these circumstances the formation of ketones from adipose tissue is inhibited, presumably as a result of increased levels of insulin (Neufeld et al., 1980). The manner in which adipose tissue is involved in AP-response is suggested by reports of an IL-1-like factor which inhibits lipoprotein lipase and thus makes available free fatty acids as sources of energy (Kawakami et al., 1981). Evidently understanding of the interrelationships of these various factors will require further investigation. At this time it is safe to state only that the metabolic pattern of patients with sepsis or inflammatory disease forms a sharp contrast with that during starvation, as was demonstrated nearly 70 years ago by Coleman and DuBois (1915). The increased release of amino acids from muscle which occurs as part of the AP-response is specially valuable because, in addition to acting as sources of energy, these molecules are required for synthesis of protein in several types of tissue. The most important are brain and cardiac muscle, essential for preservation of life, bone marrow and lymphoid tissue where extra synthesis is needed for antibody formation and finally hepatocytes in which formation of the AP-proteins takes place. In these circumstances the weight of the liver increases and undoubtedly this organ is responsible for the uptake of most of the newly available amino acids (cf. Chapter 15). The special need for amino acids during the AP-response, for protein synthesis in all these tissues provides an explanation for the fact that in infected and injured patients, the rate of release of amino acids for the muscles is increased by a factor of 3 – 5 times over normal. In patients with sepsis, up to 15 g nitrogen per day can be lost from the muscles. Despite the increase in liver weight, this organ cannot metabolise phenylalanine, tyrosine and tryptophan fast enough to prevent a degree of accumulation of these amino acids in the plasma in conditions of severe muscle wasting (Wannemacher, 1977) and also after injection of turpentine (Woloski et al., 1983 and Chapter 15).

Although the significance of IL-1, PIF and PG, with respect to muscle wasting, seems certain the roles of other mediators such as insulin and corticosteroids should not be ignored (Neufeld et al., 1980). Another unsolved aspect of the effect of IL-1 on muscle is the degree in which its action may be an indirect one due to increased rates of metabolism in all tissues caused by fever. For this to be important it would be necessary that in hyperthermia the increase in rate of degradation would exceed the increase in synthesis. While this may be true, it is of interest that rats injected with rabbit EP/IL-1 usually develop little or no fever but nevertheless manifest considerable muscle degradation.

Because the proportion of muscle greatly exceeds that of any of the other bodily tissues, the effect of IL-1/PIF on muscle, considered quantitatively, must constitute a major part of the AP-response. As already mentioned, degradation of muscle yields both extra energy and an extra supply of amino acids. These provide the units required for synthesis of AP-proteins, the response traditionally considered as the central feature of the AP-response. Only in patients in an extreme condition, with already severely wasted muscles does the supply of amino acids fall to a point at which synthesis of AP-proteins such as CRP is also reduced (Mosley et al., 1982).

References, p. 132.

Gordon/Koj (eds.) The acute-phase response to injury and infection.

CHAPTER 10

Responses of connective and other tissues and cell types to injury-derived factors

A.H. GORDON

1. Connective tissue response to IL-1

Connective tissue from different organs contains either typical fibroblasts or cells of nearly the same kind. Most information is available concerning the effect of IL-1 on fibroblasts such as those derived from guinea pig dermis and from synovial cells obtained from patients with rheumatoid arthritis. The responses of fibroblasts to IL-1 have been examined both in vivo and in culture. Experiments conducted in vivo, before the properties of IL-1 as a secreted product of macrophages had been discovered, already suggested the importance of these cells as stimulators of fibroblast proliferation (Leibovitch and Ross, 1975). This concept was based on measurements of the rate at which fibroblasts appeared in healing skin wounds of guinea pigs. Comparisons between animals depleted of macrophages by means of steroids or anti-macrophage serum and normals showed that at 5 days after wounding, the number of fibroblasts present in the healing areas of the depleted animals was less than half that in the normals. It is now known that results such as this are due to the effects of more than one factor. Among these are chemotactic factors from lymphocytes and IL-1 from the macrophages (Postlethwaite and Kang, 1982).

As methods for isolation and purification of IL-1 have been improved, some clarification of its effects on fibroblasts and fibroblast-like cells has been achieved. An initial step involved the separation of supernatants obtained from cultures of guinea pig lymph node cells using Sephadex G-50 (Nielson et al., 1982). The material, thus obtained consisting of molecules averaging 60 kDa, was found to stimulate fibroblast proliferation. On the other hand, a fraction containing smaller molecules of the size of IL-1 was slightly inhibitory. Using a similar method for estimation of fibroblast proliferation and similar cells, except that this time they were from guinea pigs which had been injected with tuberculin instead of with renal tubular antigen, Postlethwaite and Kang (1982), obtained a different result. After passage over Sephadex G-100, the fibroblast proliferation activity was found to coelute with IL-1 at the elution volume expected for molecules of 15 kDa. As shown by both of these experiments, fibroblast proliferative activity is present in supernatants from primed lymph node cells. That such activity results from molecules with properties which cannot be distinguished from one of the components of the IL-1 group of proteins, was also demonstrated by Postlethwaite and Kang (1982). By means of HPLC, six components were obtained, and as shown in Fig. 1, only one of these had fibroblast proliferation activity whereas all six were active in the thymocyte costimulator assay.

Experiments with highly purified IL-1, obtained from human blood peripheral monocytes and alternative from murine P388/D cells, also showed correspondence between fibroblast proliferative activity and positive results using the thymocyte costimulator assay. Comigration was obtained using both Sephacryl S-200 chromatography, which gave two peaks both of which were active in both assays, and isoelectric focusing (IEF) which gave 3 peaks, again all were active in both assays (Schmidt et al., 1982).

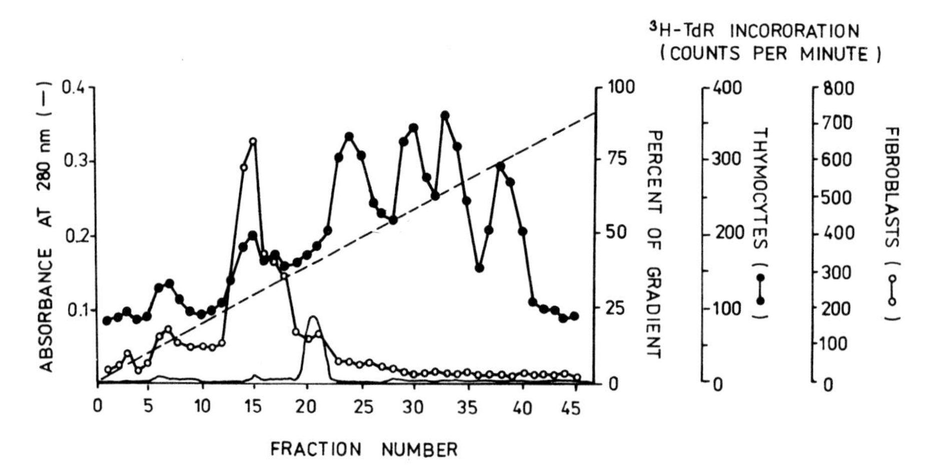

Fig. 1. Fractions (20 ml) from the 16 kDa peak of fibroblast proliferation activity obtained by fractionating supernatants from PPD immune-stimulated LNC cultures on Sephadex G-100 were pooled and concentrated 50-fold by ultrafiltration in an Amicon chamber equipped with a YM 5 membrane, dialysed against starting buffer, and applied to a Syn-Chropak AX 300 anion exchange-HPLC column. The eluted fractions were tested for IL-1 and fibroblast proliferation activities. (Postlethwaite and Kang, 1982.)

2. *Effects of IL-1 on fibroblast and synovial cells in culture*

Owing to the presence of more than one cell type in healing wounds, and in the synovium of inflamed joints, meaningful investigations have had to be conducted in vitro with cultures consisting almost exclusively of a single cell type, e.g. synovial cells (Dayer et al., 1980; Mizel et al., 1981). Under these conditions, the presence of either supernatants from stimulated macrophages or of highly purified IL-1, leads to greatly increased formation of both collagenase and PGE_2. However, study of the effects of IL-1 on proliferation of dermal fibroblasts has led to a different result (Schmidt et al., 1982). For this purpose, mononuclear cells from normal unrelated volunteers were allowed to react under the conditions of the secondary mixed lymphocyte reaction. Supernatants from these cells which were shown to contain IL-1 as indicated by the thymocyte costimulator assay, also led to fibroblast proliferation. Because the two activities were found to comigrate it follows that IL-1, at least as thus prepared, stimulates fibroblast growth. Extending the work of Postlethwaite and Kang (1982), referred to above, to highly purified IL-1, Postlethwaite and Kang (1983), using human foreskin fibroblasts stimulated by highly purified IL-1 compared collagenase activities and results in the costimulator assay. As shown in Fig. 2, dilution of IL-1 led to similar decreases in activity in both assays. The possible use of collagenase formation by fibroblasts as an assay for IL-1 is considered in Chapter 23, Section 3.4.

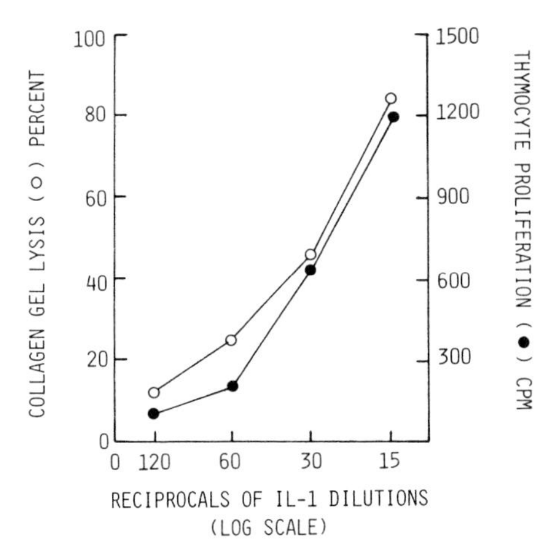

Fig. 2. Human IL-1 was partially purified (diafiltration, ultrafiltration and isoelectric focusing) from supernatants of cultures of monoblasts-monocytes from a patient with acute monocytic leukemia. IL-1 was added to cultures of mouse thymocytes and foreskin fibroblasts at the dilutions indicated. Collagenase production by fibroblasts and thymocyte proliferation were measured. (Postlethwaite et al., 1983.)

Taken together, the above results indicate that IL-1 acting on fibroblasts can lead to increased synthesis of collagen and, presumably under slightly different conditions, to increased formation of collagenase. Whether this depends on the cell type from which the stimulator has been obtained and/or the balance between different forms of IL-1, only certain of which have fibroblast stimulating activity (cf. p. 117 and Fig. 1), remains to be discovered. Also important is whether, and if so in what conditions, fibroblast growth and release of collagenase into the medium can occur simultaneously. The importance of the balance between these two activities for fibroblasts involved in wound healing, cannot be overemphasised. That increased synthesis of collagenase by fibroblasts, which have been stimulated by IL-1, does not necessarily involve growth of the cells has also been shown by Postlethwaite et al. (1983). As already mentioned, when IL-1 acts on fibroblasts or synovial cells, increased synthesis of collagenase is accompanied by a large increase in formation of PGE_2. A similar effect follows when IL-1 is incubated with brain tissue (Chapter 5, Section 1) and skeletal muscle (Chapter 9, Section 1). When either of these two tissues are directly stimulated by addition of PGE_2, in absence of IL-1, similar metabolic events to those which would have occurred with IL-1 are induced. This does not occur when fibroblasts are cultured in presence of PGE_2. Thus as PGE_2 does not induce collagenase production from fibroblasts, it cannot be an intermediate between IL-1 and collagenase. Indeed, as shown in Table 1, more collagenase was formed by synovial cells when stimulation was brought about by

TABLE 1
Collagenase production by synovial cells (ASC) treated with monocyte-macrophage supernatants (MCF): effect of indomethacin[a]

Blood monocyte-macrophages			ASC
Addition			
Fc fragments[b]	Indomethacin	PGE_2 (ng/ml)	Collagenase (units/10^6 cells)
None	None	1.1	0.20 ± 0.10
None	14 μM	0.2	0.20 ± 0.10
50 μg/ml	None	11.4	16.80 ± 2.50
50 μg/ml	14 μM	1.3	27.20 ± 3.80
No MCF			0.02 ± 0.01

[a] ASC in the second passage were plated at 1×10^5 cells/wells 3 days before bioassay. ASC were then incubated for 3 days with various media from blood monocyte-macrophages that had previously been incubated for 3 days in 0.5 ml/well containing 0.5×10^6 cells/well. Monocyte culture media were diluted 1/20 with DMEM, 10% FCS before adding to ASC in a final volume of 0.1 ml/well. Each value is the mean ± SEM for 3 wells.

[b] Fc fragments from human immunoglobulin.

monocyte supernatant produced in the presence of indomethacin (Dayer et al., 1980). The relationship between fibroblast and PGE_2 formation has been investigated by the use of supernatant from human monocytes, either incubated without any stimulant or in presence of PHA. Such supernatants were found both to suppress fibroblast growth and to induce formation of large amounts of PGE_1 (Korn et al., 1980). As under these conditions monocytes provide very small amounts of IL-1, the presence of a fibroblast growth inhibitor predominated and may well have masked any effect that might have been caused by the trace amounts of IL-1 which will have been present. Experiments in which indomethacin was included indicated that growth suppression required the presence of the PG formed by fibroblasts. The function of PGE_2 as a growth suppression was clear because when it was added, growth suppression returned after its reversal by indomethacin. The full significance of IL-1, in relation to connective tissue cells, must await further work on the contribution of other factors which have been shown to either stimulate or to inhibit fibroblast proliferation.

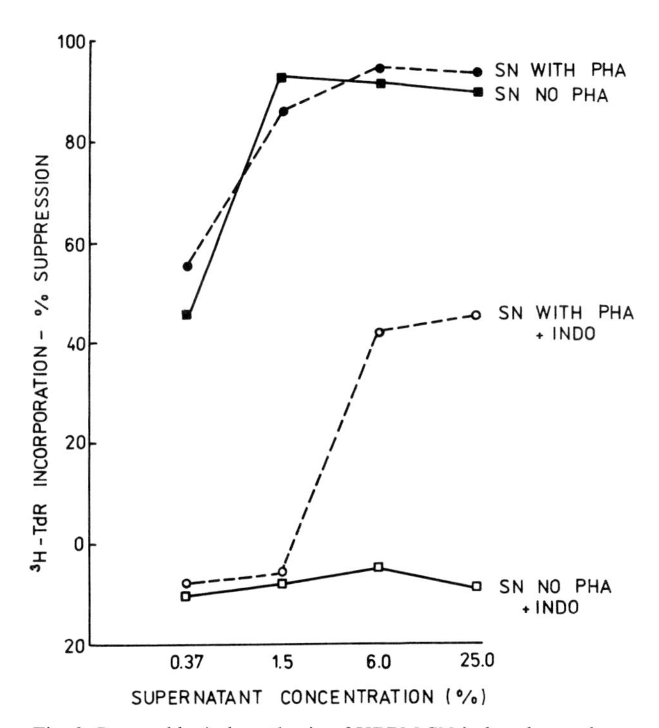

Fig. 3. Reversal by indomethacin of HPBM-SN-induced growth suppression. Fibroblasts were incubated with indomethacin (+ Indo) or buffer, as previously noted, before addition of unstimulated or PHA-stimulated SN preparations. Both SN preparations were derived from HPBM obtained from a single donor on the same date. Indomethacin in the absence of HPBM-SN gave 1% suppression of [^{3}H]TdR incorporation. (Wahl et al., 1977.) HPBM-SN = Human peripheral blood monocytes supernatant.

Particular interest will attach to the role of PG because of their inhibitory effect on the growth of fibroblasts. In the healing wound, macrophages appear before fibroblasts and are known to be producers of collagenase when stimulated by endotoxin, as is likely in healing wounds, collagenase synthesis by macrophages is greatly increased if PG are also present (Wahl et al., 1977). Later, when fibroblasts make their appearance, PG begin to have an opposite effect. As shown in Fig. 3, growth of fibroblasts is almost completely suppressed when they are cultured with as little as 1.5% of a supernatant obtained from human peripheral monocytes. As mentioned above, this is almost certainly due to the presence of PG (Korn et al., 1980). As fibroblasts replace macrophages in the healing wound collagenase synthesis will continue due to the effect of IL-1. But as this stage the PG will act to limit the rate of cell growth and thus at the same time the amount of collagenase formed. If this interpretation of the interrelationships of IL-1, macrophages and fibroblasts, collagenase and PG in healing wounds proves ultimately to be correct, the role of the latter substances in wounds will be that of controllers rather than intermediates, as seems to be in the case in muscle and perhaps also in certain areas of the brain.

3. Possible relationship of IL-1 to chondrocytes and cartilage breakdown

Degradation of the main cartilage matrix components, e.g. collagen and proteoglycans, can be brought about by proteases secreted by chondrocytes. However, normal rabbit articular chondrocytes secrete very small amounts of such enzymes when in culture. The cells are obtained from knee or hip joint cartilage of young rabbits by serial digestion with hyaluronidase, trypsin and collagenase and then grown to confluency in Ham's F12 medium containing 10% FCS. With such cells, addition of supernatants from rabbit peritoneal macrophages which have been activated by endotoxin leads, after 2 – 3 days, to significant levels of collagenase and other neutral proteases (Deshmukh-Phadke et al., 1980). Gel filtration of this macrophage supernatant has identified a factor of 13 – 15 kDa which is responsible for the production of the collagenase. There is also evidence for cartilage degrading activity by a factor from blood mononuclear cells which have been stimulated with PHA (Jastin and Dingle, 1981). They point out that the matrix degradation cannot be due to macrophage proteases because it only occurs in presence of living chondrocytes.

A protein now known as catabolin with similar activity has recently been isolated from supernatants of pig monocytes (cf. Chapter 3, Section 5.2). This has been shown to be a form of IL-1 because not only is it active in the thymocyte costimulator assay (cf. Chapter 23, Section 3.1) but it induces chondrocytes and fibroblasts to form PGE_2 and collagenase. Although it leads to proteoglycan breakdown in explants of articular cartilage, collagen is not affected (Saklatvala et al., 1984). Further strong evidence that catabolin is a form of IL-1 is that fever follows injection of as little as 5 ng ICV into a rabbit and SAA increase occurs in mice (Dr. J. Saklatvala, personal communication).

4. *Effects of macrophage-derived mediators on adipose tissue*

Supernatants obtained by treating peritoneal macrophages with endotoxin lead not only to reduced lipoprotein-lipase formation by adipocytes but also to suppression of fatty acid biosynthesis (Kawakami et al., 1982; Pekala et al., 1983). Whether these effects are due to the IL-1 which must be present in such supernatants or to some other macrophage-derived mediator, is not yet known. Evidence from preliminary investigations of molecular size and instability to heating do not rule out IL-1 as the responsible factor (Kawakami et al., 1982).

Whatever the nature of this mediator, its effects on adipose tissue and plasma very low density lipoprotein, form an important and only recently recognised aspect of the AP-response. The initial observation which led to the above results was that of hypertriglyceridaemia associated with *Trypanosoma brucei brucei* infection in rabbits (Rouzer and Cerami, 1980). Further work on the response of rabbits to en-

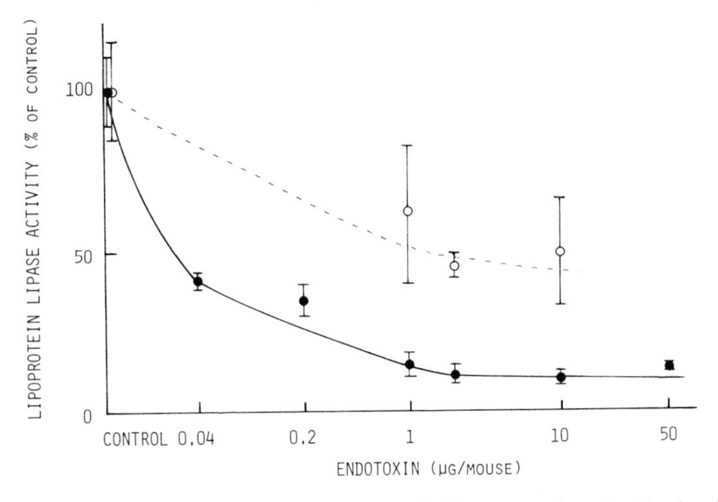

Fig. 4. Response of adipose tissue lipoprotein lipase to endotoxin injection in endotoxin-sensitive and endotoxin-resistant mice. As indicated, amounts of endotoxin dissolved in 0.2 ml of saline were injected into C3H/HeN (●) or C3H/HeJ (○) mice. Adipose tissue was obtained 16 h after the injection. Data are presented as the mean (± SEM) of percentage activity in five animals for each dose compared with those of control C3H/HeN mice (7) or control C3H/HeJ mice (10). (Kawakami and Cerami, 1981.)

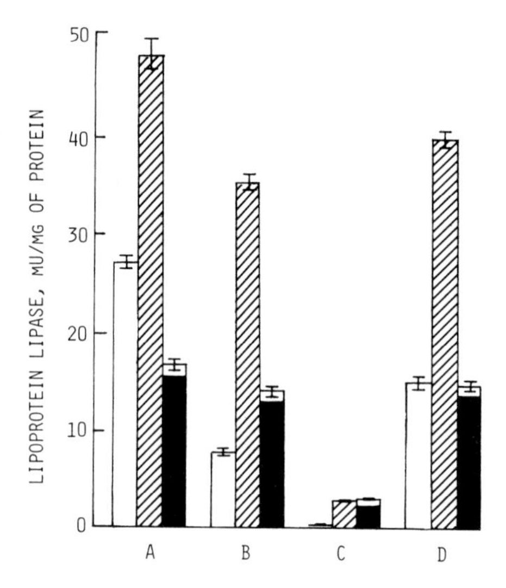

Fig. 5. Effect of conditioned medium from endotoxin-treated mouse peritoneal exudate cells on lipoprotein lipase activity of 3T3-L1 cells. One millilitre of either RPMI 1640 medium (A) or conditioned medium from exudate cells incubated in the absence (B) or presence of endotoxin (5 μg/ml) (C) or RPMI 1640 medium with endotoxin (22.5 μg/ml) (D) was added into a confluent culture of 3T3-L1 cells in 6.0 cm dishes containing 3.5 ml of DMEM medium. After 20 h of incubation, lipoprotein lipase activity in the medium (open bars), in the cell surface (hatched bars) and in cell sap (solid bars) was measured. Data are expressed as mean ± SEM ($n = 4$). (Pekala et al., 1983.)

dotoxin showed the hyperlipidaemia to be the result of inhibition of lipoprotein lipase. Using mouse strains C3H/HeN and C3H/HeJ, known to be sensitive and insensitive to endotoxin, the same effect was shown to be indirect (Fig. 4). Thus at a time after endotoxin injection, sufficient to ensure almost complete clearance of endotoxin, serum from the endotoxin-sensitive strain injected into the endotoxin-resistant mice led to a 55% decrease in lipoprotein lipase activity. Fatty acids liberated from plasma triglycerides by lipoprotein lipase are resynthesised into tissue triglyceride and stored in this form. In view of the restriction on build-ups of such stores, which is caused by these macrophage supernatants, the possibility that synthesis of fatty acids might also be inhibited was realised and has been investigated by Pekala et al. (1983). The 3T3-L1 murine cell line of pre-adipocytes was used. These cells, like adipocytes in general, utilise lipoprotein lipase and also contain acetyl-CoA carboxylase and fatty acid synthetase. As shown in Fig. 5, conditioned medium from endotoxin-treated mouse peritoneal exudate cells led to considerable reduction of both these enzymes in the cytoplasm of 3T3-L1 cells. Inhibition of synthesis of fatty acids, due to reduced activity of these two enzymes, together with reduced transfer of fatty acid from plasma to adipose tissue, taken together indicate that as part of the AP-response, a switch from anabolism to catabolism must occur. As a result of this, extra energy from glucose, which otherwise would have been transformed into and stored as triglyceride fat, becomes available. The existence of this glucose sparing effect is shown in Table 2. In presence of the mediator there was a decrease of more than 40% in the incorporation of glucose into triglyceride by 3T3-L1 cells.

TABLE 2
Effect of macrophage on incorporation of glucose into triglyceride by 3T3-L1 cells[a,b]

Additions	cpm Glucose incorporated per mg
Control medium	109 ± 10
Mediator	60 ± 17

[a] [U-^{14}C]glucose incorporation into triglyceride after 90 min incubation at 37°C.
[b] The cells at 2 days post-confluence were cultured in medium containing 10% FCS, 0.5 mM isobutylmethylxanthine, 1 μm dexamethasone, 10 μg insulin/ml. After 2 days, the medium was replaced with medium supplemented only with FCS and 50 ng/ml insulin. To determine the effect of the mediator on uptake of [^{14}C]glucose the cells were used after a further 24 h.

References, p. 132.

Gordon/Koj (eds.) The acute-phase response to injury and infection.

CHAPTER 11

Other injury-mediated metabolic changes

A.H. GORDON

1. Hormonal changes induced by injury and occurring during inflammation

The AP-response can be divided into the increased concentrations in the plasma of the AP-proteins themselves and changes such as those of certain hormones and other changes including clotting which alter the function of the blood itself. Also involved are fibrinolysis, the complement system and the changes which lead to the release of kinins (Sundsmo and Fair, 1983). Altered hormonal concentrations must necessarily be considered in relation to their effects on target organs. Evidence as to the degree in which the various hormones are involved during inflammation or after trauma has been summarised by Beisel (1981).

As originally emphasised by Selye (1946), the predominant initial change involves the adrenal cortex. Following injury, cortisol levels rise but return to normal values within hours. Selye's view that this change could be considered to be the main feature of the so-called stress response is now considered to be a serious oversimplification. Thus volunteers, given cortisol in a sequence of doses to mimic the actual pattern of glucocorticoid output during an acute infection, failed to show the expected loss of body nitrogen (Beisel et al., 1973). Plasma cortisol levels increase less during infection than after trauma or surgery and certainly are not responsible for the other multi-hormone changes actually observed.

However, the importance of the adrenocorticoid hormones is evident from the extreme sensitivity to shock seen in animals after adrenalectomy. This fact which carries the implication that these hormones form part of the body's defense mechanism, was known before 1940 when the anti-inflammatory effect of glucocorticoids was discovered. After 1940 it was necessary to take into account that acting as anti-inflammatory agents glucocorticoids can inhibit extremely important defense mechanisms as well as themselves showing protective effects, e.g. in shock. A current hypothesis which seeks to reconcile these apparently opposite effects of glucocorticoids has been suggested by Munck et al. (1984). This is that their influence on the numerous facets of bodily defense is not to prevent or impair the function of the immunological or other defense mechanisms, but to limit their action and thus prevent damage due to a too vigorous response.

2. Local changes at an injury site leading up to formation of IL-1

Before consideration of the influence of glucocorticoids on each of the more important defense mechanisms, it is necessary to outline briefly the changes which occur locally at an injury site at the time when increased production of glucocorticoids is occurring. As a result of these events, increased synthesis of AP-proteins also occurs and it is against this background that the function of the AP-proteins must be viewed. As is well known, release of histamine from mast cells is the primary local event which follows an injury. Its effect, together with contributory effects of 5-hydroxytryptamine, certain globulin permeability factors and catecholamines which also rapidly reach an injury site, is a vigorous vasodilation. This is followed by infiltration of leukocytes, neutrophils being followed by macrophages. If, as normally occurs, bacteria or other invaders are present, the macrophages become activated and IL-1 is produced. The increased rates of synthesis of particular AP-proteins which then occur will be determined by the balance of hormones and other mediators present. Certain of these of which adrenaline is an example can act both directly and via their capacities to stimulate synthesis of other hormones. This has been demonstrated for α_2-macroglobulin of rats by means of subcutaneous injections of adrenaline which led to greatly increased plasma concentrations of corticosteroids. For comparison, corticosterone was given orally (Van Gool et al., 1984). The effect of simultaneous treatment of adrenalectomised rats with both substances was also investigated. This led to the maximum plasma concentrations of α_2-macroglobulin thus indicating that the two hormones can act synergistically and that the effect of adrenaline in normal rats is not only indirect via stimulation of the cortico-adrenal axis.

3. Effects of glucocorticoids on AP-protein synthesis

The requirement for IL-1 for increased synthesis of AP-proteins and also its requirement as part of the immune system, have both been described already (cf. Chapter 7, Section 3 and Chapter 15). Further augmentation of the immune response may also occur due to the increased concentration of CRP brought about by IL-1 (Wicher et al., 1983). The function of CRP as a protective agent against pneumococcal infection in mice has been shown by Mold et al. (1981). However, insofar as these events in vitro and in vivo depend on IL-1, they must be presumed to be limited by glucocorticoids. In Fig. 1, production of IL-1 by thymocytes is shown to be inhibited by hydrocortisone. On the other hand, the rate of synthesis of certain AP-proteins, e.g. α_2-macroglobin, fibrinogen, haptoglobin and α_1-acid glycoprotein are increased by glucocorticoids (Gordon and Limaos, 1979). Evidence that this occurs due to induction of genes, the effects of which are either undetectable in hepatocytes from unstimulated rats or are present only in low abundance, has been obtained by Feinberg et al. (1983). This subject is further discussed in Chapter 18.

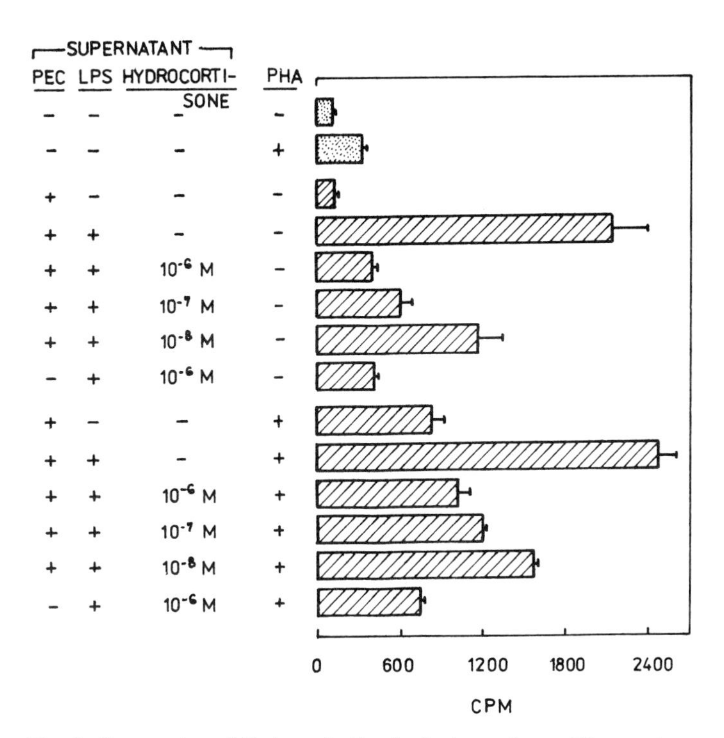

Fig. 1. Suppression of IL-1 production by hydrocortisone. Thymocytes were cultured for 2 days with or without phytohaemagglutinin, then pulsed with [^{3}H]thymidine. Data represent mean ± SEM of cpm of incorporated [^{3}H]thymidine from triplicate samples. Control of supernatants of dialysed medium containing LPS and hydrocortisone incubated for 24 h without macrophages did not suppress the basal level of thymocyte proliferation. Figure shows the results of supernatants at a 10% (v/v) concentration, where the response of thymocytes is linear to the amount of mitogen. (Snyder and Unanue, 1983.)

It is important to note that work with isolated hepatocytes has also shown that rates of synthesis of plasma proteins other than AP-proteins, e.g. albumin is also maintained at a relatively high level in presence of a glucocorticoid (Koj et al., 1984). Presumably the exact balance between the stimulatory and inhibitory effects of glucocorticoids and their differential effect on particular AP-proteins make possible the control function of glucocorticoids postulated by Munck et al. (1984). The central feature of this hypothesis is that the true role of glucocorticoids is to limit and prevent overreaction by a defense mechanism.

4. The inflammatory response and the role of macrocortin

Turning now to the inflammatory response, a defence mechanism which involves at least 13 mediators, including histamine, serotonin, bradykinin, complement fragments, PGE_2, leukotrienes B_4 and D_4, platelet-activating factor, interferon and IL-1. For details concerning the effects of glucocorticoids on several of these substances see Munck et al. (1984) and Larsen and Henson (1983). Also relevant are the functions of macrophage-inhibitory factor and the regulation of cell traffic by glucocorticoids (Fauci, 1979).

A further means by which glucocorticoids inhibit inflammation is via stimulation of the synthesis of macrocortin. This mediator inhibits synthesis of phospholipase A_2 and thus the amount of arachidonic acid available for formation of PG. Although there is no doubt that glucocorticoids act to limit inflammation, the relative importance and timing of this inhibition in respect to the numerous mediators of inflammation remains to be worked out.

Among the most important effects of glucocorticoids is their role in relation to

TABLE 1

Plasma glucagon, insulin and glucose levels[a] at time of admission in non-diabetic patients with acute infection of varying severity

Group[b]	Glucagon (pg/ml)	Insulin (U/ml)	Insulin-glucagon molar ratio	Glucose (mg/100 ml)
Severe infection (6)	409 ± 129	23 ± 9	2.48 ± 1.4	115 ± 9
Moderate infection (11)	185 ± 31	12 ± 2	2.34 ± 0.8	111 ± 6
Mild infection (5)	69 ± 16	10 ± 2	3.99 ± 1.2	96 ± 5
Normal (59)	75 ± 4	10 ± 1	3.27 ± 0.2	96 ± 1

[a] Means ± SEM.

[b] Figures in parentheses represent number in group.

carbohydrate metabolism. As shown in Table 1, during infection a doubling of insulin concentration can occur. According to Munck et al. (1984) the function of the glucocorticoids in stress is to prevent increased concentrations of insulin leading to dangerous hypoglycaemia. Thus because glucocorticoids are able to stimulate hepatic gluconeogenesis and also glucose uptake in peripheral tissue, the likelihood of hypoglycaemia is reduced.

The direct hyperglycaemic effects of glucagon and epinephrin and the indirect inhibitory effect of glucocorticoids against insulin all tend towards increased blood glucose concentrations after injury. The special function of the glucocorticoids is to prolong this effect (De Fronzo et al., 1980).

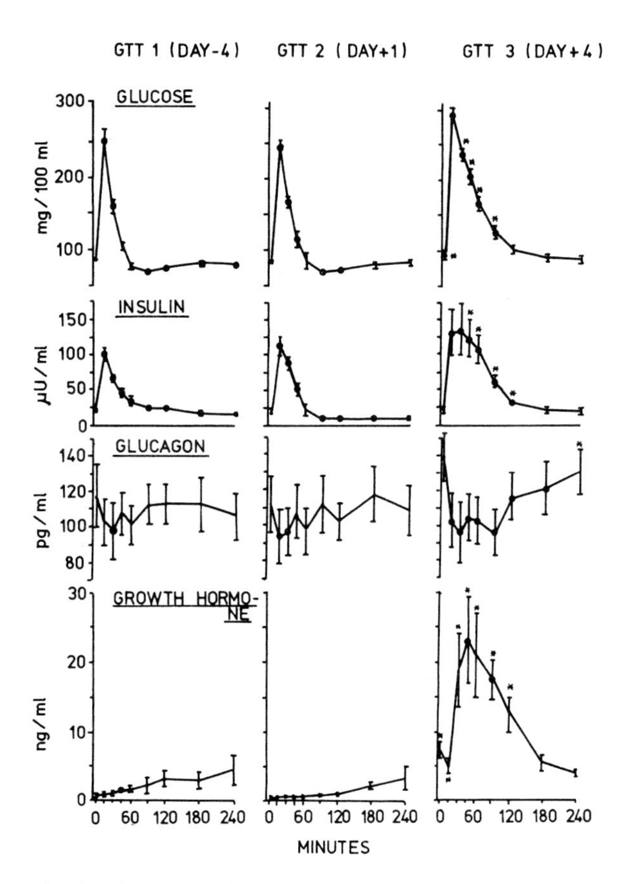

Fig. 2. Glucose, insulin, glucagon and growth-hormone responses after intravenous glucose tolerance tests (GTT) during Sandfly fever in man (mean ± SEM). The large dots represent statistically significant ($p<0.05$) differences in relation to time 0, the asterisks statistically significant differences in relation to GTT-1 at a given time.

As described in Chapter 10, Section 4, a stress-induced factor with properties similar to IL-1 has an energy-sparing effect, including inhibition of glucose conversion to fatty acids. The relationship of this factor to glucocorticoids still needs investigation.

Examples of hormonal changes observed during acute inflammatory conditions are as follows:

(1) The effects of a viral infection in patients with Sandfly fever (Rayfield, 1973). By the fourth day after infection a glucose-tolerance test showed that glucose clearance had fallen to half its normal rate. In the same circumstances, plasma insulin concentration was increased, as was that of growth hormone, to 4 times its normal concentration.

During Sandfly fever, glucagon at the fourth day reached about 130 pg/ml, an increase to 120% of normal. The response to glucose in the tolerance test reduced this value to the concentration characteristic of normals (Fig. 2).

(2) In bacteria infection, much higher plasma glucagon concentrations have been reported (Rocha et al., 1973). In an experimental investigation of pneumococcal infection in dogs, glucagon increased almost 12 times over normal whereas plasma insulin concentration was unaffected (Fig. 3).

5. *Effects of endotoxin*

Endotoxin shock has been widely investigated in order to identify the various metabolic changes which can lead rapidly to the terminal state. Injection of endotoxin, even in minute doses, leads to the complex of events known as the AP-response with which this monograph is concerned. Undoubtedly altered hormonal concentrations, such as result from injection of endotoxin, have considerable effects on other aspects of the AP-response, e.g. the requirement for cortisol for liver synthesis of α_2-macroglobulin (cf. Chapter 18, Section 2). As endotoxin injection is a relatively simple means for induction of the AP-response, some description of the hormonal changes which follow its use are worthy of note. Unfortunately, in only a few cases have changes in acute phase plasma proteins been carried out concurrently with measurements of hormonal changes. Possibly as a result, very little is known of the effects in vivo of hormonal changes on other aspects of the AP-response. In hepatocyte systems as described in Chapter 17, Section 2, rather more is known, at least in respect to requirements for insulin and cortisol (Jeejeebhoy et al., 1975).

In the absence of data relating the normal changes which occur in endotoxin shock to other aspects of the AP-response, e.g. changes in concentration of the AP-proteins, it will suffice to mention that in animals given endotoxin there is an initial hypoglycaemia at which time both insulin and glucagon concentrations increase slightly. This has been shown to occur in several species, i.e. man, monkey and rat, but extreme caution is necessary before any attempt at interpretation can be attempted because in baboons the effect is opposite (Cryer et al., 1972). This finding may indicate that it is unwise to assume, as is tempting, that all hormones are more active in the AP-response. As a first approximation to the truth, hormones may be divided into those which constitute the majority that have been shown to be more active after trauma and during infection, i.e. those of the anterior and posterior pituitaries, adrenal cortex, thyroid and endocrine pancreas and those, the activities of which remain unchanged. The hormones of the parathyroid and gonads fall into

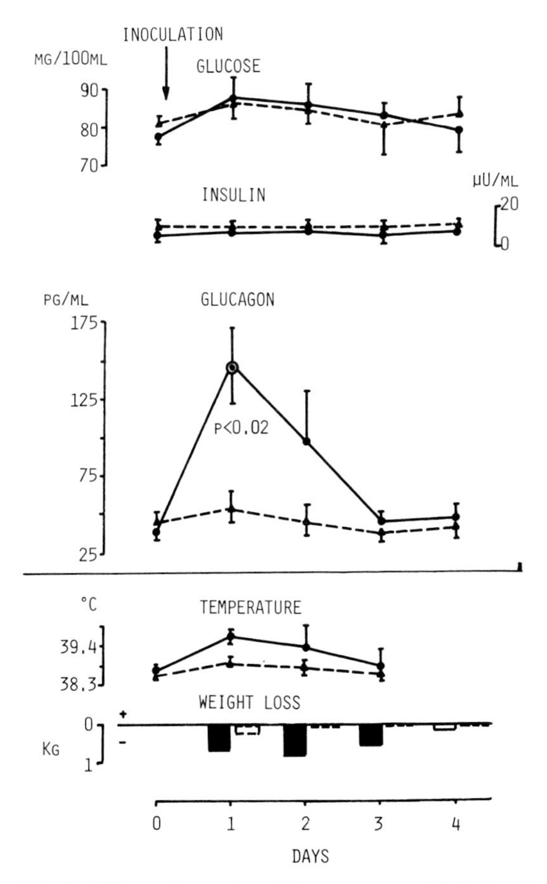

Fig. 3. Effects of experimental pneumococcal infection in dogs. ●—●, Pneumococcal inoculation, n = 3; ▲ - - - ▲, sham inoculation, n = 4.

this latter category. As already indicated, the presence or absence of only a few of the former hormones have been related to changes in synthesis rate of particular AP-proteins. This limited degree of metabolic control, exercised by the glandular hormones, contrasts with the important general effect of IL-1, the monocyte-macrophage hormone which stimulates the synthesis of almost all members of the acute phase group of plasma proteins.

6. *Involvement of the complement, kinin, coagulation and fibrinolytic systems in the response to injury and inflammation*

For further understanding of the AP-response, the nature and properties of the numerous plasma proteins which together constitute the complement, kinin, coagulation and fibrinolytic systems must be considered from several further aspects. The two most important are firstly whether each plasma protein, be it enzyme or substrate, is itself an AP-protein, and secondly how far the final outcome of each system is influenced by the increased concentration of each individual AP-protein as it occurs during inflammation and after trauma. As shown in Fig. 4, one proenzyme, prothrombin in certain species but not in rats, and one substrate, fibrinogen, in the pathway to coagulation have been shown to be AP-proteins. Likewise, in the kinin-release system, one enzyme and one substrate, kallikrein and kininogen, are AP-proteins (Borges and Gordon, 1976). Owing to the nature of the cascade system which leads to clotting and the complexity of the kinin system, it is as yet impossible to evaluate quantitatively the effects of the changed concentrations of these AP-proteins on the final products, i.e. fibrin and bradykinin. Even more difficult to assess quantitatively are the effects of changes in one system on the final result occurring in another. This is particularly so if the complement system, several

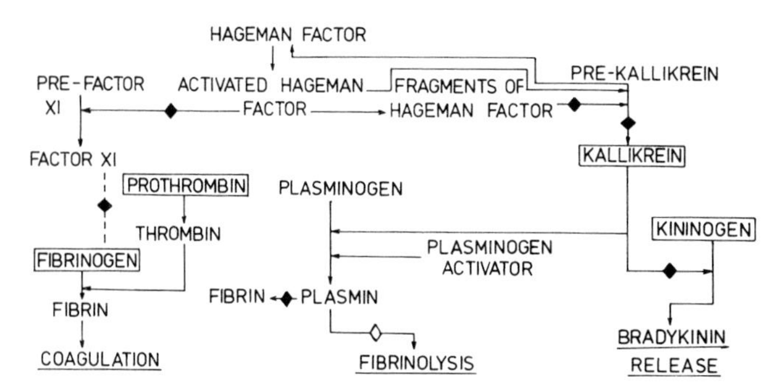

Fig. 4. Involvement of plasma proteins in coagulation, fibrinolysis and kinin-release systems. –◇–, Inhibition by α_2-antiplasmin, which is an AP-protein; –◆–, inhibition by C1 inactivator, which is also an AP-protein; □, AP-proteins.

components of which are themselves AP-proteins, is also included. 'In a few cases purified proteins have been employed and molar concentrations of enzyme required to cleave a substrate in an opposing system seem to be exceedingly large'. Although, 'This might suggest that such interactions between systems are not physiologically meaningful, the data presently available does not permit this conclusion' (Sundsmo and Fair, 1983). Most of the complement components including C2, C3, C4, C5b C9, Factor B and C3Ina have been shown to rise in concentration after injury and must thus be included as AP-proteins. However, as the changes in concentration which occur do not usually reach to much above double those in normal blood these proteins can be classified as weak AP-proteins (cf. Chapter 12, Table 2).

In an important sense, however, classification of the complement constituents, if based solely on their increase in concentration in the plasma, is inadequate because following any stimulus sufficient to initiate complement activation rapid transformation and finally removal from the plasma of a certain proportion of each of the complement constituent takes place. Thus increased plasma concentration, if it occurs, will be the result of an increased synthesis rate sufficient to more than compensate for such loss. Evidently a more satisfactory classification of complement constituents and other AP-proteins must be based on their increased rates of synthesis following a particular stimulus. So far, only a few complement constituents have been investigated in sufficient detail to provide data on which such comparison might be made. An example of such work is the investigation of C4 synthesis by murine macrophages, in which pool size and steady state synthesis rate as well as the nature of the precursor molecule were all determined (Fey et al., 1980).

Those complement components which increase in concentration after activation can be said to differ from other AP-proteins, primarily in respect to the nature of the stimulus which leads to the increase in concentration in question. Thus, complement activation results from a rather narrower set of stimuli than are required for initiation of the AP-response of the non-complement proteins. The degree of overlap between the two responses has not been thoroughly investigated and for the present whether a slight degree of complement activation always occurs as part of the AP-response, must remain an open question. Considered in the most general terms, complement activation, the release of kinins, coagulation and fibrinolysis and the synthesis of immunoglobulins are all parts of the AP-response and are all different aspects of the body's defence and repair mechanism. The degree of interdependence of the immune response and the AP-response has been described (cf. Chapter 13). Looked at from this point of view the question inevitably arises as to whether all three responses have developed from a single primitive defence mechanism. While this concept has been attractive theoretically for some time, compelling data in its support have been absent (cf. Chapter 18, Section 3 for genetic evidence).

Recently, with the cloning and sequencing of the whole of human α_2-macroglobulin (Sottrup-Jensen et al., 1984), a part of rat α_2-macroglobulin (Kan et al., manuscript in preparation) and a similar study of human C3 cDNA, the situation has changed abruptly. Thus, sufficient sequence homology between human α_2-macroglobulin and murine pro C3 (Fey et al., 1984), have been revealed to indicate that both proteins must have originated from a common ancestral protein (cf. also Chapter 14).

Strongly favouring this concept is evidence that the proteinase binding property of α_2-macroglobulin and the haemolytic activity of C3 depend on the presence of a common structural feature, an activatable internal β-cysteinyl-γ-glutamyl thiol ester (cf. Chapter 13, Section 1). This and other common features of α_2-macroglobulin and complement components C3 and C4 are discussed by Sottrup-Jensen et al. (1984).

7. *Effects of C5a on monocytes*

Although the observed changes in concentration of complement constituents are much less than those shown by certain other AP-proteins, their physiological importance may be considerable for one or more of several reasons. One of these derives from the fact that unlike other AP-proteins, small proportions of certain of the

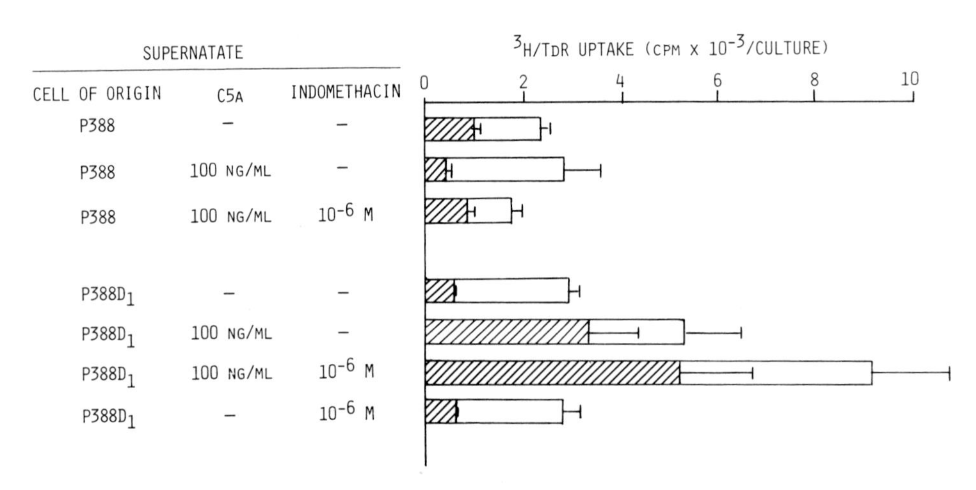

Fig. 5. IL-1 induction by C5a. Thymocytes from C3H/St mice, 5 – 7 weeks of age, were cultured at 5 × 10^5/ml of medium containing 10% heat-inactivated FCS and 50 μM 2-mercaptoethanol. Supernatants were assayed at a 1:4 dilution and Con A was added at 3 μg/ml. Cultures were pulsed with 1 μCi [^{3}H]thymidine for the final 6 h of the 3-day incubation period. Results are presented as the arithmetic means of five replicate cultures ± SEM. (Wiegle et al., 1983.)

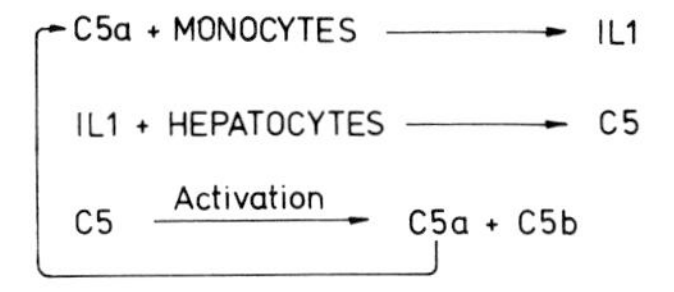

Fig. 6. Diagrammatic representation of how C5 activation may lead to stimulation of IL-1 synthesis by positive feedback.

complement proteins are synthesised in extra hepatic sites. These sites have been shown to include spleen, lymphoid cells and tissues of the reticuloendothelial system as well as blood monocytes and a few epithelial organs (Fey and Colten, 1981). Investigation of cells derived from murine monocytes, i.e. P388/D_1 cells and ^{125}I-labelled C5a, has revealed the existence of as many as 10^5 C5a receptors on each of these cells (Wiegle et al., 1983). As shown in Fig. 5, C5a uptake by these cells leads to secretion of IL-1. Thus, as indicated in Fig. 6, as a result of positive feedback, increased synthesis of C5 from both monocytes and hepatocytes must be presumed to occur. Owing to the rather similar properties of C3a, similar control of C3 synthesis may exist. These findings are of interest also in relation to the clotting, fibrinolysis and kinin-release systems, because as already mentioned, certain of their constituent proteins are themselves AP-proteins and thus will be affected by IL-1 released by C5a. The importance of the small proportion of C5 formed by monocytes may also be due to changes in its local concentration of IL-1.

The existence of a degree of interdependence between coagulation, fibrinolysis and bradykinin release is evident from Fig. 4 and the existence of a relation between these systems and the complement cascade due to the ability of thrombin, kallikrein and plasmin to cleave complement proteins is also known (Sundsmo and Fair, 1983). Because of the enzymes leading to coagulation and to kinin release, and also of the first constituent of the complement cascade, the role of C1 inactivator is of special interest (cf. Chapter 13, Table 1). The activity of this protease inhibitor tends to retard conversion by Hageman factor of profactor XI into factor XI with the final result that coagulation is slowed down. Similarly, C1 inactivator tends to retard conversion by fragments derived from Hageman factor of pre-kallikrein into kallikrein and thus ultimately the release of bradykinin. Thus when, as a result of injury, all three pathways, clotting, fibrinolysis and kinin release are activated, the degree to which they function may be influenced by the concentration of C1 inactivator present. Consequently, as C1 inactivator is an AP-protein, its concentration in the plasma will depend on the severity of the injury and at each concentration of C1 inactivator there will be an appropriate degree of inhibition to keep in check over activity of each of the three systems.

Just as C1 inactivator inhibits clotting, fibrinolysis and kinin release the concentrations of factor X11 and plasma kallikrein must affect the functioning of the complement cascade. This is because these components, being proteases, will compete with C1 for C1 inactivator.

The function of C1 inactivator as an inhibitor of all four systems will be counteracted as each system is activated; in this way certain intermediates of each system will promote the functioning of the other systems. Although the interrelations of these systems have been considered in great detail, e.g. by Bennett and Ogston (1981), both in physiological normality and states of disease, the consequences of changes in concentration of AP-proteins has attracted little or no attention. Even in the case of high molecular weight kininogen which, since 1975, has been known to play a central role in reactions which involve contact activation, acting as a cofactor stimulating the activation of pre-kallikrein by factor X11a (Griffin and Cochrane, 1976; Heimark et al., 1980) evidence is lacking as to effects which may follow increases in its concentration in plasma after injury. In the absence of relevant data it may be permissible to speculate that just as C1 inactivator may act as an inhibitor for all three systems, high molecular kininogen tends to increase, at least coagulation and kinin release.

As mentioned above, the fact that release of C5a leads to formation of IL-1 and thus presumably to increased synthesis of those AP-proteins which are involved in clotting, fibrinolysis and kinin formation emphasises the degree of interdependence of the four systems. However, despite the fact that synthesis rates of AP-proteins may in certain circumstances be thus affected, experiments involving decomplementation, i.e. depletion of circulating C3 by injection of cobra venom factor have shown that this procedure has little or no effect on the AP-response (Pepys and Rogers, 1980). This ambiguity illustrates how limited the present understanding is of this extremely complex field.

References (Chapters 5 – 11)

Allison, E.S., Cranston, W.I., Duff, G.W., Luff, R.H. and Rawlins, M.D. (1973) Clin. Sci. Mol. Med. *45*, 449 – 459.

Atkins, E. and Wood, W.B. (1955) J. Exp. Med. *102*, 499 – 516.

Bailey, P.T., Abeles, F.B., Hauer, E.C. and Mapes, C.A. (1976) Proc. Soc. Exp. Biol and Med. *153*, 419 – 423.

Baracos, V., Rodemann, H.P., Dinarello, C.A. and Goldberg, A.L. (1983) N. Engl. J. Med. *308*, 553 – 558.

Barker, J.L. and Carpenter, D.O. (1970) Science *169*, 597 – 598.

Beisel, W.R. (1981) in Infection, The Physiologic and Metabolic responses of the Host (Powanda, M.C. and Canonico, P.G., eds.) pp. 147 – 172, Elsevier/North-Holland Biomedical Press, Amsterdam.

Beisel, W.R., Sawyer, W.D., Ryll, E.D. and Crozier, D. (1973) Ann. Intern. Med. *67*, 744 – 779.

Bendtzen, K. (1983) Allergy *38*, 219 – 226.

Bendtzen, K., Baeck, L., Berild, D., Hasselbach, H., Dinarello, C.A. and Wolff, S.M. (1984) N. Engl. J. Med. *310,* 596.
Bennett, I.L. and Beeson, P.B. (1953) J. Exp. Med. *98*, 493 – 508.
Bennett, B. and Ogston, D. (1981) in Haemostasis and Thrombosis (Bloom, A.L. and Thomas, D.P., eds.) pp. 236 – 251, Churchill Livingstone, Edinburgh.
Benson, M.D., Aldo-Benson, M.A., Shirahama, T., Borel, Y. and Cohen, A.S. (1975) J. Exp. Med. *142*, 236 – 241.
Blatteis, C.M., Bealer, S.L., Hunter, W.S., Llanos, Q.J., Ahokas, R.A. and Mashburn, T.A. (1983) Brain Res. Bull. *11*, 519 – 526.
Borges, D.E. and Gordon, A.H. (1976) J. Pharm. Pharmacol *28*, 44 – 48.
Brightman, M.W., Prescott, L. and Reese, T.S. (1975) in Brain-Endocrine International II. The Ventricular System, 2nd Int. Symp. Shizuoka 1974, pp. 146 – 165. Karger, Basel.
Broadwell, R.D., Balin, B.J., Salcman, M. and Kaplan, R.S. (1983) Proc. Natl. Acad. Sci. USA *80*, 7352 – 7356.
Cannon, J.C. and Dinarello, C.A. (1984) Fed. Proc. *43*, 462.
Cannon, J.C. and Kluger, M.J. (1983) Science *220*, 617 – 619.
Chedid, L.A., Audibert, F.M. and Johnson, A. (1978) Prog. Allergy *25,* 63 – 105.
Chedid, L.A., Parant, M.A., Audibert, F.M., Riveau, G.J., Parant, F.J., Lederer, E. Choay, J.P. and Le Francier, P.L. (1982) Infect. Immun. *35,* 417 – 424.
Chu, E., Rosenwasser, L.J., Dinarello, C.A., Loreau, M. and Geha, R.S. (1984) J. Immunol. *132*, 1311 – 1316.
Clowes, G.H.A., George, B.C., Villee, C.A. and Saravis, C.A. (1983) N. Engl. J. Med. *308*, 545 – 552.
Coleman, W. and DuBois, E.F. (1915) Arch. Int. Med. *15,* 887 – 938.
Conlon, P.J., Henney, C.S. and Gillis, S. (1982) J. Immunol. 128, 797 – 801.
Cooper, K.E., Cranston, W.I. and Honour, A.J. (1967) J. Physiol. (London) *191*, 325 – 337.
Cranston, W.I., Hellon, R.F. and Mitchell, D. (1975) J. Physiol. (London) *253*, 583 – 592.
Cranston, W.I., Duff, G.W., Hellon, R.F. and Mitchell, D. (1976) J. Physiol. (London) *256*, 120P – 121P.
Cranston, W.I., Dawson, N.J., Hellon, R.F. and Townsend, Y. (1978) J. Physiol (London) *285*, 35.
Cranston, W.I., Hellon, R.F. and Townsend, Y. (1980) J. Physiol. (London) *305*, 337 – 344.
Cranston, W.I., Hellon, R.F. and Townsend, Y. (1982) J. Physiol. (London) *322*,, 441 – 445.
Cryer, P.E., Coran, A.G., Sode, J., Hennen, C.M. and Horwitz, D.L. (1972) J. Lab. Clin. Med. *79*, 622 – 638.
Dayer, J.M. (1981) FEBS Lett. *124*, 253.
Dayer, J.M., Passwell, J.H., Schneeberger, E.E. and Krane, S.M. (1980) J. Immunol. *124*, 1712 – 1720.
Defronze, R.A., Sherwin, R.S. and Felig, P. (1980) Acta Chir. Scand. Suppl. *498*, 33.
Desmukh-Phadke, K., Nanda, S. and Leek, K. (1980) Eur. J. Biochem. *104*, 175 – 180.
Dinarello, C.A., (1982) in Hormones Drugs (Gueriguian, J.L., Bransome, E.D. and Outschoorn, A.D., eds.) pp. 36 – 47. U.S. Pharm. Convention Inc., Rockville USA.
Dinarello, C.A., (1984) Rev. Infect. Dis. *6*, 51 – 94.
Dinarello, C.A., Rosenwasser, L.J. and Wolff, S.M. (1982) J. Immunol. *127*, 2517 – 2519.
Dinarello, C.A. and Bernheim, H.A. (1981) J. Neurochem. *37*, 702 – 708.
Dinarello, C.A., Marnoy, S.O. and Rosenwasser, L.J. (1983) J. Immunol. *130*, 890 – 895.
Dinarello, C.A., Bodel, P.T. and Atkins, E. (1968) Trans. Assoc. Am. Physicians *81*, 334 – 344.
Dinarello, C.A., Clowes, G.H.A., Gordon, A.H., Saravis, C.A. and Wolff, S.M. (1984) J. Immunol. *133*, 1332 – 1338.
Ellouz, F., Adam, A., Ciovaru, R. and Lederer, E. (1974) Biochem. Biophys. Res. Commun. *59*, 1317 – 1325.

Fantone, J.C. and Ward, P.A. (1982) Am. J. Pathol. *107*, 397 – 418.
Farrar, J.J. and Hilfiker, M.L. (1982) Fed. Proc. *41*, 263 – 268.
Farrar, W.R., Mizel, B. and Farrar, J.J. (1980) J. Immunol. *124*, 1371 – 1377.
Fauci, A.S. (1979) J. Immunopharmacol *1*, 1 – 25.
Feinberg, R.F., Sun, L.-H.K., Ordahl, C.P. and Frankel, F.R. (1983) Proc. Natl. Acad. Sci. USA *80*, 5042 – 5046.
Feldberg, W. and Milton, A.S. (1978) in Handbook of Experimental Pharmacology (Vane, J.R. and Ferriera, S.H., eds.) pp. 617 – 656, Vol. 50, Springer-Verlag, Berlin.
Feldberg, W., Gupta, K.P., Milton, A.S. and Wendlandt, S. (1972) Br. J. Pharmacol. *46*, 550 – 551P.
Feldberg, W., Gupta, K.P., Milton, A.S. and Wendlandt, S. (1973) J. Physiol. (London) *234*, 279 – 293.
Fey, G.H. and Colten, H.R. (1981) Fed. Proc. *40*, 2099 – 2104.
Fey, G.H., Lundwall, A., Wetsel, R.A., Tack, B.F., DeBruijn, M.H.L. and Domdey, H. (1984) Philos. Trans R. Soc. London, Ser. B., in press.
Fontana, A., Kristensen, F., Dubs, R., Gemsa, D. and Weber, E. (1982) J. Immunol. *129*, 2413 – 2419.
Fontana, A., McAdam, K.P.W.J., Kristensen, F. and Weber, E. (1983) Eur. J. Immunol. *13*, 685 – 689.
Forsmann, W.G. and Ito, S. (1977) J. Cell. Biol. *74*, 299 – 313.
Gery, I., Gershon, R.K. and Waksman, B.H. (1972) J. Exp. Med. *136*, 128 – 142.
Goodwin, J.S., Bankhurst, A.D. and Messner, R.P. (1977) J. Exp. Med. *146*, 1719 – 1733.
Gordon, A.H. and Limaos, E.A. (1979) Br. J. Exp. Pathol. *60*, 441 – 446.
Gordon, A.H. and Parker, J.D. (1980) Br. J. Exp. Pathol. *61*, 534 – 539.
Griffin, J.H. and Cochrane, C.G. (1976) Proc. Natl. Acad. Sci. USA *73*, 2554 – 2558.
Heimark, R.L., Kurachi, K., Fujikawa, K. and Davie, E.W. (1980) Nature (London) *286*, 456 – 460.
Hellon, R.F., Cranston, W.I., Townsend, Y., Mitchell, D., Dawson, N.J. and Duff, G.W. (1980) in Fever (Lipton, J.M., ed.) pp. 159 – 164, Raven Press, New York.
Hoffman, M.K. (1980) J. Immunol. *125*, 2076 – 2081.
Hori, T., Nakashima, T., Hori, N. and Kiyohara, T. (1980) Brain Res. *186*, 203 – 207.
Jandinski, J., Cantor, H., Tadakuma, T., Peavy, D.L. and Pierce, C.W. (1976) J. Exp. Med. *143*, 1382 – 1390.
Jankovic, B.D. and Isakovic, K. (1973) Int. Arch. Allergy Appl. Immunol. *45*, 360 – 372.
Jastin, H.E. and Dingle, J.T. (1981) J. Clin. Invest. *68*, 571 – 581.
Jeejeebhoy, K.N., Ho, J., Greenberg, G.R., Philips, M.J., Bruce-Robertson, A. and Sodtke, U. (1975) Biochem. J. *146*, 141 – 155.
Kampschmidt, R.F., Upchurch, H.F. (1977) Proc. Soc. Exp. Biol. Med. *155*, 89 – 93.
Kampschmidt, R.F., Upchurch, H.F. and Worthington, M.L. (1983) Infect. Immun. *41*, 6 – 10.
Kawakami, M. and Cerami, A. (1981) J. Exp. Med. *154*, 631 – 639.
Kawakami, M., Pekala, P.H., Lane, M.D. and Cerami, A. (1982) Proc. Natl. Acad. Sci. USA *79*, 912 – 916.
Klempner, M.S., Dinarello, C.A. and Gallin, J.I. (1978) J. Clin. Invest *61*, 1330 – 1336.
Klempner, M.S., Dinarello, C.A., Henderson, W.R. and Gallin, J.I. (1979) J. Clin. Invest. *64*, 996 – 1002.
Koj, A. Gauldie, J., Regoeczi, E., Sauder, D.N. and Sweeney, G.D. (1984) Biochem. J. *224*, 505 – 514.
Korn, J.H., Halshuka, P.V. and Le Roy, E.C. (1980) Int. J. Dermatol. *19*, 487 – 495.
Kristensen, F., Walker, C., Bettens, F., Joncourt, F. and de Weck, A.L. (1982) Cell. Immunol. *74*, 140 – 149.
Krueger, J.M., Pappenheimer, J.R. and Karnovsky, M.L. (1982) Proc. Natl. Acad. Sci. USA *79*, 6102 – 6106.
Laburn, H., Mitchell, D. and Rosendorff, C. (1977) J. Physiol. (London) *267*, 559 – 570.
Lachman, L.B. and Maizel, A.L. (1983) in Contemporary Topics in Molecular Immunology (Inman, F.P. and Kindt, T.J., eds.) Vol. 9, pp. 147 – 167, Plenum Press, New York.

Larsen, G.L. and Henson, P.M. (1983) Ann. Rev. Immunol. *1*, 335 – 359.
Leibovich, S.J. and Ross, R. (1975) Am. J. Pathol. *78*, 71 – 100.
Lipton, J.M. and Ticknor, C.B. (1979) J. Physiol. (London) *287*, 535 – 543.
Lipton, J.M., Welch, J.P. and Clark, W.G. (1973) Experientia *29*, 806 – 808.
Lipton, J.M., Whisenant, J.D., Gean, J.T. and Ticknor, C.B. (1979) Brain Res. Bull. *4*, 297 – 300.
McAdam, K.P.W.J., Foss, N.T., Garcia, C., Delellis, R., Chedid, L., Rees, R.J.W. and Wolff, S.M. (1983) Infect. Immun. *39*, 1147 – 1154.
Milton, A.S. and Sawhney, V.K. (1981) Br. J. Pharmol. *74*, 786P.
Milton, A.S. and Wendland, S. (1968) Br. J. Pharmacol. *34*, 215 – 216P.
Milton, A.S. and Wendlandt, S. (1970) J. Physiol. (London) *207*, 76 p.
Milton, A.S. and Wendlandt, S. (1971) J. Physiol. (London) *218*, 325 – 336.
Mizel, S.B. and Ben-Zvi, A. (1980) Cell. Immunol. *54*, 388 – 389.
Mizel, S.B. and Mizel, D. (1981) J. Immunol. *126*, 834 – 837.
Mizel, S.B., Oppenheim, J.J. and Rosenstreich, D.L. (1978) J. Immunol. *120*, 1497 – 1503.
Mizel, S.B., Dayer, J.M., Krane, S.M. and Mergenhagen, S.E. (1981) Proc. Natl. Acad. Sci. USA *78*, 2474 – 2477.
Mold, C., Nakayama, S., Holzer, T.J., Gewurz, H. and Du Clos, T.W. (1981) J. Exp. Med. *154*, 1703 – 1708.
Morley, J.J. and Kushner, I. (1982) Ann. N.Y. Acad. Sci. *389*, 406 – 417.
Munck, A., Guyne, P.M. and Holbrook, N.J. (1984) Endocrinol. Rev. *5*, 25 – 44.
Neilson, E.G., Phillips, S.M. and Jiminez, S. (1982) J. Immunol. *128*, 1484 – 1486.
Oppenheim, J.J. and Gery, I. (1983) in T Lymphocytes Today (Inglis, J.R., ed.) pp. 89 – 95, Elsevier, Amsterdam.
Parant, M., Riveau, G., Parant, F., Dinarello, C.A., Wolff, S.M. and Chedid, L. (1980) J. Infect. Dis. *142*, 708 – 715.
Pekala, P.H., Kawakami, M., Angus, C.W., Lane, M.D. and Cerami, A. (1983) Proc. Natl. Acad. Sci. USA *80*, 2743 – 2747.
Pepys, M.B. and Rogers, S.L. (1980) Br. J. Exp. Pathol. *61*, 156 – 159.
Poole, S., Gordon, A.H., Baltz, M. and Stenning, B.E. (1984) Br. J. Exp. Pathol. *65*, 431 – 439.
Postlethwaite, A.E. and Kang, A.H. (1982) Cell. Immunol. *73*, 169 – 178.
Postlethwaite, A.E., Lachman, L.B., Mainardi, C.L. and Kang, A.H. (1983) J. Exp. Med. *157*, 801 – 806.
Puri, J. and Lonai, P. (1980) Eur. J. Immunol. *10*, 273 – 281.
Puri, J., Shinitzky, M. and Lonai, P. (1980) J. Immunol. *124*, 1937 – 1942.
Rayfield, E.J., Curnow, R.T., George, D.T. and Beisel, W.R. (1973) N. Engl. J. Med. *289*, 618 – 621.
Resh, K., Heckmann, B., Schober, I., Barlin, E. and Gemsa, D. (1981) Eur. J. Immunol. *11*, 120 – 126.
Rocha, D.M., Santeusanio, F., Faloona, G.R. and Unger, R.H. (1973) N. Engl. J. Med. *288*, 700 – 703.
Rodermann, H.P. and Goldberg, A.L. (1982) J. Biol. Chem. *257*, 1632 – 1638.
Rouzer, C.A. and Cerami, A. (1980) Mol. Biochem. Parasitol, *2*, 31 – 38.
Ruwe, W.D. and Myers, R.D. (1979) Brain Res. Bull. *4*, 741 – 745.
Saklatvala, J., Pilsworth, L.M.C., Sarsfield, S.J., Gavrilovic, J. and Heath, J.K. (1984) Biochem. J. *224*, 461 – 466.
Sauder, D.N., Carter, C.S., Katz, S.I. and Oppenheim, J.J. (1982) J. Invest. Dermatol. *79*, 34 – 39.
Scala, G. and Oppenheim, J.J. (1983) J. Immunol. *131*, 1160 – 1166.
Schmidt, J.A., Mizel, S.B., Cohen, D. and Green, I. (1982) J. Immunol. *128*, 2177 – 2182.
Schoener, E.P. and Wang, S.C. (1975) Am. J. Physiol. *229*, 185 – 190.
Selye, H. (1946) J. Clin. Endocrinol. *6*, 117 – 230.
Sheagren, J.N. and Wolff, S.M. (1966) Nature (London) *210*, 539 – 540.

Sipe, J.D., Johns, H., DeMaria, A., Cohen, A.S. and McCabe, W.R. (1984) Fed. Proc. *43,* 1542.
Smith, K.A. (1984) Immunol. Today *5*, 83 – 84.
Smith, K.A., Baker, P.E., Gillis, S. and Ruscetti, F.W. (1980) Mol. Immunol. *17*, 579 – 589.
Snyder, D.S. and Unanue, E.R. (1983) J. Immunol. *129*, 1803 – 1805.
Sobrado, J., Moldawer, L.L., Bistrian, B.R., Dinarello, C.A. and Blackburn, G.L. (1983) Infect. Immun. *42*, 997 – 1005.
Sottrup-Jensen, L., Stepanik, T.M., Kristensen, T., Wierzbicke, D.M. Jones, M., Lonblad, P.B., Magnasson, S. and Petersen, T.E. (1984) J. Biol. Chem. *259*, 8318 – 8327.
Splawinski, J.A., Zacny, E. and Zbigniew, G. (1977) Pfluegers Arch. *368*, 125 – 128.
Staruch, M.J. and Wood, D.D. (1983) J. Immunol. *130*, 2191 – 2194.
Stein, M., Schivia, R.C. and Camarino, M. (1976) Science *191*, 435 – 440.
Stewart, W.E. (1979) in The Interferon System, pp. 196 – 222, Springer-Verlag, Berlin.
Stobo, J.D., Kennedy, M.S. and Goldyne, M.E. (1979) J. Clin. Invest. *64*, 1188 – 1195.
Sundsmo, J.S. and Fair, D.S. (1983) Springer Semin. Immunpathol. *6*, 231 – 258.
Tocco, R.J., Khan, L.L., Kluger, M.J. and Vander, A.J. (1983) Am. J. Physiol. *244*, R368 – R373.
Townsend, Y., Gourine, V.N. and Cranston, W.I. (1981) IRCS Med. Sci. *9,* 89.
Turchick, J.B. and Bornstein, D.L. (1980) Infect. Immun. *30*, 439 – 444.
Vane, J.R. (1971) Nature (London) New Biol. *231,* 232 – 235.
Van Gool, J., Boers, W., Sala, M. and Ladiges, N.C.J.J. (1984) Biochem. J. *220*, 125 – 132.
Veale, W.L. and Cooper, K.E. (1975) in Temperature Regulation and Drug Action (Lomax, P., Schonbaum, G. and Jacobs, J., eds.) pp. 218 – 226, Karger, Basel.
Wahl, L.M., Olsen, C.E., Sandberg, A.L. and Mergenhagen, S.E. (1977) Proc. Natl. Ac. Sci. USA *74*, 4955 – 4958.
Wannemacher, R.W. (1977) Am. J. Clin. Nutr. *30*, 1269 – 1280.
Wetzel, G.D., Swain, L. and Dutton, R.W. (1982) J. Exp. Med. *15*, 306 – 311.
Wicher, J.T., Newhouse, Y.G. and Mortensen, R.F. (1983) J. Immunol. *130*, 248 – 253.
Wiegle, W.O., Goodman, M.G., Morgan, E.L. and Hugli, T.E. (1983) Springer Semi. Immunopthol. *6*, 173 – 194.

PART III

Liver response to inflammation and synthesis of acute-phase plasma proteins

Gordon/Koj (eds.) The acute-phase response to injury and infection.

CHAPTER 12

Definition and classification of acute-phase proteins

A. KOJ

Liver response to injury occupies a central position in the acute-phase reaction (AP-reaction) but is intimately linked to other adaptative measures such as fever, leucocytosis and enhanced protein degradation in muscles. As early as 2 h after surgery or injection of bacterial endotoxin there occurs an increased uptake of amino acids by the liver. This is followed by accumulation of some metals (iron and zinc) changes in the activities of several enzymes (glycosyltransferases, fatty acid synthetase, tyrosine aminotransferase), profound alterations in RNA metabolism (increased formation of rRNA and maturation of ribosomes; enhanced transcription and processing of many specific mRNAs), and finally, by increased, or reduced, synthesis of several plasma proteins. AP-proteins have been defined (cf. Introduction) as those plasma proteins synthesized predominantly in liver parenchymal cells whose concentration in the blood is swiftly and markedly increased after injury, as it occurs with C-reactive protein (CRP), α_1-acid glycoprotein (α_1-AGP) or fibrinogen. Later it became clear that reduced concentration of albumin, transferrin or α_2-HS-glycoprotein is also a common feature of the AP-response and the term 'negative AP-reactants' was introduced (Lebreton et al., 1979). Unfortunately this term is in many ways ambiguous and perhaps should be replaced by one such as 'concentration depressed' acute-phase plasma proteins, especially in the studies in vivo based on measurements of protein concentration in the blood. Until recently it was possible to believe that the fall in plasma protein concentration of albumin or transferrin after injury might be solely explained by their transfer to the extravascular space or by increased catabolism (cf. Chapters 20 and 24). However, experiments both in vivo and in vitro (cf. Chapter 17) have shown that during the AP-response the synthesis of albumin may be significantly reduced in parallel with a drop in the level of its mRNA in the liver cell while at the same time 'positive' AP-proteins show increased synthesis and increased abundance of mRNA. This inhibition of albumin synthesis after injury has not been observed in all animal species and in all experimental models. By using [^{14}C]carbonate method, Koj and McFarlane (1968) showed increased synthesis of not only fibrinogen but also of albumin in rabbits injected with endotoxin. Balegno and Neuhaus (1970) observed a transient, insulin-dependent rise in the incorporation of [^{14}C]glycine into albumin in rats subjected to laparatomy. In case of transferrin, the situation is also complicated since reduced plasma concentrations of this protein during the AP-response are not accompanied by a drop in the level of corresponding liver mRNA (cf. Chapter 15, Table 1 and Chapter 17). Clearly the 'negative' or 'concentration depressed' acute-phase plasma proteins represent a heterogenous group and more research is needed to elucidate the underlying mechanisms.

In the early stages of the inflammatory reaction some synthesis-stimulated AP-proteins show a transient fall in plasma concentration probably resulting not only from increased vascular permeability but also from their enhanced consumption or catabolism (e.g. haptoglobin – Lombart et al., 1968; Koj, 1970; proteinase inhibitors – Fritz, 1980). The subsequent rate of increase and maximum concentrations of particular AP-proteins differ widely: change in the level of CRP, α_1-antichymotrypsin (α_1-ACh), α_1-AGP, fibrinogen and haptoglobin is much faster in comparison with ceruloplasmin (Aronsen et al., 1972). In addition, the kinetics of these changes depends on the type of injury or disease, on the previous occurrence of inflammatory episodes (cf. Chapter 18) and even is influenced by such poorly defined phenomena as individual responsiveness. For all these reasons the term 'acute-phase proteins' is rather vague and analysis of individual well defined plasma proteins in specific pathological states is much to be preferred. The picture is complicated by the fact that apart from the principal blood constituents (occurring in micromolar concentrations) such as albumin, fibrinogen, α_1-AGP etc., plasma contains also a great variety of trace protein components (occurring in nanomolar concentrations) which are detectable mainly by the virtue of their biological activities, e.g. certain clotting factors, enzymes or hormones. Some of these are synthesized in the liver cells at rates which may be significantly affected by injury and thus fall into the class of AP-reactants but in principle they will not be discussed here.

A brief review of physicochemical properties and structure of human AP-proteins (Table 1) shows that they are highly heterogenous in respect to molecular weight, isoelectric points, carbohydrate content etc., hence no general rule can be drawn. Formerly three groups of AP-proteins were distinguished depending on their presence in normal and foetal serum (Koj, 1974):

TABLE 1

Some physicochemical and structural properties of principal positive (synthesis stimulated) and negative AP-plasma proteins listed in order of increasing molecular weight

Protein	Concentration (mg/ml)		M_r	No. of polypeptide chains	Carbohydrate contents (%)	Complexes with or binds	Ref.
	Normal plasma	AP-plasma					
Serum amyloid A (SAA)	ca. 0.01	1.0 – 2.0	11 – 14,000 (subunit)	1[a]	0	HDL_3 lipoprotein	B
α_1-AGP	0.6 – 1.2	1.5 – 2.5	40,000	1	42	–	A, C, H, O
Prealbumin	0.3 – 0.4	0.1 – 0.3	52,000	4	0	Thyroxine vitamin A	A, K, O
α_1-proteinase inhibitor (α_1-PI)	1.5 – 3.0	4.0 – 6.0	54,000	1	12	–	A, H, J, O
α_2-HS-glycoprotein	0.4 – 0.8	0.2 – 0.4	56,000	1 – 2[b]	13 – 14	Ba (Ca)	A, D, H
Albumin	35 – 45	20 – 30	66,000	1	0	Anions, drugs	A, H, O
α_1-ACh	0.2 – 0.6	1.5 – 2.0	68,000	1	24	–	I, J
α_2-antiplasmin (α_2-API)	0.06 – 0.1	0.1 – 0.2	68,000	1	11	–	G, J
Transferrin	2.0 – 3.0	1.5 – 2.0	77,000	1	6	2 Fe	A, H, O
Haptoglobin (1 – 1, 2 – 1, 2 – 2)	1.0 – 2.0	5.0 – 6.0	85 – 400,000[c]	4	20	–	A, H, O
C1-inactivator	0.15 – 0.23	0.3 – 0.4	100,000	1	34	–	E, F, G
C-reactive protein	0.001	0.3 – 0.5	105,500	5	0	(Ca)	B
Ceruloplasmin	0.3 – 0.4	0.4 – 0.6	132,000	1	9	8 Cu	A, H, O
C3 component of complement	1.0 – 1.6	2.0 – 2.5	180,000	2	3	–	A, L, O, P
Fibrinogen	1.9 – 3.3	7.0 – 9.0	340,000	6	5	–	A, M, O

[a] Isolated after denaturation.

[b] Conflicting reports.

[c] Depending on phenotype.

A, Allen et al. (1977); B, Pepys and Baltz (1983); C, Charlwood et al. (1976); D, Gejyo and Schmid (1981); E, Nilsson and Wiman (1982); F, Heimburger (1975); G, Hedner et al. (1983); H, Lebreton et al. (1979); I, Bowen et al. (1982); J, Travis and Salvesen (1983); K, Aronsen et al. (1972); L, Alper (1974); M, Regoeczi (1974); N, Giblett (1974); O, Putnam (1975); P, Fey et al. (1983).

TABLE 2
Classification of plasma proteins based on changes of their concentration in response to trauma

A	B	C	D	E
CRP (man, rabbit) 1, 2, 3	α_1-AGP (many species) 1, 6, 7	Ceruloplasmin (many species) 1,11	α_2-M (man, dog) 6, 19, 20	Albumin (many species) 6, 19, 20
SAA protein (man, mouse) 3, 4, 5	Fibrinogen (many species) 1, 6, 7	α_1-PI (rat, rabbit) 9, 10	α_1-M (rat, dog) 7, 21	Transferrin (many species) 6, 19
α_2-M (rat) 1,7	Haptoglobin (many species) 1, 6, 7	C3 component of complement (man, mouse) 1, 12, 32	Antithrombin III (AT III) (rat, rabbit) 10, 7, 22	α_2-HS-glycoprotein (man) 20, 28
	α_1-AP-globulin (rat) 1,7	α_2-API (man, rat) 13	Haemopexin (man) 23	α_1-Lipoprotein (man) 6, 25
	α_1-ACh (man) 6	Kininogen (rat) 14	SAP component (man, rat) 2, 24	Prealbumin (man, rat) 6, 29
	α_1-PI (man) 1, 6	Prekallikrein (rat) 15	Immunoglobulins (many species) 6, 25	α_1-Inhibitor$_3$ (rat) 30
	SAP component (mouse) 2	Haemopexin (rat) 16, 17	Prothrombin (rat, man) 6, 27	Inter-α-antitrypsin (man) 12, 31
	α_1-M (rabbit) 8	C1-inactivator (man) 18		
		CRP (rat) 26		

A = very strong or spectacular AP-reactants, usually increase 20 to 1000-fold; B = strong AP-reactants, usually increase 2 to 5-fold; C = weak AP-reactants, usually increase by 30 – 60%; D = 'neutral' proteins, show no regular and prompt change in concentration; E = 'negative' or synthesis-inhibited AP-reactants, usually decrease by 30 – 60%. For the sake of brevity references are limited to review articles rather than original papers. 1, Koj (1974); 2, Pepys (1981a); 3, Kushner et al. (1981); 4, Sipe (1978); 5, McAdam et al. (1978); 6, Aronsen et al. (1972); 7, Koj et al. (1982); 8, Lebreton et al. (1970); 9, Koj et al. (1978b); 10, Koj and Regoeczi (1978); 11, Voelkel et al. (1978); 12, Pepys (1979); 13, Matsuda et al. (1980); 14, Borges and Gordon (1976); 15, Limaos et al. (1981); 16, Abd-el-Fattah et al. (1981); 17, Merriman et al. (1978); 18, Hedner et al. (1983); 19, Smith et al. (1977); 20, Lebreton et al. (1979); 21, Ohlsson (1971); 22, Owen and Miller (1980); 23, Kushner and Müller-Eberhard (1972); 24, Pontet et al. (1981); 25, Fleck (1976); 26, de Beer et al. (1982); 27, Koj et al. (1984b); 28, Gejyo and Schmid (1981); 29, Scherer et al. (1977); 30, Gauthier and Ohlsson (1978); 31, Frank and Pedersen (1983); 32, Fey et al. (1983).

(1) 'New' or induced proteins, which appear only in pathological states, such as CRP in man, but are absent in foetal and neonatal blood (cf. Chapter 21).
(2) 'Foetal' proteins, such as α_2-macroglobulin (α_2-M) of the rat named also α-macrofoetoprotein. Its synthesis stops soon after birth but may be resumed after injury.
(3) 'Normal' or constitutive proteins, such as fibrinogen or α_1-AGP, which are always present in plasma but their synthesis increases in inflammatory states.

However, as the widespread use of radioimmunoassay and other techniques enabling detection of trace amounts of plasma constituents has demonstrated that all these 'new' or 'foetal' proteins can always be detected in the blood of apparently healthy individuals (cf. Chapters 21 and 24) the difference between constitutive and induced proteins has become arbitrary.

When the magnitude of the relative changes in the concentrations of individual proteins occurring within the first 2 – 3 days after injury is taken into account, at least five classes of plasma proteins may be distinguished (Table 2). 'Spectacular' or very strong AP-reactants which occur in healthy individuals in trace amounts correspond to 'new' or 'foetal' proteins. Their rise to a concentration of 0.5 mg/ml may represent a 50 to 500-fold increase (α_2-M in rat, CRP in man). If, however, changes in plasma protein concentration after injury are reported in absolute terms the leading position will be occupied by such proteins as fibrinogen and haptoglobin with increases of 1 – 5 mg/ml, or by albumin which decreases by 5 – 10 mg/ml. Hence, spectacular proteins are not necessarily the major products of the stimulated liver cell and both relative and absolute changes should always be compared when evaluating the magnitude of the response of individual acute-phase plasma proteins.

Any attempt at classification of AP-proteins is bound to be difficult because of their species-related variability. CRP is one of many known examples since it is a spectacular AP-protein in man, monkey, rabbit and dog, but a weak AP-reactant in rat. It is probably relevant that healthy rats have a rather high concentration of CRP in plasma. Even such closely related species as rat and mouse differ greatly in respect to serum amyloid P (SAP) component and α_2-M (cf. Table 2). Differences have also been found between various strains of rats in respect of induced synthesis of α_2-M (Weimer et al., 1972) and between various strains of mice in respect to plasma levels and changes after injury of haptoglobin (Peacock et al., 1967; Baumann et al., 1983b), ceruloplasmin (Meier and MacPike, 1968) and SAP component (Pepys et al., 1979a). Finally, the magnitude of the AP-response of a particular protein often depends on the sex of the animal. Bosanquet et al. (1976) reported considerable differences in the changes of plasma level of α_1- and α_2-M of male and female rats after injection with cortisol and turpentine. Probably the most striking example of the sexual dimorphism in the AP-response has been described in hamsters by Coe and Ross (1983) for a protein belonging to the pentraxin family and resembling CRP or SAP. Hamster females show a high level of this protein in their plasma (approximately 1.2 mg/ml) which is reduced by 50% during the inflammatory episode, while in males the initial low level (approximately 0.025 mg/ml) is increased approximately 3-fold (cf. Chapter 14, Table 2). All these data taken together suggest that classification of a protein in respect to the AP-response must be relative and will depend not only on species but also on the particular genotype and sex of the animal.

References, p. 232.

Gordon/Koj (eds.) The acute-phase response to injury and infection.

CHAPTER 13

Biological functions of acute-phase proteins

A. KOJ

The AP-proteins represent a highly heterogenous group of plasma proteins not only in respect to their physicochemical properties (cf. Chapter 12, Table 1) but also to their biological functions. They may be found among almost all the main groups of plasma proteins including proteinase inhibitors, coagulation proteins, enzymes, transport proteins and modulators of the immune response (Geisow and Gordon, 1978). In some cases their functions are well established, in others less so. However, all available evidence points to the fact that all AP-proteins have at least one common function: they are engaged in various physiological processes which tend to restore the delicate hemeostatic balance disturbed by the injury, tissue necrosis or infection (Table 1).

1. Inhibition of proteinases

Bacterial invasion, injury and cell death lead invariably to local release of lysosomal hydrolases (Davies and Allison, 1976) which then digest and remove necrotic tissues. This process is greatly enhanced by infiltrating polymorphonuclear leucocytes and macrophages equipped with a set of neutral proteinases of various specificities: collagenase, elastase and cathepsin G (for references see Chapter 1, and also Havemann and Janoff, 1978; Lorand, 1981). Excessive and unrestricted action of these proteinases may lead to secondary tissue damage. At the same time injury causes activation of zymogens of serine proteinases involved in clotting, comple-

TABLE 1
Homeostatic functions of AP-proteins

Function	AP-proteins engaged
Inhibition of proteinases	Macroglobulins, α_1-PI, α_1-ACh, ITI, α_1-AP-globulin, haptoglobin
Blood clotting and fibrinolysis	Fibrinogen, α_2-API, C1-inactivator, α_1-AGP, SAP, CRP
Removal of foreign materials from the organism	CRP, SAA, SAP, C3 complement, fibrinogen
Modulation of immunological response of the host	Proteinase inhibitors, CRP, C3 complement, α_2-HS-glycoprotein, α_1-AGP, fibrinogen, haptoglobin
Anti-inflammatory properties	Proteinase inhibitors, fibrinopeptides, haptoglobin, ceruloplasmin
Binding and transport of metals and biologically active compounds	Haptoglobin, haemopexin, transferrin, ceruloplasmin, prealbumin, albumin, α_1-AGP, macroglobulins

TABLE 2
Concentration in plasma of principal proteinase inhibitors of human blood (based on the data from: Heimburger, 1975; Collen and Wiman, 1978; Fritz, 1980; Daniels et al., 1974; Macartney and Tschesche, 1983; Travis and Salvesen, 1983)

Inhibitor	Concentration (μM)	(%)	AP-response
α_1-PI	53.7	72.0	+ +
α_1-ACh	7.0	9.4	+ + +
AT III	3.6	4.9	0
α_2-M	3.6	4.9	0
ITI	3.1	4.1	−*
C1-Inactivator	2.2	2.9	+
α_2-API	1.0	1.3	+
β_1-AC	0.4	0.5	?
Total	72.6	100.0	

The inhibitors are listed in decreasing order of proteinase-binding capacity in plasma (based on their M_r and assuming that 1 μmol of the inhibitor can inactivate 1 μmol of a proteinase). Changes in plasma concentration during the AP-response are expressed in relative terms: − = reduced concentration; 0 = no change; + = slight increase ($\leqslant$ 50%); + + = significant increase (2 to 3-fold); + + + = very strong increase (3 to 5-fold); * = decreased concentration of native inhibitor due to its degradation. (Frank and Pedersen, 1983.)

ment and kinin-forming cascades; processes, which if uncontrolled could lead to very adverse physiological consequences (Chapter 11; Kaplan et al., 1982; Sundsmo and Fair, 1983). Only the presence in the blood and tissue fluids of numerous proteins with antiproteolytic activity prevents widespread tissue damage during acute inflammation (cf. Heimburger, 1975 and Fig. 1).

The concept that the biological integrity of a variety of tissues is intimately linked with the availability and proper functioning of serum proteinase inhibitors is relatively new. Well known examples in support of this idea are the high incidence of chronic obstructive lung disease and liver cirrhosis in inborn α_1-PI deficiency (for review see Kueppers and Black, 1974; Carrell et al., 1982; Travis and Salvesen,

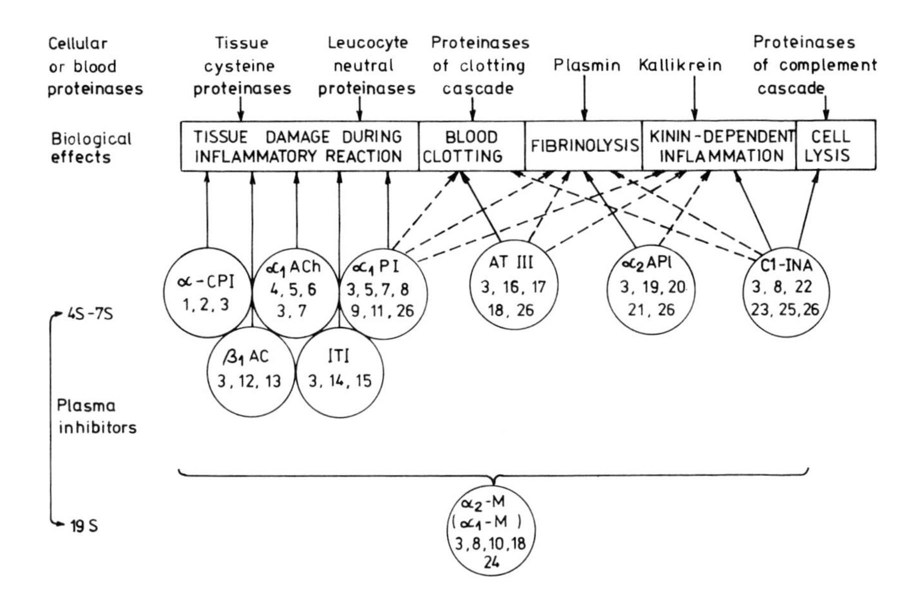

Fig. 1. Homeostatic functions of some plasma antiproteinases. →, Main targets of a given proteinase inhibitor; - - - →, secondary targets (less effective inhibition); α_1-CPI, α_1-cysteine proteinase inhibitor; β_1-AC, β_1-anticollagenase; α_1-ACh, α_1-antichymotrypsin; ITI, inter-α-trypsin inhibitor; α_1-PI, α_1-proteinase inhibitor (formerly known as α_1-antitrypsin); AT III, antithrombin III; α_2-APl, α_2-antiplasmin; Cl-INA. Cl-esterase inactivator. Numbers in circles are principal references to literature. 1, Ryley, 1979; 2, Esnard and Gauthier, 1983; 3, Travis and Salvesen, 1983; 4, Ohlsson and Akesson, 1976; 5, Travis et al., 1978; 6, Moroi and Travis, 1983; 7, Ohlsson, 1978; 8, Heimburger, 1975; 9, Kueppers and Black, 1974; 10, Ohlsson, 1975; 11, Hirano et al., 1984; 12, Wolley et al., 1976; 13, Macartney and Tschesche, 1983; 14, Albrecht et al., 1983; 15, Jochum and Bittner, 1983; 16, Jesty, 1978; 17, Stead et al., 1976; 18, Steinbuch and Audran, 1974; 19, Collen and Wiman, 1978; 20, Moroi and Aoki, 1977; 21, Müllertz and Clemensen, 1976; 22, Nilsson and Wiman, 1982; 23, Daniels et al., 1974; 24, Starkey and Barrett, 1977; 25, Schapira et al., 1982; 26, Scott et al., 1982.

1983), the disturbances in the clotting system accompanying AT III abnormalities (Marciniak et al., 1974; Sas et al., 1974; Koide et al., 1984), severe haemorrhagic tendency in patients with congenital α_2-APl deficiency (Aoki et al., 1979; Kluft et al., 1979), cystic fibrosis related to disturbances of α_2-M function (Shapira et al., 1977) and hereditary angioedema associated with C1 esterase and its inhibitor (reviewed by Steinbuch and Audran, 1974).

Human plasma contains at least seven distinct and well characterized proteins responsible for the inhibition of various proteinases (Table 2). They share certain common features such as being glycosylated and being of liver origin but differ considerably in their molecular weights and plasma concentrations. According to the majority of authors all these inhibitors inactivate proteinases in 1:1 molar ratio (except α_2-M which can bind up to two molecules of a proteinase – Starkey and Barrett, 1977; Travis and Salvesen, 1983) hence it is justified to express their relative abundances as a fraction of total plasma antiproteinase activity. In this respect, α_1-PI (formerly known as α_1-antitrypsin, α_1-AT) dominates, being responsible for over 70% of human plasma antiproteolytic activity (Table 2). However, the situation is complicated due to the preferential reactivity of individual plasma inhibitors for certain proteinases. As shown in Fig. 1, β_1-anticollagenase (β_1-AC), inter-α-trypsin inhibitor (ITI) and α_1-ACh show the narrowest specificity which is limited to either collagenase or cathepsin G. The opposite pole is occupied by α_2-M as it rapidly binds not only almost all known serine proteinases but also some cysteine- and metallo-proteinases (for references see Starkey and Barrett, 1977; Travis and Salvesen, 1983). The broad specificity and high reactivity of α_2-M are together responsible for the fact that any proteinase appearing in the blood is easily entrapped by this large inhibitor molecule and the resulting complex quickly removed by the cells of the reticuloendothelial system. However, the large molecular weight of α_2-M (approximately 725,000) prevents its penetration into tissue fluids where local release of some proteinases may occur, e.g. in the inflamed areas. Fortunately other inhibitors of lower molecular weight (4S – 7S range in Fig. 1) escape more easily from the blood (e.g. in the rabbit over 50% of α_1-PI is estimated to be present in the extravascular compartment; Koj and Regoeczi, 1978). Some authors postulated the existence of a two-step mechanism of clearance of proteinases locally released in tissues. They assumed that at first the enzymes are bound to 4S – 7S antiproteinases and then transported to the blood where they are transferred to α_2-M and finally removed and degraded. Indeed, Ohlsson (1975) demonstrated transfer of trypsin and granulocyte elastase from α_1-PI to α_2-M in dog and human plasma. However, this mechanism does not operate in all species and in many, removal of proteinases complexed with inhibitors other than α_2-M is almost as efficient as removal of α_2-M-proteinase complexes (Travis and Salvesen, 1983). A more important limitation of the biological efficiency of a given inhibitor is the rate of proteinase inactivation. As pointed out by Bieth (1980) only those inhibitors may be regarded as physiologically important in vivo which can remove a proteinase from body fluids within less than a few seconds (cf. also Travis and Salvesen, 1983).

Since the reactions between proteinases and inhibitors lead in most cases to irreversible loss of the inhibitors the enhanced proteolytic activity in acute inflammation could quickly deplete the body reserves of these antiproteinases. As demonstrated by the elegant studies of Ohlsson (1974) and Fritz (1980) such situations occur in certain cases of acute pancreatitis and in septic shock or endotoxaemia. Severe depletion of plasma level of proteinase inhibitors, and especially of α_2-M, has a bad prognostic significance. However, the AP-response usually stimulates a prompt replenishment of proteinase inhibitors due to their enhanced liver synthesis. In man, the level of α_1-ACh begins to rise in plasma 8 h after injury and is followed by rise of α_1-PI (Aronsen et al., 1972; Kueppers and Black, 1974). A different pattern is observed in the rat: α_1-PI is only a little affected (Koj et al., 1978b) but α_2-M belongs to the spectacular AP-reactants. Rats appear to be unique in their ability to produce during the inflammatory reaction considerable amounts of the inhibitor of cysteine proteinases. Thus Esnard and Gauthier (1983) showed inhibition of papain by one of the major rat AP-proteins – α_1-AP-globulin. Lysosomal cysteine proteinases were shown to be inhibited competitively by haptoglobin: rat liver cathepsin C by human Hp (Koj, 1972) and cathepsins B and L by homologous native Hp, asialo-Hp and Hp-Hb complex (Kalsheker et al., 1981; Pagano et al., 1982). However, these results should be treated with caution since Snellman and Sylven (1974) demonstrated that the inhibitory activity of haptoglobin against cathepsin B depends on a low molecular weight component associated with haptoglobin.

The conclusion indicated by the above findings is that despite considerable species-dependent variability the AP-response always leads to increased proteinase-binding capacity of the blood due to the augmented synthesis of some of the proteinase inhibitors in the liver.

2. *Blood clotting and fibrinolysis*

As bleeding is one of the common effects of trauma, clot formation and subsequent clot lysis form part of the principal defense mechanisms of the injured organism. The clotting cascade includes several zymogens – precursors of highly specific serine proteinases – as well as a range of phospholipid modulators and protein inhibitors, many of these components being associated also with plasma kinin and complement systems (see Chapter 11 and also Eisen, 1977; Jackson and Nemerson, 1980; Sundsmo and Fair, 1983). Already in 1966 Miller and co-workers (Olson et al., 1966) demonstrated that not only fibrinogen (factor I) but also prothrombin (factor II) as well as factors V, VII and X are synthesized by the perfused rat liver. Subsequent studies showed that the last four factors belong to a group of proteins requiring vitamin K during post-translational carboxylation essential for their biological activity (Suttie and Jackson, 1977; Jackson and Nemerson, 1980).

Among clotting factors only fibrinogen is definitely established as a strong AP-reactant and in some inflammatory states its synthesis may exceed that of albumin (for references see Koj, 1968, 1974; Mosesson and Doolittle, 1983). The synthesis of the other clotting factors appears to be less affected by injury. Thus the plasma level of factor XII is slightly reduced in post-operative states (Hedner et al., 1983). The prothrombin level is similarly decreased in patients with disseminated intravascular coagulation (McDuffie et al., 1979) and in rats injected with endotoxin but is slightly increased in rats with turpentine-induced inflammation (Koj et al., 1984b); all these changes are insufficient to qualify factors XII and II as AP-proteins. On the other hand, some evidence indicates that typical AP-proteins, such as α_1-AGP, CRP, or mouse SAP component may affect the clotting mechanism. Thus Andersen et al. (1980, 1981) demonstrated that highly purified α_1-AGP shows an antiheparin effect and contributes to the increased heparin tolerance regularly observed during the acute phase of various diseases. Other authors (Snyder and Coodley, 1976; Anderson and Eika, 1979; Costello et al., 1979) reported inhibition of platelet aggregation by native and desialylated α_1-AGP. However, the physiological significance of this effect has been questioned (Coller, 1980). Certain functions for SAP during clotting have been suggested by Ku and Fiedel (1983). Thus supraphysiological concentrations of human SAP (0.15 – 0.3 mg/ml) were shown to reduce release of fibrinopeptide A from fibrinogen, an effect which is potentiated by heparin. At the same time spontaneous polymerization of fibrin monomers is significantly delayed. Moreover, according to the latter authors, SAP appears to inhibit platelet activation thus acting oppositely to aggregated or complexed CRP (Fiedel et al., 1982). Direct involvement of SAP in coagulation has been critically evaluated by Pepys et al. (1980) who found on the other side that CRP interferes with the inhibition of factor Xa in the presence of heparin.

In the fibrinolytic pathway only the inhibitors α_2-APl (Matsuda et al., 1980; Högstrop et al., 1981; Hedner et al., 1983) and C1 inactivator (Hedner et al., 1983; Daniels et al., 1974) are regularly increased after injury although Aronsen et al. (1972) reported augmented levels of plasminogen in patients at 5 – 10 days after surgery.

3. *Removal of 'foreign' materials from the organism*

Injury and tissue necrosis lead to the formation of breakdown products of endogenous origin. As this is usually accompanied by bacterial infection, foreign products and materials of pathogenic origin become involved. The immunological response of the injured or infected organism is not instantaneous and before production of specific antibodies can occur the first line of defense includes increased synthesis of CRP and to some extent also SAP, SAA, fibrinogen and the complement system.

It is well established that CRP in the presence of calcium ions binds phosphorylcholine residues with very high affinity comparable to that of specific antibody-hapten interaction (Pepys, 1981). This is important because materials containing phosphorylcholine are widely distributed in nature, both in mammalian cell membranes and in polysaccharides of bacterial, fungal, parasitic and plant origin. Moreover, CRP can directly precipitate certain polycations (polylysine, protamine) the reaction being modified by calcium or heparin (Uhlenbruck et al., 1981; Potempa et al., 1982) as well as some galactans, invertebrate haemocyanins (Uhlenbruck et al., 1982) and neuraminidase-treated cells (Uhlenbruck et al., 1981). Robey et al. (1984) have demonstrated a high affinity binding of rabbit CRP to chromatin and nucleosome core particles. At saturation there is approximately one CRP-binding site for every 160 base pairs in chromatin. The authors suggest that CRP may help in removal of chromatin fragments from the body after cell death at sites of tissue damage.

Kushner and Kaplan (1961) were the first to demonstrate that CRP is selectively deposited on necrotic cells that have phospholipids exposed on damaged membranes (Volanakis, 1982). This may lead to activation of complement, generation of chemotactic and opsonic activities resulting in enhanced phagocytosis, resorption and repair of the lesion (Uhlenbruck et al., 1981; Volanakis, 1982; Pepys and Baltz, 1983). During the course of tissue damage, some cellular constituents are released into the circulation where they can complex with CRP (Pepys, 1981). Thus another major role of CRP is to bind and detoxify such abnormal autologous materials or facilitate their clearance (Pepys and Baltz, 1983). Simple binding of these products may reduce their toxicity or pathogenic properties while involvement of secondary processes such as phagocytosis and complement activation may lead to their final removal. If indeed the evidence for the involvement of CRP in the removal of modified host tissues is still partly conjectural, the role of this protein in the defence against microbial infection is firmly established. Having become bound to the bacterial cell, CRP can mediate precipitation, agglutination, capsular swelling of bacteria, or activation of the classical complement pathway (for references see Pepys, 1981). This protective function of CRP in bacterial infection has been clearly demonstrated in animal models such as in mice infected with pneumococci (Mold et al., 1982; Yother et al., 1982).

Although SAP can bind to galactans, to some microbial cell wall polysaccharides and to the fixed complement component C3b (Pepys, 1981; Uhlenbruck et al., 1981), its role in the removal of foreign materials appears to be less important than that of CRP. On the other hand, it has been firmly established that SAA (a spectacular AP-reactant in man and mouse) is associated with plasma lipoproteins of the HDL_3 class which are cleared rapidly from the blood (Benditt et al., 1982). Thus a possible function of SAA would be to increase clearance of some HDL molecules and in this way bring about elimination of toxic or foreign materials complexed with lipoproteins (Pepys and Baltz, 1983). Evidence supporting this hypothesis is provided by the observation that lipopolysaccharides added to rabbit serum complex with HDL (Ulevitch et al., 1981) and that in AP-rabbit serum the HDL complex contains SAA (Tobias et al., 1982; Pepys and Baltz, 1983).

Clumping of staphylococci or streptococci is facilitated by another AP-protein, fibrinogen (Duthie, 1955; Lipinski et al., 1967). The binding sites for staphylococci, and for blood platelets, have been localized on C-terminal portions of gamma chains of human fibrinogen (Hawiger et al., 1982, 1983). Since two segments of gamma chains are present at each end of the elongated fibrinogen molecule (cf. Chapter 14, Fig. 4) the divalency requirements for cell agglutination are fulfilled. This function of fibrinogen to interact with foreign particles appearing in the blood is phylogenetically an old phenomenon and resembles behaviour of other clottable proteins in lower animals (cf. Chapter 14).

Finally, CRP complexes may participate in activation of the complement cascade leading to destruction of foreign cells. Although there is some controversy in respect to which complement components can be reckoned among AP-proteins (Pepys, 1981; Kushner, 1982) the level of C3 in plasma increases after cardiac infarction, in pneumonia and in inflammatory diseases (for references see Koj, 1974; Fey and Colten, 1981). Activated C3b shows opsonin and chemotactic activity facilitating phagocytosis and removal of foreign materials (Sundsmo and Fair, 1983). Further, it is also well established that activated enzymes of the complement system can kill bacteria, protozoa, fungi and viruses even in the absence of specific antibodies (Volanakis, 1982).

4. Modulation of the immunological response

The literature abounds in reports describing either activation or inhibition by various AP-proteins of cellular or humoral components of the immune response. However, many of the studies carried out in tissue culture are difficult to interpret in terms of the whole organism, while the results of others are mutually contradictory. A good example of the latter is CRP binding by and stimulation of lymphocytes reported by several laboratories (Hornung and Fritschi, 1971; Mortensen et al., 1975; Williams et al., 1978) but questioned in other studies (James et al., 1982; Pepys and Baltz, 1983). Li and co-workers (1982) suggest that CRP, SAP and SAA are involved with down regulation of the human-immune response after exposure to antigens, mitogens and inflammatory stimuli. Even if CRP does not affect B and T-lymphocytes directly, it certainly has indirect modulatory effects on the immune reactions through the complement cascade and activation of macrophages. The complement proteins C3, C4, C5 and their active fragments are all involved in immune adherence, chemotaxis and release of various cytokines and thus in general propagation of the inflammatory reaction (Volanakis et al., 1982; Sundsmo and Fair, 1983) (cf. Chapter 11, Section 6). Liposome-associated human CRP may activate macrophages and generate tumoricidal activity (Barna et al., 1984). Opsonic and phagocytosis-enhancing activity has been reported for human α_2-HS-glycoprotein (references in Gejyo and Schmid, 1981) and for peptides obtained by digestion of human fibrinogen or fibrin by plasmin (Kopeć et al., 1982; Saldeen, 1983). On the other hand, these peptides were found to be immunosuppressive both in vitro and in vivo: they inhibited thymidine uptake by mitogen-stimulated lymphocytes in concentrations that had no cytotoxic effect (Gerdin et al., 1980), depressed cell-mediated immunoreactivity in vivo and promoted tumour growth and metastases in mice (Roszkowski et al., 1981).

The immunoregulatory properties of human and murine macroglobulins are rather well documented. Thus these proteins were reported to be responsible for the inhibition of antibody-induced lysis of erythrocytes by neutrophils (Ganrot and Schersten, 1967; Cordier and Revillard, 1980) and the decrease of the mixed lymphocyte response (Tunstal et al., 1975), inhibition of DNA synthesis in certain murine tumour cells (Koo, 1982), suppression of chemotactic responses of monocytes and modulation of the responses of lymphocytes in vitro to mitogenic and antigenic stimuli (reviewed by James, 1980). All these effects are dose dependent in such a manner that low concentrations often exert stimulation while high doses are inhibitory (Goutner et al., 1976). Some of the biological effects of macroglobulins are directly related to inactivation of proteinases while others are mediated through complement components, the kinin generating system or certain chemotactic responses. In addition, murine macroglobulins have been shown to be involved in binding and transport of various cytokines capable of reacting with lymphoreticular cells (Koo, 1982 and Section 6).

In 1981, Hudig et al. provided evidence that macromolecular antiproteases from blood plasma suppress natural cytotoxicity of human cells. These observations have been elaborated and extended by Ades and co-workers who demonstrated that human α_1-PI and α_2-M (Ades et al., 1982), as well as α_1-ACh (Gravagna et al., 1983) inhibit in a dose-dependent fashion two cytolytic events of the immune response: activity of natural killer cells and antibody-dependent cell-mediated cytotoxicity. The results suggest that antiproteases do not influence binding of effector to target cells but eliminate lytic potential. Although Matsumoto and colleagues (1981) reported that injection of human ACh into mice enhances the plaque-forming cell response to sheep erythrocytes the reaction is probably unspecific since it was not altered by heat treatment of α_1-ACh sufficient to destroy its antiproteolytic activity (Matsumoto et al., 1982).

An interesting observation concerning the immunosuppressive action of α_1-PI has been made by Willoughby and co-workers (Simon et al., 1983). Commercial preparation of human α_1-PI added to culture of mouse thymocytes at physiological concentrations (2 mg/ml) suppressed IL-1 dependent and PHA induced cell proliferation as well as proliferation of PHA-induced mouse spleen cells. This inhibition of DNA synthesis in both types of cells could be reversed by the addition of purified IL-1 (Table 3). Although the mechanism of antagonistic action of IL-1 and α_1-PI is not clear, these findings, if confirmed, may be important for an understanding of the regulation of the AP-response. One may speculate that α_1-PI competes with IL-1 for binding sites on target cells thus reducing certain effects mediated by this cytokine. Bata and Revillard (1981) suggested that immunoregulatory activity of α_1-PI may be related to blocking of lymphocyte surface protease.

Immunosuppressive activities of AP-proteins are not limited to proteinase inhibitors. Van Oss et al. (1974) reported that α_1-AGP inhibits not only phagocytosis but also proliferation of lymphocytes stimulated by mitogen or by allogenic cells. These findings have been confirmed by Chiu et al. (1977), Cheresh et al. (1984) and Bennett and Schmid (1980). The latter authors found that native human α_1-AGP and its asialo- agalacto-derivatives added to cultured mouse spleen cells inhibit an-

TABLE 3
Effect of IL-1 and α_1-PI on thymus cell proliferation (after Simon et al., 1983)

IL-1 (units/ml)	α_1-PI (mg/ml)	[^{3}H]thymidine incorporation CPM ± SE
0	0	463 ± 29
3	0	10,273 ± 141
6	0	12,161 ± 129
12	0	12,346 ± 762
3	2	1,275 ± 83
6	2	5,601 ± 246
12	2	9,376 ± 125

The results are means of quadriplicate cultures of 1.5 × 10^6 mouse thymus cells/0.2 ml medium. IL-1 was obtained from rabbit alveolar macrophages and quantitated in arbitrary units. Human α_1-PI was from Sigma. PHA was added to all cultures (1 μg/ml).

tibody formation, the mitogenic responses to Con A and lipopolysaccharide, the mixed lymphocyte reaction and the induction of cytolytic lymphocytes reactive to alloantigens. On the other hand, α_1-AGP did not inhibit lysis of tumour target cells by natural killer lymphocytes and did not suppress proliferation of tumour cells. Whether α_1-AGP and its deglycosylated derivatives indeed may function to regulate various immune responses in vivo evidently requires further studies.

Samak et al. (1982) described inhibition of blastogenic responses of PHA-stimulated human blood lymphocytes by supraphysiological concentrations of homologous haptoglobin and fibrinogen, or alternatively by mixtures of α_1-AGP, α_1-PI and ceruloplasmin, while chemotactic responses of human blood monocytes to casein were suppressed by haptoglobin, fibrinogen and α_1-AGP. These results correspond rather well to the observations of Baskies et al. (1980) who reported suppression of cellular immunity in 147 patients with solid tumours, all of whom had high levels of α_1-AGP, Hp and α_1-PI in their plasmas.

Many years ago the assumption that carbohydrate-rich glycoproteins might have immunosuppressive properties prompted Apffel and Peters (1969) to propound the hypothesis of 'symbodies', i.e. liver-produced antagonists to immunoglobulins. They suggested that cells of the inflamed areas might share with cancer cells the capacity to stimulate the synthesis of symbodies in the liver. Using current terminology this would correspond to stimulation of AP-protein synthesis by IL-1. The symbodies, i.e. AP-proteins, were then supposed to localize on the surface of injured cells, or on cancer cells, covering their abnormal antigens and thus preventing development of autoimmune diseases as well as impeding immunological attack by the host on the tumour. Although the present state of immunology may be supposed to have rendered these ideas rather naive, they should not be forgotten in the light of the new evidence reviewed above. As pointed out by Samak et al. (1982) at least two mechanisms may be responsible for the inhibition of the effector-cell function by AP-proteins: firstly, binding of these proteins to surface receptors of lymphocytes and monocytes (perhaps on the principle of carbohydrate-lectin interaction), or secondly, by generation of suppressor cells. The first mechanism is somewhat supported by the observations of Mannick et al. (1977) who showed that the response of lymphocytes from cancer patients to PHA in vitro improves after washing of the cells, and by Cheresh et al. (1984) who observed α_1-AGP-induced displacement of an anionic probe from liposome membranes.

While immunosuppression may be beneficial for the injured organism by quenching the excessive or prolonged inflammatory response (including in some cases autoimmune aggression) it is certainly deleterious in respect to immunological combat against cancer. However, this function of AP-proteins is still highly conjectural and much research is required to elucidate the mechanism of the immunosuppressive properties of some AP-proteins as well as their importance in immunological reactions in vivo.

5. Anti-inflammatory properties

The inflammatory reaction is highly complex (cf. Chapter 1; Movat, 1985) and AP-proteins may affect some of its mechanisms, as described below. Neutrophils, macrophages and other migrating cells which accumulate at the site of tissue injury are the source not only of proteolytic enzymes but also of free radicals and superoxide ions (Weissmann, 1980; Fantone and Ward, 1982) as well as prostaglandins (Humes et al., 1977) and various other mediators (Chapters 1, 7, 9 and 11).

The homeostatic functions of proteinase inhibitors have been presented above (Section 1) but direct evidence for specifically anti-inflammatory properties of these proteins is rather limited. Van Gool and co-workers described the inhibitory effect of rat α_2-M on carrageenin oedema (1974) and polymorphonuclear chemotaxis (1982). The inhibition of carrageenin inflammation by α_2-M in rat has been confirmed by Nakagawa et al. (1984). This effect may be indirect since α_2-M forms easily complexes with nerve-growth factor (cf. Section 6) which had been shown to possess strong anti-inflammatory activity in the rat hind paw oedema (Banks et al., 1984). The human α_2-M-trypsin complex inhibits secretion by murine macrophages of such inflammation-promoting compounds as neutral proteinases, cytolytic factor and plasminogen activator (Johnson et al., 1982). As reviewed by Lewis (1977), some of the symptoms of experimental inflammation such as oedema formation, granuloma tissue deposition, delayed hypersensitivity reactions and adjuvant arthritis may be reduced by synthetic and natural proteinase inhibitors including specific anti-inflammatory protein synthesized in the liver of injured rats (Billingham et al., 1971). Nakagawa et al. (1983) reported that development of granulation tissue and the exudate during carrageenin-induced inflammation in rats were markedly suppressed by some synthetic proteinase inhibitors.

A formerly unknown function of fibrinopeptides A and B is inhibition of carrageenin-induced inflammatory rat paw swelling (Ruhenstroth-Bauer et al., 1978, 1981). Although the molecular mechanism of this effect is unkown it appears to be specific since fibrinogen-degradation products derived by plasmin digestion were found to be ineffective.

Early reports on inhibition of prostaglandin synthesis by haptoglobin (Shim, 1976; Saeed et al., 1977) have not been confirmed. However, relevant work showing that haptoglobin does indeed inhibit prostaglandin synthetase by restricting available haem group which is required for the enzyme activity is now published (Jue et al., 1983). For this reason, the significance of haptoglobin as an inhibitor of prostaglandin synthesis in vivo appears to be rather limited and in need of further study.

The involvement of oxygen-derived free radicals in leucocyte and macrophage dependent inflammatory reactions is well documented (for references see Goldstein et al., 1979, 1982; Fantone and Ward, 1982). The superoxide anion ($0_2^{-}\cdot$), a highly reactive free radical formed by the univalent reduction of molecular oxygen, can be generated in a variety of biological systems either by auto-oxidative processes or by enzymes involved in aerobic metabolism of phagocytic leucocytes (Goldstein et al., 1982). As shown in Fig. 2, this ion or other reactive molecules (such as hydroxyl radicals or singlet oxygen) can directly damage cell constituents, lipid membranes (Lynch and Fridovich, 1978; Sacks et al., 1978; Goldstein et al., 1982) and connective tissue of the joints (Greenwald and Moy, 1979). In addition, it can inactivate α_1-PI (Carp and Janoff, 1980; Matheson et al., 1982) thus enabling unrestricted action of neutral proteinases. An interesting feed-back regulatory mechanism has been demonstrated by Pizzo and co-workers (Pizzo et al., 1983) who showed that superoxide production by activated mouse peritoneal macrophages can be inhibited by proteolytically modified 'fast' form of α_2-M.

Cytosolic and mitochondrial superoxide dismutases are involved in the intracellular removal of these radicals (Fridovich, 1975) while the AP-protein, ceruloplasmin, fulfills a similar function in blood and body fluids (Dendo, 1979; Goldstein et al., 1979, 1982; Gutteridge, 1983). Goldstein et al. (1982) demonstrated that ceruloplasmin, at concentrations similar to those in plasma, inhibits several superoxide-mediated enzymatic reactions and thus may function as a circulating anti-inflammatory protein. Although some authors have reported that ceruloplasmin removes superoxide by a dismutase reaction (Plonka and Metocheva, 1979) more recent evidence suggests that ceruloplasmin scavenges superoxide in a non-catalytic, stoichiometric fashion (Goldstein et al., 1982). However, ceruloplasmin may also remove hydroxyl radicals and singlet oxygen by an, as yet, undefined mechanism (Dendo, 1979). The antioxidant properties of ceruloplasmin also result from its ferro-oxidase activity: since Fe^{2+} ions may initiate lipid perox-

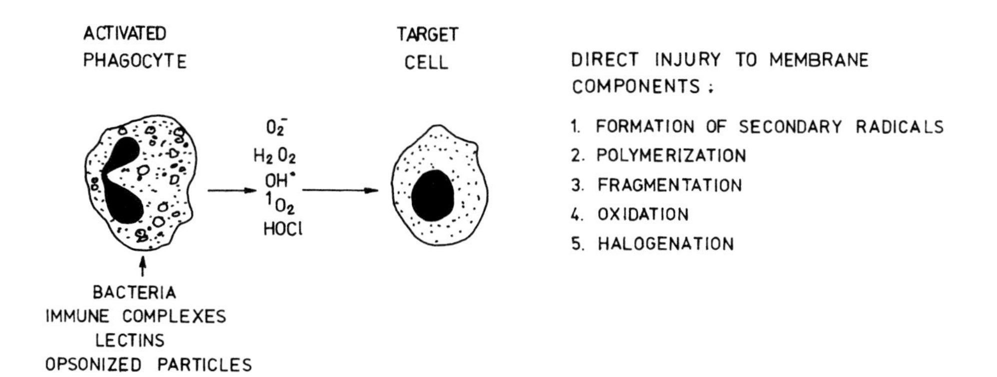

Fig. 2. Mechanism of cytotoxic effects of phagocyte-derived free radicals. (After Fantone and Ward, 1982.)

idation through their reaction with superoxide, thus ceruloplasmin could inhibit these reactions by the enzymic oxidation of Fe^{2+} (Frieden, 1973; Williams et al., 1974; Loustad, 1981). It has also been suggested that apotransferrin may have antioxidant activity due to its ability to bind iron (Fantone and Ward, 1982). Some further relationships between ceruloplasmin and transferrin are discussed below (Section 6).

6. Binding and transport of metals and biologically active compounds

Injury and infection are often accompanied by haemolysis which may lead to the loss of valuable iron. Haptoglobin, a typical AP-protein, binds haemoglobin stoichiometrically and irreversibly. Several amino acid residues, mainly in the alpha chain, are involved in the interaction (Dobryszycka and Bec-Katnik, 1975; Kazim and Atasn, 1983; Lustblader et al., 1983). The complex which exhibits true peroxidase activity (Jayle, 1951) and considerable inter-species variations in respect of this activity (Dobryszycka and Krawczyk, 1979) is rapidly cleared from the plasma. It accumulates in the liver due to the presence of a specific receptor (Kino et al., 1980; Lowe and Ashwell, 1982). This action of haptoglobin as a very effective and specific scavenger is important because iron haemoglobin is known to support bacterial growth (Weinberg, 1974). Indeed, Eaton and co-workers (1982) suggested that haptoglobin may serve as a natural bacteriostat since rats injected intraperitoneally with pathogenic E.coli and small amounts of haemoglobin all died unless haptoglobin was administered simultaneously.

On degradation, haemoglobin yields haem for which haptoglobin has no affinity. However, haem reacts in equimolar proportions with haemopexin, an AP-protein in the rat. Haemopexin also forms complexes with certain cellular haemoproteins including myoglobin and cytochrome C (Allen et al., 1977) that may enter the blood during tissue destruction. Finally, iron released from haem is trapped by transferrin which is a negative AP-reactant in all species and which can bind 2 atoms of Fe^{3+}. Since circulating transferrin is normally only partly saturated with iron, it represents an iron buffer system and may increase resistance to infection by reducing availability of iron for microorganisms (Aisen and Listowsky, 1980). Its main function is to deliver iron from the reticuloendothelial cells of the liver where the metal is stored in a complex with ferritin, to haemopoietic cells and other tissues (Fig. 3). Konijn et al. (1982) demonstrated that hepatic synthesis of apoferritin is increased during the first hours of inflammation. Incorporation of iron into apotransferrin and apoferritin is accelerated by another AP-protein, ceruloplasmin, which acts as a ferroxidase (Frieden, 1973; Frieden and Hsieh, 1976; Boyer and Schori, 1983). Hence ceruloplasmin appears to be essential for both deposition of iron in the form of fer-

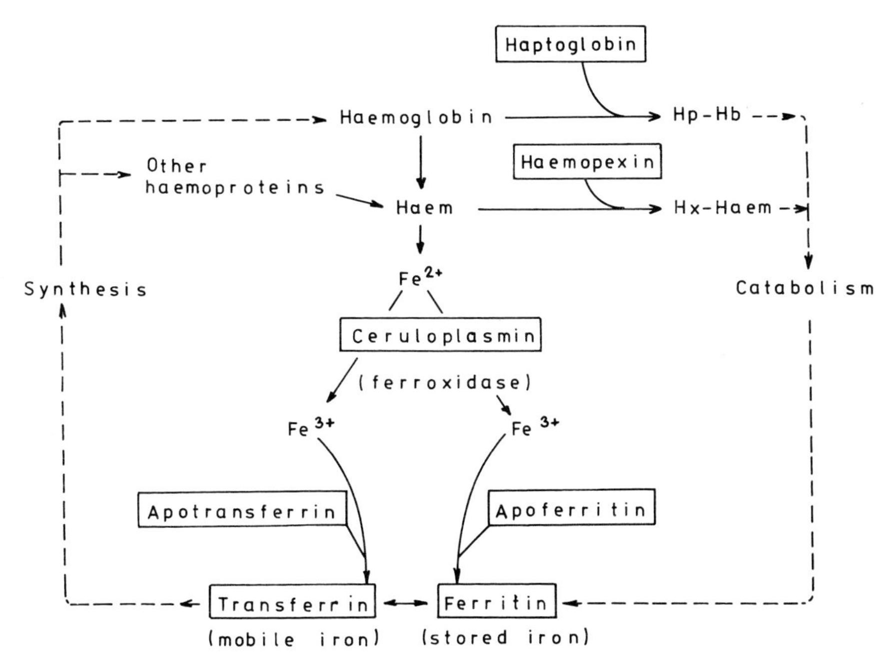

Fig. 3. Involvement of AP-proteins in binding, removal, storage and transport of iron in the animal organism.

ritin and for movement of iron from reticuloendothelial cells to plasma transferrin and then to haemopoietic cells (Williams et al., 1974). In addition, ceruloplasmin may oxidise numerous substrates (reviewed by Frieden, 1979; Goldstein et al., 1982) and functions as the main copper transport protein delivering this metal to tissue enzymes, such as cytochrome oxidase (Hsieh and Frieden, 1975; Linder and Moor, 1977; Frieden, 1979).

As reviewed by Geisow and Gordon (1978), a significant proportion of low molecular weight compounds transported by blood, including vitamins and hormones, are bound to plasma proteins. Albumin binds a variety of amino acids, fatty acids and drugs. Prealbumin binds not only thyroxine and vitamin A but also aspirin and barbiturates. Concentrations of albumin and prealbumin in plasma are significantly reduced during the AP-response. On the other hand, α_1-AGP, the classical positive AP-protein has been implicated in binding and transport of progesterone (Ganguly and Westphal, 1968), pteridine (Ziegler et al., 1982), phenothiazine neuroleptics and various basic drugs (Piafsky et al., 1978; Javaid et al., 1983; Müller et al., 1983). Thus changes in plasma concentration of α_1-AGP in acute inflammatory diseases and cancer patients may alter kinetics and efficacy of those drugs (Abramson et al., 1982).

More recent evidence reviewed by James (1980) and Koo (1982) suggests that macroglobulins are important transport proteins. In addition to carrying inactivated (or partly active) proteinases, macroglobulins were shown to bind reversibly certain other enzymes (L-asparaginase and cationic aspartate aminotransferase), lectins and mitogenic substances (Con A, endotoxin), and cytokines (nerve-growth factor, macrophage activating factor, lymphocyte activating factor). By employing acid dialysis Koo (1982) was able to dissociate from purified mouse macroglobulin two cytokine fractions with M_r of about 12,000 and 3,000. These components were found to inhibit growth of murine tumour cells but to stimulate B and T-lymphocyte reactivities to mitogens in vitro. Huang et al. (1984) identified human plasma α_2-M as a specific protein binding platelet-derived growth factor (PDGF) by a disulphide-exchange reaction. Although the PDGF-α_2-M complex retains mitogenic activity, it appears that α_2-M is the main protein responsible for rapid clearance of PDGF, either locally or from systemic circulation.

References, p. 232.

Gordon/Koj (eds.) The acute-phase response to injury and infection.

CHAPTER 14

Phylogenetic aspects of the acute-phase response and evolution of some acute-phase proteins

A. KOJ

The AP-response represents an adaptative reaction of the organism to infection or injury (cf. Introduction and Chapter 13). Fever is a typical sign of the AP-response and has been regarded as beneficial until antipyretic drugs became widely available. Experimental and clinical evidence reviewed by Kluger (1981) suggests that moderate fever indeed has survival value in various bacterial and viral infections. This may be true not only for mammals but also in lower vertebrates: increased body temperature of infected lizards and fish significantly reduced their mortality (Kluger, 1981). Fever may improve host defence in many ways: by increasing mobility of leucocytes or their phagocytic and bactericidal activities (Roberts and Steigbigel, 1977; Sebag et al., 1977), by accelerating blastic transformation of lymphocytes (Smith et al., 1978; Roberts, 1979), and even by direct action on bacterial growth, especially in iron-poor medium, such as blood plasma during the AP-response (Kluger and Rothenburg, 1979).

Studies of Hanson and co-workers (1983) and Duff and Durum (1983) indicate that murine T-cell proliferation in response to purified IL-1 in vitro is greatly increased when the cells are exposed to a temperature typical of fever, and that injection of the same IL-1 causes fever in mice. Hyperthermia resulted in an increase in the magnitude of T-cell proliferation response to IL-1 and Con A, and not only an acceleration (Duff and Durum, 1983). Temperature-sensitive events occurred early in the course of T-cell culture and the effect could be explained, at least in part, by increased proliferation in response to IL-2, although an additional augmentation of IL-1-induced IL-2 production at the higher temperature was not excluded. If this relationship exists also in vivo, the resulting facilitation of T-cell dependent immune response may well confer survival value and contribute to evolutionary conservation of fever which is energetically expensive. However, the nonspecific resistance to infection and injury provided by leucocyte endogenous mediator (LEM), endogenous pyrogen (EP) and interleukin 1 (IL-1) cannot be limited to fever since LEM did not increase body temperature of rats (Kampschmidt and Upchurch, 1969) although it protected them from a usually fatal dose of *Salmonella typhimurium* (Kampschidt and Pulliam, 1975). In addition, IL-1, and epidermal cell-derived thymocyte activating factor (ETAF) was shown to be directly mitogenic to human skin fibroblasts (Schmidt et al., 1982; Luger et al., 1983) and to exert chemotactic activity for polymorphonuclear leucocytes and monocytes in vitro (Luger et al., 1983; Sauder et al., 1984). The immunostimulatory activity of IL-1 has been demonstrated both in vitro and in vivo where it enhances secondary antibody responses in mice to a protein antigen (Staruch and Wood, 1983). Involvement of IL-1 in the defence mechanism based on the immunological response is reviewed in Chapter 6.

Although the occurrence of EP/IL-1 has been demonstrated in only a few species of lower vertebrates (Bernheim and Kluger, 1977; Kluger, 1979) the indirect evidence suggests that the AP-response is phylogenetically an old phenomenon. Live bacteria and endotoxin were shown to induce fever or febrile-like behaviour in poikilothermic animals, and antipyresis could be obtained by sodium salicylate or acetaminophenon (Kluger, 1979; Reynolds and Casterlin, 1982), although some contradictory results were reported in a teleost fish (Marx et al., 1984). Early evolutionary origin of the AP-response is supported by comparative studies of some AP-proteins belonging to three functional classes: proteinase inhibitors, clotting proteins and proteins involved in removal of foreign materials. It appears that in lower animals, these proteins are produced constitutively and occur in relatively high concentrations in body fluids. When more sophisticated defense mechanisms came into being, synthesis of these proteins in healthy organisms was either reduced (clotting proteins, complement components) or completely shut off (CRP in man and rabbit), but in all cases can easily be elevated in emergency states. The signal, IL-1, generated by macrophages represents the first line of defense. IL-1 has multiple targets within the body and exhibits truly pleiotropic effects in a variety of adaptative reactions (cf. Chapter 4). It represents a link between cellular and humoral defense on one side, and between non-specific resistance and highly specific immune response on the other. Elucidation of the evolutionary origin of IL-1 will throw light on many obscure aspects of the AP-response in higher animals.

1. Phylogeny of proteinase inhibitors

Structural analysis of proteinase inhibitors from blood plasma revealed the existence of two main protein families: one including α_1-PI and AT III and the other, macroglobulins. In the case of α_1-PI and AT III, not only the reactive sites recognized by serine proteinases are homologous (Carrell et al., 1980; Owen et al., 1983) but also the amino acid sequence of polypeptide chains show considerable similarities; an additional and surprising homology was found to exist between these two inhibitors and chicken ovalbumin (Hunt and Dayhoff, 1980; Leicht et al., 1982). It appears that genes for all three proteins were present in the mammal-bird ancestral line and can still be expressed in birds; in mammals the ovalbumin gene may have been lost, or be present but not expressed by any type of cell (Hunt and Dayhoff, 1980). α_1-PI is often a positive AP-reactant in distinction to AT III being a neutral protein (cf. Chapter 12, Table 2). In several mammalian species, α_1-PI occurs in multiple molecular forms resulting from variations in the structure of both the carbohydrate components and the polypeptide chains: man (Carrell et al., 1982; Travis and Salvesen, 1983), rabbit (Koj et al., 1978; Regoeczi et al., 1980; Koj and Regoeczi, 1981), or horse (Pellegrini and von Fellenberg, 1980; Kurdowska et al., 1982). Molecular heterogeneity of α_1-PI may reflect instability of a gene in the course of evolutionary divergence into related but distinct inhibitors. Genetic polymorphism of human α_1-PI is responsible for the occurrence of functionally disabled variants Z and S showing single amino acid substitutions. Individuals with α_1-PI Z and S variants are prone to develop lung emphysema due to reduced antiproteolytic capacity of the blood (cf. Chapter 13 and Carrell et al., 1982; Travis and Salvesen, 1983).

More is known about the evolution of macroglobulins which appear to be primordial proteinase inhibitors. All species of mammals contain in blood plasma at least one, and often two, macroglobulins designated according to their electrophoretic mobility as α_1-M and α_2-M. The most extensively investigated human α_2-M is a glycoprotein composed of four identical subunits of M_r 185,000 linked in pairs by disulphide bonds (Starkey and Barrett, 1977 and Fig. 1). Proteolytic attack at a 'bait' region near the middle of the primary sequence of one or more subunits leads to a conformational change of α_2-M and 'trapping' of the proteinase which is then unable to digest macromolecular substrates but still can hydrolyze low M_r synthetic substrates (Starkey and Barrett, 1977). Ensuing conformational change (Fig. 1) affects electrophoretic mobility of the protein: macroglobulin modified by a proteinase (or by nucleophilic reagent such as methylamine) migrates faster (F-α_2-M) than the native form (S-α_2-M) (Barrett et al., 1979). In addition to the trapping reaction specific for proteinases, we now recognize two other binding reactions: one is the 'adherence' reaction seen with anhydrotrypsin or various cationic proteins

(Sayers and Barrett, 1980; Chapter 13, Section 6), the other is a covalent 'linking' reaction through a thiol ester bond exposed by attack of a proteinase on α_2-M (Salvesen and Barrett, 1980; Salvesen et al., 1981; Sottrup-Jensen et al., 1981). A reactive thiol ester is responsible not only for covalent binding but also for internal peptide bond cleavage under denaturing conditions. This unusual property is shared by macroglobulins with complement proteins C3 and C4 (Sim and Sim, 1981).

Rat α_1-M (and perhaps some α_1- and α_2-M of other species) may have slightly different molecular weights and subunit structure compared with human α_2-M (Gordon, 1976; Nelles and Schnebli, 1982) but such a conclusion is uncertain due to the possibility of autolytic cleavage of macroglobulin polypeptide chains in hot alkaline sodium dodecyl sulphate (SDS) solutions (Barrett et al., 1979; Salvesen and Barrett, 1980). Despite similarities in amino acid composition of α_1- and α_2-M (Gordon, 1976) they are immunologically distinct and already in 1972 Berne and co-workers (cited after Starkey and Barrett, 1977) distinguished two groups: those closely related to human α_2-M were rabbit α_1-M, rat α_2-M and both dog macroglobulins, while the second group was represented by rabbit α_2-M, rat α_1-M and mouse α_2-M. Out of these proteins only rat α_2-M and rabbit α_1-M are regarded as strong AP-reactants (cf. Chapter 12, Table 2). Thus it is clear that the response to injury cannot be predicted neither from electrophoretic mobility nor from immunological affiliations. The immunological relationships between some macroglobulins reinvestigated by Weström et al. (1983) are shown in Fig. 2.

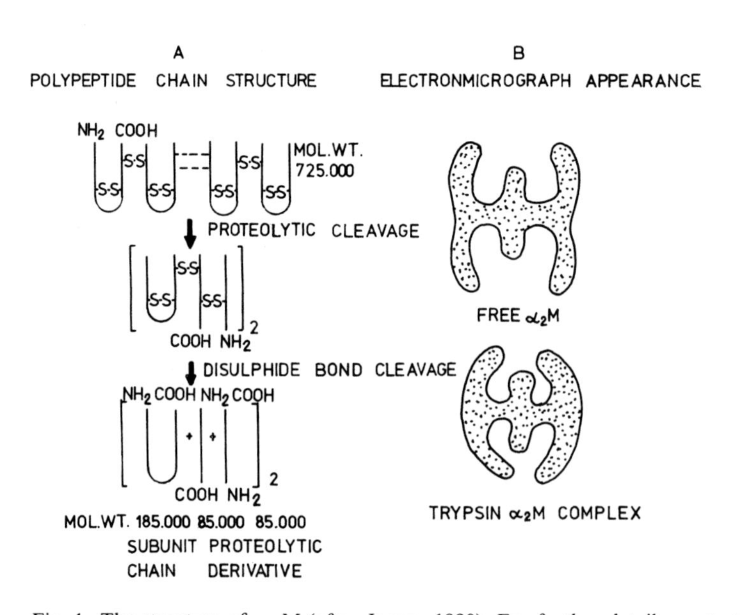

Fig. 1. The structure of α_2-M (after James, 1980). For further details see text.

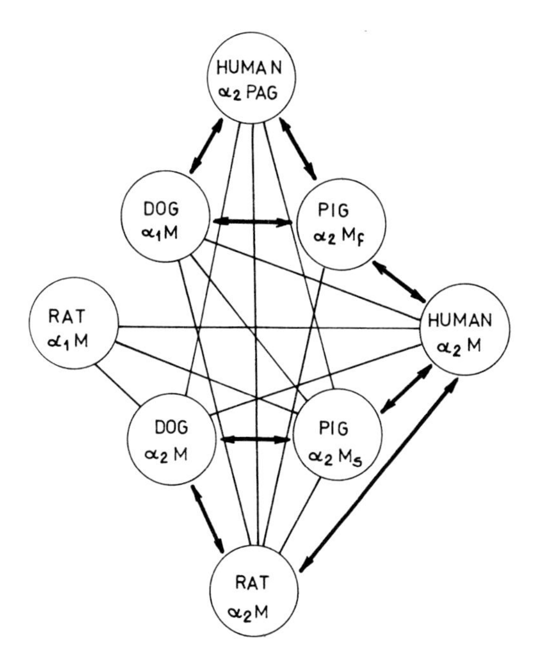

Fig. 2. Immunological relationships between some mammalian macroglobulins (after Weström et al., 1983). —, antigenic cross-reactivity; ⟷, strong reciprocal cross-reactivity, i.e. when a pair of macroglobulins both cross-react with corresponding antisera.

Although such tests have limited value since they reflect the characteristics of not only antigens but also particular antisera, again at least two groups of macroglobulins may be distinguished. Moreover, dog α_1-M and pig α_2-M (fast form) appear to be related to human pregnancy-associated α_2-glycoprotein (α_2-PAG, consisting of two subunits and showing no antiproteolytic activity). All these data suggest that macroglobulins of mammalian plasmas (and possibly also α_2-PAG) derive from a common ancestral protein. Due to the studies of Starkey and Barrett (1982a, b) its occurrence can be traced back to invertebrates. The authors used an original approach based on the property of α_2-M-proteinase complex in which the bound proteinase preserves a significant proportion of its activity toward low molecular weight substrates. As shown in Table 1, such a papain-binding protein is present in the plasma of all vertebrates so far examined, although in fish (including dipnoans) and in cyclostomes the protein has M_r of 360,000 corresponding to half of the mammalian molecule. This 'small' macroglobulin shares other properties of mammalian macroglobulins, such as susceptibility to methylamine, but under reducing and denaturing conditions can be dissociated into four subunits: two with M_r of 105,000 (subunit I) and two with 90,000 (subunit II) (Starkey et al., 1982; Starkey and Barrett, 1982b). Studies of other authors point to considerable similarities in the structure and some biochemical properties of macroglobulins and complement proteins C3 and C4 (Sim and Sim, 1981). Thus, as stated by Starkey and Barrett (1982b), 'all three proteins undergo a conformational change as a result of specific proteolytic attack, all are inactivated by amines, all undergo an autolytic cleavage in denaturing conditions that is prevented by prior inactivation by methylamine, and all are capable of forming covalent links with other proteins'. Moreover, the same sequence of seven amino acids occurs around the reactive thiol ester bond in C3, C4 and α_2-M, and additional sequence analogies have also been reported (Mortenson et al., 1981; Fey et al., 1983). In view of this evidence Starkey and Barrett (1982b) proposed that the α_2-M subunit of M_r 185,000 present in mammals arose from subunits I and II by loss of the post-translational cleavage site; this mutation probably occurring in an early tetrapod

TABLE 1
Distribution of a papain-binding protein in the serum (or plasma) of some vertebrate species (modified from Starkey and Barrett, 1982a)

Classification	Species	Size of papain-binding protein
Mammals	Man (*Homo sapiens*)	L
	Chimpanzee (*Pan troglodytes*)	L
	Hedgehog (*Erinaceus europaeus*)	L
	Squirrel (*Sciurus vulgaris*)	L
	Asian elephant (*Elephas maximux*)	L
	Camel (*Camelus ferus*)	L
	Wallaby (*Wallabia rufogrisea*)	L
Birds	Canada goose (*Branta canadensis*)	L
Crocodiles and alligators	Chinese alligator (*Alligator sin.*)	L
Lizards and snakes	Nile monitor (*Varanus niltoticus*)	L
	Boa constrictor (*Boa constrictor*)	L
Amphibians	American bullfrog (*Rana catesb.*)	L
	European frog (*Rana temporaria*)	L
Dipnoans	African lungfish (*Protopterus amphibius*)	S
Actinopterygians	*Reinhardtius hippoglossoides*	S
	Halibut (*Hippoglossus hippoglossus*)	S
	Plaice (*Pleuronectes platessa*)	S
Elasmobranchs	Guitar fish (*Rhinobatus cerniculas*)	S
	Dogfish (*Squalus acanthus*)	S
Cyclostomes	Hagfish (*Myxine glutinosa*)	S
	Lamprey (*Petromyzon marinus*)	S

The size of the papain-binding protein is indicated by L (M_r approximately 725,000) or S (M_r approximately 360,000).

(Fig. 3). The large single-type subunit enabled formation of the tetrameric molecule of M_r 725,000 with apparently greater efficiency for trapping proteinases than that of 'small' macroglobulins.

The findings of Starkey and Barrett have been extended by Quigley and Armstrong (1983) who detected an endopeptidase inhibitor similar to vertebrate macroglobulins in the haemolymph of the horseshoe crab, *Limulus polyphemus*. This inhibitor blocks digestion of casein (but not low M_r substrates) by a variety of prokaryotic and eukaryotic endopeptidases and is also sensitive to mild acidification or methylamine treatment. Since the evolutionary lineages of *Limulus* and vertebrates diverged at least 550 million years ago, the presence of an α_2-M-like molecule in the horseshoe crab suggests that the property of certain macroglobulins as protease inhibitors or protease carriers is an ancient and evolutionary conserved protein function (Quigley and Armstrong, 1983).

2. *Phylogeny of clotting proteins*

Fibrinogen, the clotting protein of vertebrates, is a typical AP-reactant in all mammalian species (cf. Chapter 12, Table 2) and in birds (Pindyck et al., 1983). It is not yet clear whether its synthesis after injury is increased also in lower

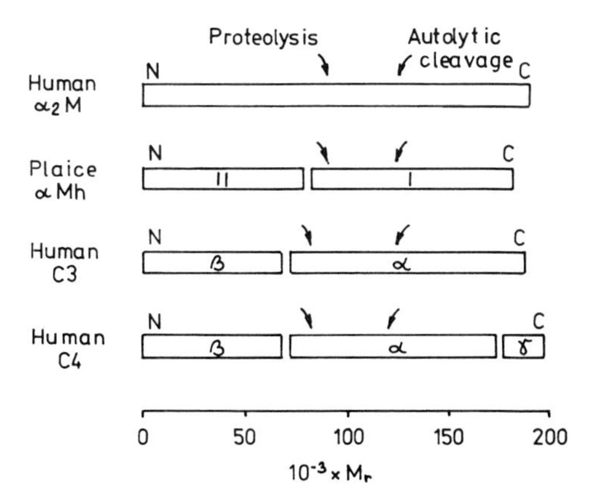

Fig. 3. Comparison of the subunits of human α_2-M, plaice α-Mh and human complement components C3 and C4 (after Starkey and Barrett, 1982b). The figure depicts the quarter-subunit of human α_2-M and the polypeptide chains of αMh and human complement proteins C3 and C4 drawn to the scale on the basis of the M_r values indicated by SDS/polyacrylamide gel electrophoresis. The polypeptide chains of proteins C3 and C4 are aligned as they occur in the single chain precursor molecules, and they, and the quarter-subunit of α_2-M are orientated with the *N-terminus* at the left. The orientation and order of the αMh subunits and its putative precursor molecule is speculative. The position of the sites of specific proteolytic cleavage and autolytic cleavage are indicated by arrows.

vertebrates. However, enhancement of fibrinogen synthesis by adrenal steroids in birds (Grieninger et al., 1983) and frogs (Waugh et al., 1983) has been reported. Fibrinogen has a M_r of approximately 340,000 and is composed of three pairs of non-identical chains which are converted to fibrin by thrombin. This is accompanied by release of fibrinopeptides A and B from the amino terminal segments of the Aα and Bβ chains (Fig. 4). The arrangement of interchain disulphide linkages is such that the amino terminals of all six chains are clustered in the central zone of the molecule (domain E) which is separated by a three-stranded bridge of coiled-coils from the second domain (D) near the C-terminal. Direct comparison of the amino acid sequences of the three polypeptide chains reveals that they are homologous and must have evolved from a common ancestor at least one million years ago (Doolittle, 1983). The origin of the fibrinogen gene is unclear since invertebrates have clottable proteins of an entirely different structure. As described by Doolittle (1975) some crustaceans, including the lobster, have a protein in their blood which is referred to as 'fibrinogen' (but is unrelated structurally to vertebrate fibrinogen) and which can be 'clotted' by a calcium-dependent enzyme found in coagulocytes (Fig. 5). This type of gelation is achieved by direct introduction of covalent bonds due to transamidation analogous to cross-linking of vertebrate fibrin by factor XIIIa. On the other hand, the haemolymph cells of the horseshoe crab, *Limulus polyphemus,* contain a coagulable protein, coagulogen, M_r 23,000 (Solum, 1973). Gelation of coagulogen is the result of limited proteolysis carried out by an enzyme which is indirectly stimulated by endotoxin and leads to a series of reactions (Fig. 5). These show a certain resemblance to the clotting cascade in higher vertebrates. As demonstrated by Nakamura and Levin (1982) proteinase N in amebocyte lysates

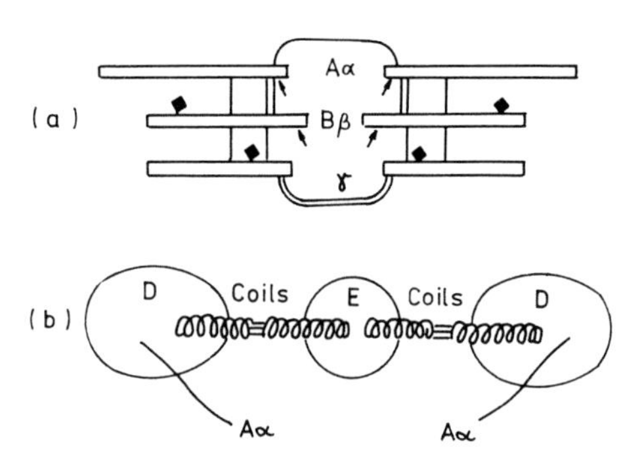

Fig. 4. Schematic models of human fibrinogen molecule. (a) Primary structure (after Henschen et al., 1983). The chains have been aligned according to homology, the *N-termini* pointing to the middle. The connecting lines represent interchain disulphide bridges, the arrows – thrombin cleavage sites, and the diamonds – carbohydrate side chains. (b) Simplified trinodular model of fibrinogen (after Plow et al., 1983). The domainal identifications E and D, the coiled coils and the carboxyterminal of the Aα chain have been superimposed on the model originally proposed by Hall and Slayter.

when exposed to endotoxin (LPS) attacks a proactivator (M_r 50,000) thus yielding a serine proteinase. This enzyme can hydrolyze synthetic substrates of thrombin and factor Xa but its main function is to activate the proclotting enzyme by limited proteolysis. The proclotting enzyme is a single-chain protein (M_r 150,000) containing in the N-terminal region γ-carboxyglutamic acid residues (Tai and Liu, 1977). In this respect it is strikingly similar to mammalian prothrombin. The clotting enzyme splits Arg-Gly and Arg-Thr bonds in the N-terminal portion of the coagulogen molecule thus liberating peptide C (Nakamura et al., 1976). It should be noted that an Arg-Gly bond is also cleaved by thrombin during conversion of mammalian fibrinogen to fibrin. Activated coagulogen (M_r 17,000) spontaneously polymerizes to a soft clot which can be dissolved in 0.87 M acetic acid (Solum, 1973) – similarly to the primary fibrin clot of vertebrates. We do not know yet whether serine proteinases in the *Limulus* coagulation system can be regarded as ancestors of zymogens in the vertebrate clotting cascade, or whether they are examples of convergent evolution. It appears, however, that endotoxin-induced clotting in *Limulus,* first described by Levin and Bang (1968), has a different physiological

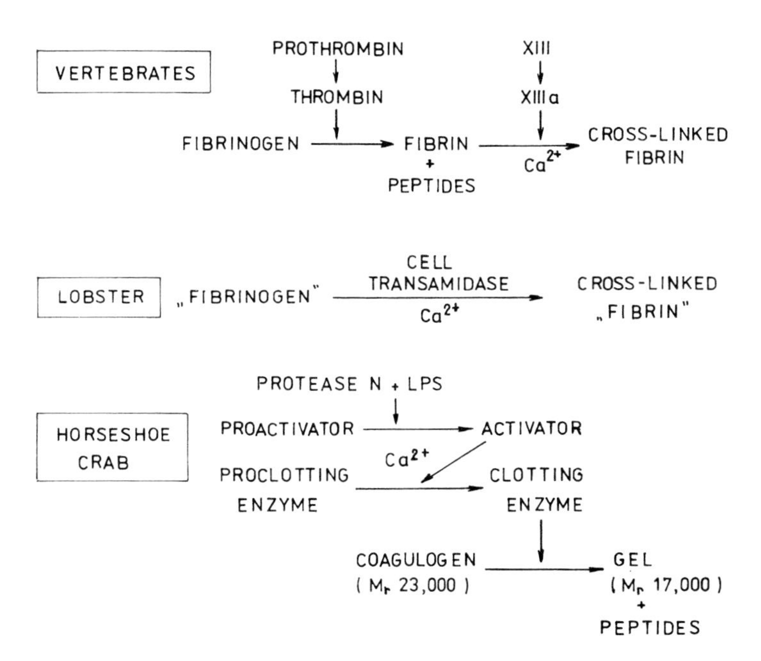

Fig. 5. Comparison of vertebrate and lobster blood coagulation system (after Doolittle, 1975) and endotoxin-induced coagulation mechanism in the haemolymph of *Limulus polyphemus.* The complex sequence of events leading to the transformation of prothrombin to thrombin in vertebrates has been omitted. Similarly, the events provoking the disruption of coagulocytes containing the transamidase which clots lobster 'fibrinogen' have not been included. For details of *Limulus* 'clotting cascade' see text.

significance to blood clotting in vertebrates, as it is aimed primarily at sequestration and removal of invading bacteria and their products. This function in vertebrates has largely been taken over by more sophisticated inflammatory and immunologic mechanisms and by AP-proteins such as those from the pentraxin family.

3. Phylogeny of pentraxin proteins

CRP was initially detected in sera of patients with acute pneumococcal pneumonia due to its ability to precipitate bacterial C-polysaccharide in the presence of calcium ions (cf. Introduction; Pepys and Baltz, 1983). Later, the characteristic pentameric structure of CRP was elucidated and its occurrence in the blood of many animal species described (reviewed in Baltz et al., 1982; Pepys and Baltz, 1983). Considerable structural similarities were found to exist between CRP and another trace protein of human plasma named variously 9.5 S α-glycoprotein or C1t complement but finally identified as SAP component, (Pepys et al., 1977). CRP and SAP are now recognized as members of a unique plasma protein family termed 'pentraxins' (Osmand et al., 1977), or 'pentaxins' (Pepys and Baltz, 1983), since their negatively stained electron micrographs show annular discs with pentameric symmetry. They share similar subunit composition, calcium-dependent ligand binding specificity and extensive homology of amino acid sequence. The subunits are held together by non-covalent interactions, except for rat CRP which may have an interchain disulphide bridge between some subunits (de Beer et al., 1982). The relative molecular mass of each subunit is around 20,000 although there are clear species-dependent differences (cf. Table 2). A complication is that the protein often behaves anomalously on SDS-PAGE gels (Pepys et al., 1978). Whole molecules of mammalian CRP or SAP in the native state are composed of five subunits but in some cases the pentameric discs may form stable pairs (e.g. human SAP, Pepys and Baltz, 1983). The majority of pentraxins are glycosylated, with the exception of human and rabbit CRP.

Early observations of Baldo and Fletcher (1973) on the occurrence of a CRP-like protein in the serum of plaice and other marine teleosts were extended by Pepys and co-workers (1978) who isolated by calcium-dependent affinity chromatography on Sepharose 4B or Sepharose-C-polysaccharide, analogues of human SAP and CRP from several species of mammals, birds, amphibia and fish (cf. Table 2). Later Liu and co-workers using calcium-dependent affinity chromatography on a phosphorylcholine matrix isolated a CRP-like protein also from the haemolymph of horseshoe crab, *Limulus polyphemus* (Robey and Liu, 1981; Liu et al., 1982). The *Limulus* CRP, called limulin, differs from its vertebrate counterpart in that it consists of 12 complex subunits, each containing glycosylated A and B chains which are non-covalently associated in a double-stacked hexamer. Limulin has broad binding

TABLE 2
Plasma concentration, subunit M_r and response to injury of SAP and CRP in some animal species

Species	Serum Amyloid P Protein			C-reactive Protein		
	Plasma conc. (mg/1)	Subunit (M_r)	AP-response	Plasma conc. (mg/1)	Subunit (M_r)	AP-response
Man[1]	30 – 40	23,650	0	0.8	21,100	+ + +
Rabbit	30 – 40	19,950	+	1.0	18,900	+ + +
Rat[2]	30 – 40	24,500	0	300 – 600	23,100	+
Mouse	5 – 100	25,300	+ + +	0.1	?	0
Hamster[3] male	?	?	?	4 – 20	?	+ +
female	?	?	?	500 – 3000	?	–
Chicken	?	?	?	?	21,100	+7
Marine toad[a]	?	27,300	?	?	?	?
Plaice[b,4]	200	25,600	0	50 – 60	18,600	+
Dogfish[c,5]	100	27,500	?	400	26,900	?
Horseshoe crab[d,6]	?	?	?	1000	A – 18,000	?
					B – 24,000	?

Concentrations of SAP and CRP in the blood of healthy individuals are reported either as the mean or range of values (to show individual or strain-dependent variability, e.g. SAP in mice). The response to injury is given in relation to normal plasma level: 0 = no significant and reproducible change; – = decreased level; + = elevation less than 2-fold; + + = elevation 2 to 10-fold; + + + = elevation of 100-fold or more; ? = not known to the author.

[a] *Bufo marinis*;

[b] *Pleuronectus platessa*;

[c] *Mustelus canis*;

[d] *Limulus polyphemus*.

The data are taken mainly from: Pepys et al., 1978; Baltz et al., 1982; Pepys and Baltz, 1983, with supplementation from the following sources: 1, Kushner et al., 1978; 2, de Beer et al., 1982; 3, Coe and Ross, 1983; 4, White et al., 1981; 5, Robey et al., 1983; 6, Robey and Liu, 1981; 7, Patterson and Mora, 1965 (cited after 4).

specificity and shows also the properties of sialic acid-binding lectin (Robey and Liu, 1981). Despite structural differences some homology of amino acid sequence was detected between human and *Limulus* CRP suggesting the possibility of a common ancestral gene and evolutionary conservation of the pentraxin family. Distant homologies have been noted with C_H2 domain of IgG and with the alpha chain of C3 complement, but they are insufficient to support the concept of a common evolutionary origin (Oliveira et al., 1979). The origin of similar C1 binding sites in human CRP and $C\gamma2$ domain of human IgG_1 is at present unclear (cf. Pepys and Baltz, 1983).

CRP and SAP bind numerous ligands: some strongly and with high specificity in the presence of calcium, others loosely and dependent on species origin of the protein (reviewed in Pepys and Baltz, 1983; see also Chapter 13, Section 3). For human CRP, specific ligands include phosphorylcholine and some microbial complex polysaccharides. Human SAP binds agarose and amyloid fibrils but when aggregated shows high reactivity toward plasma fibronectin and C4-binding protein. Loosely bound ligands of pentraxin proteins include a variety of lipids, lipoproteins, polyanions (dextran sulphate, nucleic acids), polycations (histones, myelin basic proteins, cationic proteins of leucocyte granules), components of complement, and many others. Species-dependent variations in binding specificities may be responsible for discrepancies in the classification of isolated proteins (e.g. CRP-SAP of the rat – Pontet et al., 1981; hamster female protein – Coe and Ross, 1983). There is no doubt, however, that binding of some of these ligands may aid in elimination of foreign materials from the organism (Pepys and Baltz, 1983; Chapter 13, Section 3). Thus proteins of the pentraxin family appear to represent an ancient, primitive defense mechanism before the advent of the immunological system. The concentration of CRP (limulin) in the haemolymph of the horseshoe crab – which has no immunoglobulins – is always very high whereas in the majority of species of higher vertebrates, pentraxins are trace components in the blood of healthy animals and their production increases only during the AP-response (Table 2).

References, p. 232.

Gordon/Koj (eds.) The acute-phase response to injury and infection.

CHAPTER 15

Stimulation of liver by injury-derived factors

A. KOJ

Infection, injury, tumour growth or injections of LEM/EP/IL-1 to experimental animals elicit the following principal changes in the liver listed in the approximate order of occurrence:

(a) Accumulation of iron and zinc;
(b) enhanced uptake of amino acids;
(c) increased synthesis of RNA, especially mRNA;
(d) change in the contents and activity of several cellular enzymes;
(e) altered synthesis and secretion of AP-proteins.

Some of these changes may be indirect (e.g. uptake of iron and zinc) but others have been reproduced in the primary hepatocyte cultures and thus are due to the action of IL-1 (or other mediators) on the liver cell.

1. Effects of IL-1 on trace metals in the liver

Injections of LEM/IL-1 cause redistribution in the body of at least three metals: iron, zinc and copper (reviewed by Kampschmidt, 1978, 1981) but the kinetics of this process and mechanisms involved appear to be different in each case. Iron and zinc are removed from the plasma and deposited in the liver (Pekarek et al., 1972) while the copper content of the blood increases due to enhanced synthesis and secretion of ceruloplasmin. The level of serum iron may be reduced by as much as 70% within 6 – 12 h after injection of LEM into rats (Kampschmidt and Upchurch, 1969b; Pekarek et al., 1974). This process is probably mediated by lactoferrin (van Snick et al., 1974) which is released by LEM/IL-1 from specific granules of neutrophils (Klempner et al., 1978 and Chapter 1). Lactoferrin binds plasma iron and is promptly taken up by the reticuloendothelial cells of the liver (Courtoy et al., 1984; Moguilevsky et al., 1984) where iron is stored in the form of ferritin (cf. Chapter 13). On the other hand, uptake of zinc appears to be mediated through a liver protein, metallothionein, which is synthesized at an increased rate in response to various stresses and administration of metals (Zn, Cd, Cu, Hg and Ag), while glucocorticoids and glucagon are involved in the process of induction (Brady, 1982; Sobocinski and Canterbury, 1982; Poole et al., 1984). Since, however, both hypozincaemia and synthesis of AP-proteins are evoked by intracranial injection of small doses of IL-1 that are ineffective when injected parenterally in rabbits (Turchik and Bornstein, 1980), it is possible that the central nervous system is involved in the increased uptake of zinc by the liver, especially in the early stages of the AP-response before metallothionein concentration in the liver could have risen. However, evidence against this view has been obtained from guinea pigs by Blatteis et al. (1983) (cf. also Chapter 5). The fall in plasma zinc concentration in mice following endotoxin injection was unaffected by previous treatment with indomethacin but was reduced by dexamethasone (Poole et al., 1984). It should be mentioned here that human monocyte pyrogen injected into rabbits and rats is as effective as homologous preparations in reducing plasma iron and zinc (Kampschmidt et al., 1983).

2. *Enhanced uptake of amino acids by the liver*

By using non-metabolizable amino acid analogues such as cycloleucine or α-amino isobutyric acid (AIB) it has been established that laparatomy, bacterial infection or turpentine-elicited local inflammation all lead to accumulation in the liver of free amino acids (Shibata et al., 1970; Powanda et al., 1973). A similar effect was observed when rabbit leucocyte-derived LEM/IL-1 was injected into rats (Wannemacher et al., 1972; Pekarek et al., 1974) or added to the perfusion medium of an isolated rat liver (Wannemacher et al., 1975; Fig. 1). Measurements of liver uptake of cycloleucine carried out in vivo after injection of LEM indicated log dose proportionality. The cycloleucine uptake commenced within 2 – 4 h of the LEM injection. With the perfused liver constant infusion of LEM/IL-1 was required in order to raise the uptake of [^{14}C]AIB above that in the control rat liver. With the addition of glucagon instead of LEM the effect was even stronger. Dihydroergot-

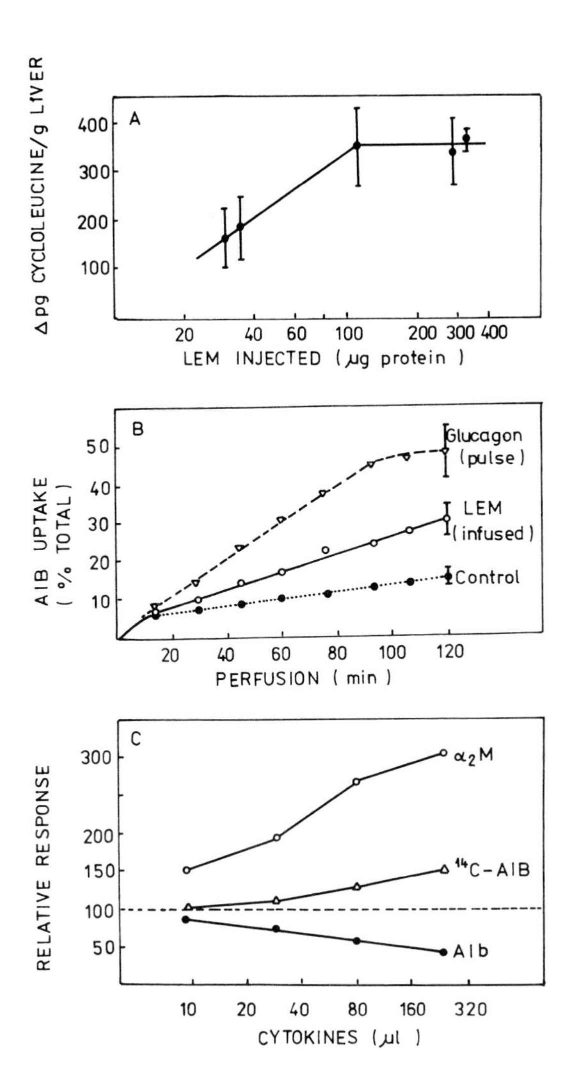

Fig. 1. Effect of LEM/HSF/IL-1 on uptake of amino acids by rat liver in vivo (A), during liver perfusion (B) and in hepatocyte culture (C). (A) Change in hepatic cycloleucine 3 h after I.P. administration of rabbit leucocyte-derived LEM preparation in comparison with saline-injected rats (after Wannemacher et al., 1972). (B) Effect of rabbit leucocyte-derived LEM and glucagon on the uptake of [^{14}C]AIB by isolated perfused rat liver. LEM was added 30 min before AIB and then constantly infused during perfusion (Wannemacher et al., 1975). (C) Effect of human monocyte-derived cytokines on plasma protein synthesis and AIB uptake by rat hepatocytes cultured for 3 days (10^6 per 1 ml of medium changed daily). The medium collected on the last day was used for determination of albumin and α_2-M. The fresh medium containing [^{14}C]AIB was added and the uptake of the label by the cells was determined after 2 h incubation. Data are expressed as percentages of the respective control values from a culture without added cytokine (Koj, Gauldie, Regoeczi, Sauder and Sweeney, unpublished observations).

amine inhibited the stimulatory action of glucagon but did not affect the response to LEM (Wannemacher et al., 1975). These observations suggest that glucagon and LEM affect the liver amino acid transport by different mechanisms. It is interesting that other hormones tested (cortisone, growth hormone, insulin, thyroxine, adrenaline, testosterone) had negligible effects on cycloleucine uptake by rat liver (Wannemacher et al., 1975).

Direct action of hepatocyte stimulating factor (HSF)/IL-1 on amino acid transport by liver cells has been confirmed in a primary culture of rat hepatocytes (Fig. 1). Addition of human monocyte-derived cytokines to the medium resulted in reduced synthesis of albumin, increased synthesis of α_2-M and enhanced uptake of [^{14}C]AIB. Stimulation of AIB uptake required prolonged exposure of the hepatocytes to crude human cytokines (monocyte supernatant) in the presence of dexamethasone (Koj, Gauldie, Regoeczi, Sauder and Sweeney, unpublished observations). Mechanisms of IL-1 action on hepatocyte amino acid transport remain obscure but it is interesting that Bayer et al. (1980) reported that alterations of amino acid transport in a rat hepatoma cell culture could be brought about by certain anti-inflammatory drugs.

Qualitative changes during increase in the liver amino acid pool have been analyzed by Woloski et al. (1983a). At 4 h after injection of turpentine into rats, the concentrations in liver of the majority of free amino acids (except taurine) were reduced but later there were significant rises. In the period from 8 to 24 h Pro, Gly, Ala, Val, Phe and His were most affected. Muscles are regarded as the main source of plasma and liver amino acids during infection and inflammation (Beisel, 1975). It has now been shown that EP/IL-1 and what may be an IL-1 degradation product (M_r 4,200), proteolysis inducing factor (PIF), can stimulate muscle protein breakdown. Lysosomal thiol proteinases and prostaglandin E_2 are both involved in this response (Baracos et al., 1983; Clowes et al., 1983; Dinarello et al., 1984; Chapter 9). The relative importance of PIF and IL-1 in vivo compared with the many other factors which are affected by injury (cortisol, glucagon, insulin, thyroxine) and are involved in the transfer of amino acids from muscles and other tissue to the liver is as yet unknown.

3. *Increased synthesis of RNA*

Already in 1966, Neuhaus and co-workers observed that the increased hepatic production of certain serum glycoproteins which follows acute injury is preceded by an accelerated rate of synthesis of RNA (Chandler and Neuhaus, 1968). Thus, by 8 h after laparatomy the incorporation of [^{14}C]orotate into nuclear and microsomal RNA was increased 3 to 5-fold. Wannemacher et al. (1975) achieved a similar stimulation of orotate incorporation (mainly into the membrane-bound ribosomal fraction) by injecting rats with rabbit-derived LEM, or by adding LEM to liver perfusate containing cortisol (Thompson et al., 1976). Increased RNA synthesis in nuclei isolated from the liver of rats undergoing an acute inflammatory reaction is essentially determined by a faster polyribonucleotide elongation rate while the number of transcribing polymerase molecules is unchanged (Schiaffonati et al., 1984). The stimulation affects both α-amanitin-resistant (polymerase I) and α-amanitin-sensitive (polymerase II) activities, suggesting that pre-ribosomal and pre-messenger RNA formation are activated at the same time. The maximum stimulation of both polymerases in rat liver nuclei was observed 10 h after injection of turpentine (Piccoletti et al., 1984).

The enhanced RNA synthesis and subsequent synthesis of AP-proteins can be inhibited by actinomycin D (Chandler and Neuhaus, 1968) or by galactosamine and ethionine, both of which are known to interfere with liver RNA metabolism by trapping cellular UTP or ATP (Koj and Dubin, 1976; Sipe, 1978; Kisilevsky et al., 1979; Kasperczyk and Koj, 1983). All these data suggested that formation of AP-proteins

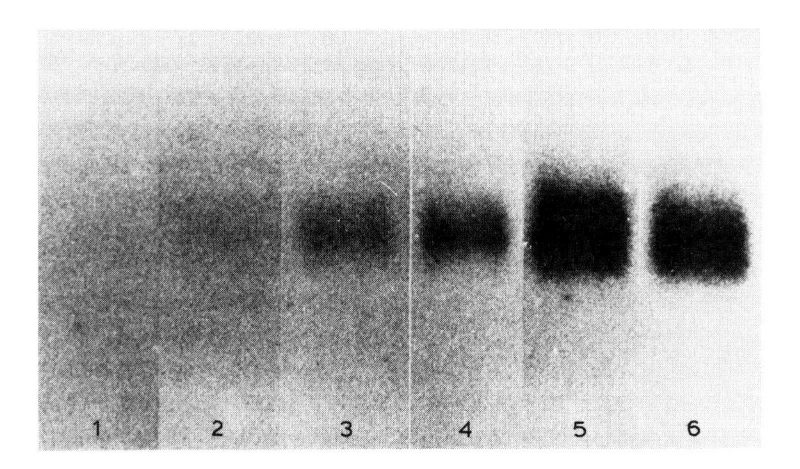

Fig. 2. Increase in the contents of α_1-AGP-mRNA in rat hepatocytes cultured with dexamethasone and human cytokines (crude IL-1 from COLO-16 cells and stimulated monocytes) (after Baumann et al., 1984b). Total RNA was extracted from cultured hepatocytes, 25 μg RNA separated on agarose gel, transferred to nitrocellulose and hybridized to ^{32}P-labelled DNA from the plasmid p10-14 containing a cDNA insert complementary to rat α_1-AGP-mRNA. The fluorographic image was exposed for 5 days. The lanes were rearranged and represent following RNA samples: 1 – 5, female rat hepatocytes: immediately after isolation (1), or cultured 48 h in control medium (2), in medium containing supernatant of squamous carcinoma cell line COLO-16 (50 lymphocyte activating factor (LAF) units/ml) (3), in medium with 1 μM dexamethasone (4), and in medium containing both COLO-16 cytokines and dexamethasone (5). Lane (6) contains RNA from male rat hepatocytes cultured 48 h in the medium containing dexamethasone and supernatant from stimulated human monocytes (170 LAF units/ml).

in the liver requires enhanced transcription of specific mRNAs. However, direct proof that this does occur was produced only recently.

Princen et al. (1981) using tritiated DNA complementary to rat fibrinogen polypeptide mRNA and albumin mRNA showed a 7-fold increase in fibrinogen mRNAs and a 2-fold decrease in albumin mRNA in the liver of rats 24 h after injecting turpentine. In separate studies, and using ^{32}P-labelled cDNA complementary to rat α_1-AGP-mRNA, Taylor and co-workers observed a progressive rise in abundance of this mRNA in rat liver following turpentine-elicited inflammation. The first detectable increase in mRNA content was observed 4 h after turpentine injection, while the well-defined maximum, corresponding to a 90-fold increase in α_1-AGP-mRNA over control values, occurred at 36 h (Ricca et al., 1981). These results have been amply confirmed by several authors and extended to other proteins. Heinrich and co-workers found over 70-fold increase in translatable α_1-AGP-mRNA at 15 h after turpentine and over 60-fold of α_2-M-mRNA 3 h later (Table 1). Moreover, Baumann et al. (1984b) using a cDNA probe were the first to demonstrate that crude human IL-1 preparations increase the content of α_1-AGP-mRNA in cultured rat hepatocytes, the effect being potentiated by addition of dexamethasone to culture medium (Fig. 2). All these data lead to the conclusion that IL-1-mediated enhanced transcription of certain genes is an obligatory step in the liver AP-response (cf. also Chapter 18).

4. *Changes in hepatocyte enzymes*

Injury and inflammation affect several liver enzymes. Thus the activities of tyrosine aminotransferase and tryptophan oxygenase are doubled at 4 – 6 h after laparatomy in the rat (Tsukada et al., 1968) or after injection of celite into mice (Koj and Dubin, 1976). The rise of glucosamine synthetase reaches its maximum (180 – 200% of control values) some 12 – 18 h after injury (Okubo and Chandler, 1974). These increases can be blocked by actinomycin D or large doses of galactosamine suggesting that enhanced transcription is required. Since, however, glucocorticoids and other stress-related hormones acting through the cAMP system, such as glucagon and adrenalin, are known to be inducers of several liver enzymes, including tyrosine aminotransferase (for references see Ernest et al., 1977; Marston et al., 1981; Augberger et al., 1983), it is not yet known whether IL-1 and other mediators of inflammation are also involved. The situation is somewhat clearer in the case of liver lysosomal β-galactosidase and *N*-acetylglucosaminidase which are reduced by half at 24 h after injection of turpentine into rats (Kaplan and Jamieson, 1977); similar, although less pronounced changes were obtained with crude cytokine from rat peritoneal cells (Woloski et al., 1983b). The same procedure resulted in elevated ac-

TABLE 1
Translatability of mRNAs for five rat plasma proteins during experimental inflammation (after Northemann et al., 1983)

	Albumin		α_1-AGP		α_1-PI		α_2-M		Transferrin	
Hours after turpentine	(%)	-fold change	(%)	-fold change	(%) × 10	-fold change	(%) × 100	-fold change	(%) × 100	-fold change
1	2.66	1.0	0.03	1.0	1.91	1.0	0.07	1.0	1.27	1.0
6	1.94	0.73	0.24	7.1	2.62	1.37	0.81	11.2	1.73	1.36
12	1.14	0.43	1.55	45.5	2.91	1.52	1.83	25.0	N.D.	N.D.
15	0.98	0.37	2.48	72.9	3.30	1.73	3.24	45.0	2.32	1.83
18	0.96	0.36	1.82	53.5	3.72	1.95	4.74	65.9	2.50	1.97
21	0.95	0.36	1.50	44.2	3.19	1.67	2.31	32.1	3.56	2.80
24	0.86	0.32	1.19	35.0	2.27	1.19	0.93	13.0	1.73	1.36
40	0.83	0.31	0.85	25.0	2.10	1.10	0.19	2.7	1.23	0.97
60	0.78	0.29	N.D.	N.D.	2. 1	1.05	N.D.	N.D.	1.13	0.89

The data are given in per cent of total trichloroacetic acid-insoluble radioactivity obtained after cell-free translation of poly (A^+) RNA in a reticulocyte lysate. For each immunoprecipitation 15 × 10^6 cpm of the total trichloroacetic acid-insoluble radioactivity were used. N.D. = not determined.

tivity of another liver enzyme, sialyltransferase, while serum levels of α_1-AGP were increased and albumin reduced, thus confirming the occurrence of a typical AP-response. Increased activity of glucosamine-6-phosphate synthase and UDG-GlcNAc epimerase, as well as expanded liver pools of UDP-GlcNAc and UDG-GalNAc, facilitate synthesis and secretion of those AP-reactants which are glycoproteins (Jamieson et al., 1983).

On the other hand, repeated injections of rabbit-cell-derived LEM/IL-1 gave a pronounced fall in rat liver catalase activity (Merriman et al., 1978) and Shutler et al. (1977) reported that inflammation suppresses adaptive synthesis of rat liver fatty acid synthetase. As a result of inflammation, a 60% decrease in liver mRNA translatable as fatty acid synthetase has also been observed (Langstaff and Burton, 1982). A similar suppression of this enzyme in cultured preadipocytes exposed to crude cytokine from mouse macrophages has been described by Pekala et al. (1983). This reduction in activity was, at least in part, due to a specific effect of macrophage-derived mediators on synthesis of this enzyme, as shown by the decreased incorporation of [^{35}S]methionine.

This brief review indicates that altered activities of several enzymes in the liver (and other tissues) elicited by injury-derived factors and stress-related hormones reflect various adaptative changes in the metabolic pathways and are caused by profound changes in the protein-synthesizing cellular machinery during the AP-response.

References, p. 232.

Gordon/Koj (eds.) The acute-phase response to injury and infection.

CHAPTER 16

Synthesis and secretion of acute-phase proteins from the liver

A. KOJ

1. Sequential recruitment of hepatocytes

Experimental inflammation leads to enhanced synthesis of several AP-proteins in the liver parenchymal cells. Thus an important question is whether this occurs due to recruitment of formerly inactive cells, or is due to augmented activity of all hepatocytes. Another crucial problem concerns possible specialization of hepatocytes as producers of one, or several AP-proteins. Early immunofluorescence studies of Barnhart and co-workers and Hamashima and associates reviewed by Koj (1974) supported the idea that a stimulated hepatocyte is able to produce the whole range of AP-proteins. Later by using the immunoenzymatic technique, Kushner and co-workers (Kushner and Feldmann, 1978; Macintyre et al., 1982) demonstrated that in rabbits with local inflammation, CRP is produced by progressively increasing numbers of cells starting from the periportal area. At the peak-serum response occurring 38 h after turpentine injection, CRP could be demonstrated in most hepatocytes. Extending these observations, Courtoy et al. (1981) showed that following turpentine injection into rats four typical positive AP-reactants (fibrinogen, α_1-AGP, α_2-M and Hp) increased simultaneously in the same cells in a very characteristic order related to the position occupied by the hepatocyte in the hepatic lobule. During the first stage of inflammatory reaction (8 – 16 h after turpentine) the synthesizing cells were detected primarily in the periportal zone; later between 24 and 36 h specific immunoenzymatic staining for these proteins extended to the midlobular and centrilobular areas. The characteristic order of hepatocyte response agrees with the concept that mediators derived from the site of injury are transported to the liver in the blood and first stimulate the cells at the periphery of the lobule adjacent to the portal triads (Kushner and Feldmann, 1978). Lack of specialization of hepatocytes in respect of protein synthesis was confirmed by Bernuau et al. (1983): by using a haemolytic plaque technique fibrinogen was detected in 93% of hepatocytes isolated from rat livers 24 h after turpentine injection. Moreover, a reciprocal modulation of fibrinogen and albumin secretion at the single cell level was observed by these authors indicating that the stimulated hepatocyte simultaneously increases synthesis of positive AP-proteins and reduces formation of albumin.

2. *Polypeptide chain synthesis and post-translational modifications of AP-proteins*

Similarly to all secretory proteins AP-reactants are synthesized on membrane-bound polysomes of the liver cell and initially contain signal peptide extensions at the NH_2 terminal ends. During the AP-response, a characteristic shift in the proportion of heavy polysomes is observed reaching its maximum 18 h after laparatomy in the rat (Liu and Neuhaus, 1968). This is related to increased efficiency of translation (for references see Koj, 1974) and abundance of mRNAs for individual AP-proteins (cf. Chapter 15). As shown by Jamieson and Ashton (1973) the amount of immunoprecipitable α_1-AGP associated with rat liver microsome fraction increased over 4-fold, 12 h after injecting turpentine. Kwan and Fuller (1977) demonstrated that whereas in control rats only 4.4% of total polysomes were involved in fibrinogen synthesis this figure was increased to 15% 24 h after turpentine stimulation. The synthesis of fibrinogen has been investigated by Kudryk et al. (1982) in some detail. Assembly of its three non-identical polypeptide chains (cf. Chapter 14) has been shown to occur in the rough endoplasmic reticulum, although their production may not yet be fully co-ordinated. Thus Yu et al. (1983) reported the existence of a large intracellular pool of gamma chains, at least in Hep-G2 cells. However, only fully assembled fibrinogen molecules are secreted in vivo and in tissue culture (Amrani et al., 1983).

In distinction to fibrinogen, haptoglobin is synthesized as a single chain precursor in the rat, rabbit and man (Haugen et al., 1981; Chow et al., 1983; Rangei et al., 1983). Thc nativc form of this glycoprotein consists of a tetrameric structure composed of two halves each containing non-identical alpha and beta subunits which are linked by intermolecular disulphide bonds (Kurosky et al., 1980). In the rat, the molecular weight of Hp has been estimated to be near 90,000, whereas the alpha subunit is around 9,500 and beta 35,000. The primary in vitro translation product, pre-prohaptoglobin (M_r 40,000) possesses an NH_2-terminal hydrophobic signal peptide of 18 amino acid residues followed directly by the alpha subunit region, with the beta subunit located in the carboxyl-terminal portion of the protein (Hanley et al., 1983). After cleavage of the signal sequence the α-α and α-β disulphide bonds are formed and the β-region undergoes core glycosylation consisting in the attachment of N-linked, mannose-rich oligosaccharide chains. While still in the endoplasmic reticulum, prohaptoglobin (M_r 45,000) is dimerized and 60 – 70% of the molecules are proteolytically processed to form the individual alpha and beta subunits in the $\alpha_2\beta_2$ tetramer. The carbohydrate chains of prohaptoglobin and the core-glycosylated β-subunit (M_r 35,000) are then converted to complex, sialylated glycoprotein in the Golgi apparatus. The secreted fully glycosylated prohaptoglobin which constitutes 30 – 40% of total synthesized Hp is then promptly processed by a plasma proteinase so that in the circulating blood only the mature tetrameric form is present. The proportion of processed and unprocessed prohaptoglobin is unaffected by either turpentine-induced acute inflammation, or by inhibition of glycosylation by tunicamycin. Moreover, it is similar both in vivo and in hepatocyte culture (Hanley et al., 1983), and remains so even after treatment with carboxylic ionophore, monensin, which is known to inhibit intracellular transport of proteins (Misumi et al., 1983). Thus it appears that the post-translational cleavage of prohaptoglobin is partly accomplished intracellularly (presumably in the endoplasmic reticulum) and partly after the protein has been secreted into the plasma. The biological significance of prohaptoglobin secretion is not understood at present, but the phenomenon appears to be unique among AP-proteins.

That proteinases are involved in the processing of polypeptide precursors of plasma proteins synthesized by the hepatocyte is indirectly supported by the effects of some proteinase inhibitors. Thus, Algranati and Sabatini (1979) observed inhibition of albumin secretion by antipain and leupeptin, while Schreiber and co-workers (Edwards et al., 1979; Schreiber et al., 1979) described a similar effect of tosyl-lysyl-chloromethyl ketone (TLCK) on secretion of albumin and transferrin. However, Mbikay and Garrick (1981) suggest that TLCK may act at the level of chain initiation rather than chain processing.

Suggestion that increased synthesis of AP-proteins is accompanied by more efficient secretion, initially based on studies with liver slices (Koj, 1980), has been recently confirmed by Macintyre et al. (1985) with isolated hepatocytes. In rabbit hepatocyte cultures from progressively more responsive animals, a decrease in the ratio of intracellular CRP content to rate of secretion ('residence time') has been observed. The authors speculate that secretion of CRP (and possibly other AP-proteins) is regulated by availability of receptors at the stage of transfer from rough endoplasmic reticulum to Golgi apparatus (Macintyre et al., 1985).

The majority of AP-reactants synthesized in liver cells are glycoproteins containing N-glycans, i.e. asparagine-linked complex oligosaccharides in the form of bi-, tri- and tetra-antennary chains (for references see Montreuil, 1980; Berger et al., 1982; Jamieson, 1983; Hatton et al., 1983). A key step in biosynthesis of these glycoproteins is the transfer of an oligosaccharide from a dolichol intermediate to a polypeptide asparagine residue located in the sequence -Asn-X-Ser(Thr). These core oligosaccharides containing glucose and mannose residues are then processed by glucosidases and mannosidases; finally, the terminal triplet sugars, N-acetyl-glucosamine, galactose and N-acetyl-neuraminic acid (or fucose), are added by glycosyl transferases in the Golgi complex (Jamieson, 1983). The intrahepatic precursors of several AP-proteins have been characterized, both in the liver (Schreiber et al., 1979; Friesen and Jamieson, 1980; Nagashima et al., 1980), and in hepatocyte culture (Andus et al., 1983a; Hanley et al., 1983; Gross et al.,1983). Due to the presence of glucose and mannose they all show high affinity to Con A.

Among mature, sialylated oligosaccharide chains the bi-antennary glycans bind most easily to Con A and thus chromatography on Con A-Sepharose (Bayard et al., 1982), or affino-electrophoresis in agarose containing Con A (Bøg-Hansen et al., 1975), are routinely used for fractionation of glycoprotein variants. Advances in structural studies of glycoproteins, including the results of NMR spectroscopy (Fournet et al., 1978; Halbeek et al., 1981) demonstrated that many plasma glycoproteins contain mixed type glycan chains, the bi-antennary being the most common (reviewed by Hatton et al., 1983). α_1-AGP is an exception having a majority of tri- and tetra-antennary chains and thus rather poor affinity for Con A. Heterogeneity of glycan structure in plasma glycoproteins may be related to some of their biological functions but it certainly demonstrates the availability of alternative biosynthetic pathways (Hatton et al., 1983).

Among the many inhibitors which can interfere with protein glycosylation (reviewed by Elbein, 1983) the most important is tunicamycin, which blocks the first step, i.e. formation of the oligosaccharide-dolichol complex. It is somewhat surprising that despite its effectiveness in preventing glycosylation, tunicamycin only partly and variably reduces synthesis and secretion of glycoproteins by cultured hepatocytes: α_2-M, α_1-AGP and α_1-PI being more affected (Carlson and Stenflo, 1982; Geiger et al., 1982; Andus et al., 1983a) than transferrin, apolipoprotein B or haptoglobin (Struck et al., 1978; Edwards et al., 1979; Hanley et al., 1983).

Incubation of liver slices from turpentine- and tunicamycin-injected rats led to detection in the medium of several non-glycosylated plasma proteins, although in vivo only one, α_1-AP-globulin, was found in plasma (Fig. 1). The half-life in the rat blood of the non-glycosylated α_1-AP-globulin has been estimated to be approximately five times shorter than that of the native form (Koj, Regoeczi, Bereta, Dubin, Kurdowska and Chindemi, in preparation). It is likely that other non-glycosylated AP-proteins secreted into the blood are quickly removed and degraded.

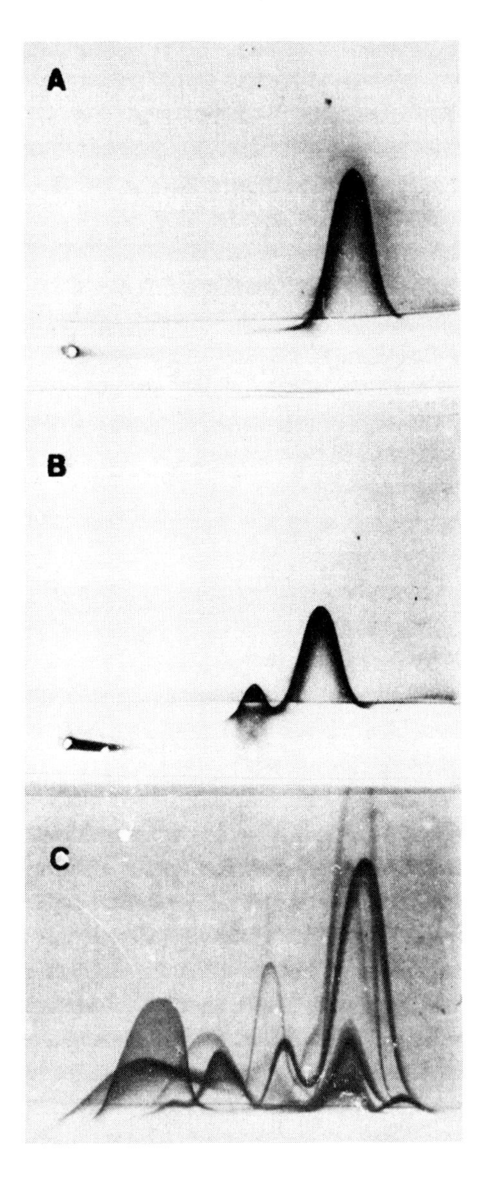

Fig. 1. Crossed immunoelectrophoresis of plasma from a turpentine-injected, or turpentine- and tunicamycin-injected rat. In each case 2 μl of plasma were first subjected to agarose electrophoresis for 2 h (anode on the right). Then agarose gels containing either monospecific antiserum to α_1-AP-globulin (A) and (B), or polyspecific antiserum to rat plasma (C), were cast and electrophoresis was carried out for 15 h (anode on the top). (A) Plasma from rat injected with turpentine 24 h earlier; (B and C) plasma from rat injected with turpentine 24 h earlier and with tunicamycin (2 μg/g body weight) 12 h earlier. In order to obtain a clearer picture, albumin was removed from plasma by means of a column of Sepharose-Cibacron blue.

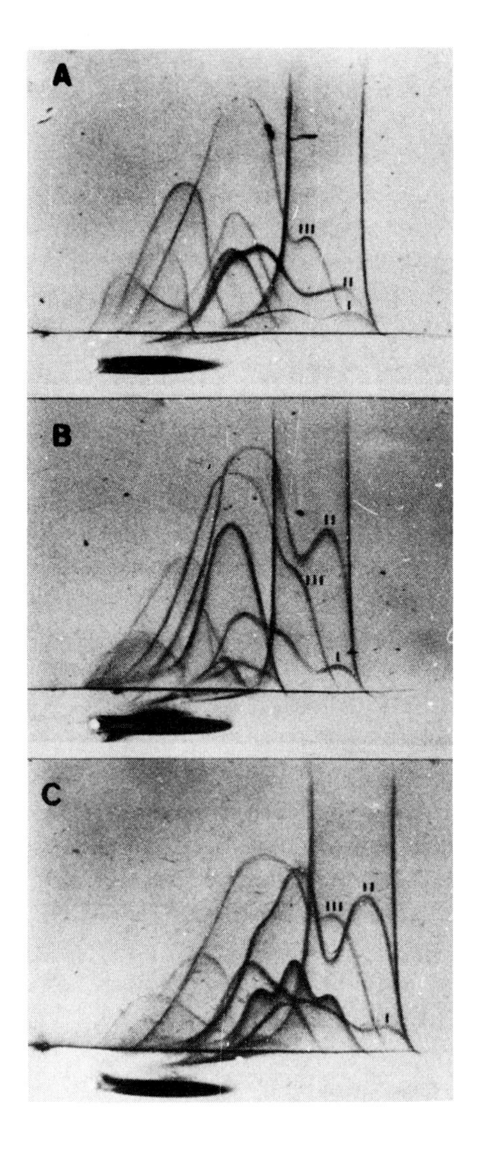

Fig. 2. Crossed immunoaffinoelectrophoresis of plasma from a control rat (A), rat injected 48 h earlier with turpentine (B) and rat carrying Morris hepatoma 7777 21 days after implantation (C). Agarose used for the first-dimension electrophoresis (run from left to right) contained Con A (0.5 mg/ml), while that used in the second dimension (run from bottom to top) contained a polyspecific rabbit antiserum to plasma from hepatoma-bearing rats. I, α_1-AGP; II, α_1-AP-globulin; III, α_1-PI.

The non-glycosylated α_1-AP-globulin has been found to retain its activity as the inhibitor of cysteine proteinases (Koj et al., in preparation) thus behaving similarly to α_1-PI which after being deprived of carbohydrate chains is still able to react with elastase (Andus et al., 1983b). All these results of experiments with tunicamycin are consistent with the hypothesis of Olden et al. (1982) who suggested that the carbohydrate moiety of some glycoproteins is not essential for their secretion from the cell, or even for mediation of specific biological activity, but is required for conformational stability and resistance to proteolytic enzymes.

3. Occurrence in the plasma of injury-induced variants of AP-proteins

Many glycosylated AP-proteins exist in plasma in multiple molecular forms differing either in the primary sequences of their polypeptide chains, or in composition and arrangement of their glycans, and it has been speculated that trauma affects only some of these components. Early studies could not provide convincing evidence supporting this idea (cf. Koj, 1984) until Nicollet et al. (1981) found a predominant increase of the Con A-reactive fraction of α_1-AGP in the blood of patients with inflammatory diseases. These observations were extended to rats with turpentine abscess and rats carrying transplantable Morris hepatomas (Koj et al., 1982). When the control and AP-plasmas were subjected to affino-electrophoresis with Con A

TABLE 1

Relative abundance of the Con A-reactive fractions in three glycoproteins of rat plasma as estimated after crossed immunoaffinoelectrophoresis (after Koj et al., 1982)

Protein	Source of rat plasma	Time (days)	Plasma level (mg/ml)	Con A-reactive fractions (%)
α_1-AGP	Control	–	0.09 ± 0.01	57.2 (53 – 61)
	Turpentine-injected	2	0.74 ± 0.08	78.1 (67 – 82)
	Hepatoma-bearing	21	0.76 ± 0.12	75.6 (64 – 80)
α_1-AP-globulin	Control	–	0.94 ± 0.11	58.6 (53 – 63)
	Turpentine-injected	2	3.85 ± 0.28	69.7 (64 – 78)
	Hepatoma-bearing	21	5.73 ± 0.36	67.6 (58 – 74)
α_1-PI	Control	–	1.10 ± 0.09	83.5 (77 – 88)
	Turpentine-injected	2	1.43 ± 0.11	85.7 (80 – 89)
	Hepatoma-bearing	21	1.48 ± 0.18	83.8 (79 – 86)

Plasma levels of the glycoproteins were measured independently by rocket immunoelectrophoresis and are given as mg/ml (± S.E.M., n = 6). The proportions reacting with Con A are in each case the mean of six electrophoretic runs (0.5 mg of Con A/ml of agarose gel) and were calculated as percentages of the whole peak area (cf. Fig. 2). Scatter of results is reported in parentheses.

followed by immunoelectrophoresis with polyspecific antiserum to rat plasma proteins the pattern shown in Fig. 2 was usually observed. Apart from albumin, which migrates as a single peak, most of the glycoproteins exhibited heterogeneity in the form of multiple peaks or shoulders, resulting from variable degrees of retardation during affinoelectrophoresis. The peaks were identified with monospecific antisera, the most conspicuous belonging to α_1-AGP, α_1-AP-globulin and α_1-PI (Fig. 2, peaks I, II and III, respectively). In hepatoma-bearing animals the additional conspicuous triple peak corresponds to α-foetoprotein (AFP).

By measuring the areas under the peaks resulting from Con A non-reactive molecules, i.e. those with the highest electrophoretic mobilities and those areas due to molecules retarded by interaction with the lectin, it was possible to evaluate their relative abundance in plasma. As depicted in Table 1, α_1-AGP and α_1-AP-globulin responded to injury or tumour growth by marked increases in the plasma level and by shifts toward the Con A-reactive forms. On the other hand, α_1-PI showed rather negligible changes in both plasma concentration and as a proportion of the components analyzed. It is tempting to speculate that during injury-induced synthesis of AP-proteins, the processing of α_1-AGP and α_1-AP-globulin in the rat liver cell is affected and either some 'unfinished' molecules, or those having predominantly bi-antennary glycan chains, are secreted. This idea is in agreement with observations of Nicollet et al. (1981, 1982) on human α_1-AGP and of Bowen et al. (1982) on human α_1-ACh. The latter authors reported that the Con A-reactive fractions of α_1-ACh are mainly responsible for the observed 5-fold increase of the plasma level of this glycoprotein during various inflammatory diseases. However, the switch to biantennary glycan structure during the AP-response may not be a general rule since Vaughan et al. (1982) reported an opposite trend with human α_1-PI.

As demonstrated by Morii and Travis (1983), human α_1-ACh isolated from the AP-plasma differs not only in glycosylation pattern but also in the structure of its polypeptide chain which is shorter by 15 amino acid residues. In normal plasma over 90% of the protein has an amino terminal sequence beginning with aspartic acid and less than 10% with arginine. In acute rheumatoid arthritis plasma, 55% of the inhibitor molecules begin with arginine and only 45% with aspartic acid. Sequence studies indicate that a 15-amino acid fragment is missing in the AP-form of α_1-ACh which despite this loss retains full antiproteolytic activity. We do not know yet whether the occurrence of a 'shorter' molecule of α_1-ACh is due to the presence of unusual processing enzymes (the cleaved peptide may also have unknown biological activity), or whether there exists a second gene for α_1-ACh which is specifically activated during the AP-response. However, the latter idea is not supported by the results of experiments with human hepatoma cell line Hep-G2 stimulated with human cytokines (Baumann et al., 1984b).

Although injury-induced molecular variants of AP-proteins are of considerable medical interest, the search for them is not easy, owing to the fact that such variants may be short-lived in the circulation, in analogy to desialylated or non-glycosylated glycoproteins.

References, p. 232.

Gordon/Koj (eds.) The acute-phase response to injury and infection.

CHAPTER 17

Hepatocyte stimulating factor and its relationship to interleukin 1

A. KOJ

1. Experiments with cytokines in vivo

Numerous experiments with intact animals, and also more recently in hepatocyte cultures, strongly suggest that the factor responsible for eliciting the characteristic AP-response of the liver cell is present in preparations of cytokines which at the same time often show IL-1/LEM/EP/LAF activities. Following the suggestion made by Ritchie and Fuller (1983) this factor will be referred to as hepatocyte stimulating factor (HSF). Stimulation of synthesis in vivo of numerous AP-proteins including α_2-M, fibrinogen, ceruloplasmin, haptoglobin, haemopexin, CRP in rats, rabbits and mice by cytokines obtained from rabbit peritoneal exudate cells was repeatedly reported by Kampschmidt and co-workers (Eddington et al., 1971; Kampschmidt and Upchurch, 1974; Merriman et al., 1978; Kampschmidt et al., 1983), by Pekarek et al. (1972b), Wannemacher et al. (1975), Thompson et al. (1976), Bornstein and Walsh (1978) and Gordon and Limaos (1979). Also mouse and human macrophage cytokines were shown to stimulate synthesis of SAA and SAP in mice (Sipe et al., 1979, 1982; Sztein et al., 1981, 1982; Le et al., 1982). These results demonstrated that the action of HSF is not species-restricted since rabbit and human preparations were active both in rats and mice, although not all signs of the AP-reaction could be elicited in animals treated with individual cytokine preparations (Fig. 1). Indeed, the lack of fibrinogen response to LEM p*I* 5 in rats and rabbits may only indicate that this cytokine is devoid of HSF activity in these species (in distinction to mice), or that it is quickly degraded or removed before it reaches

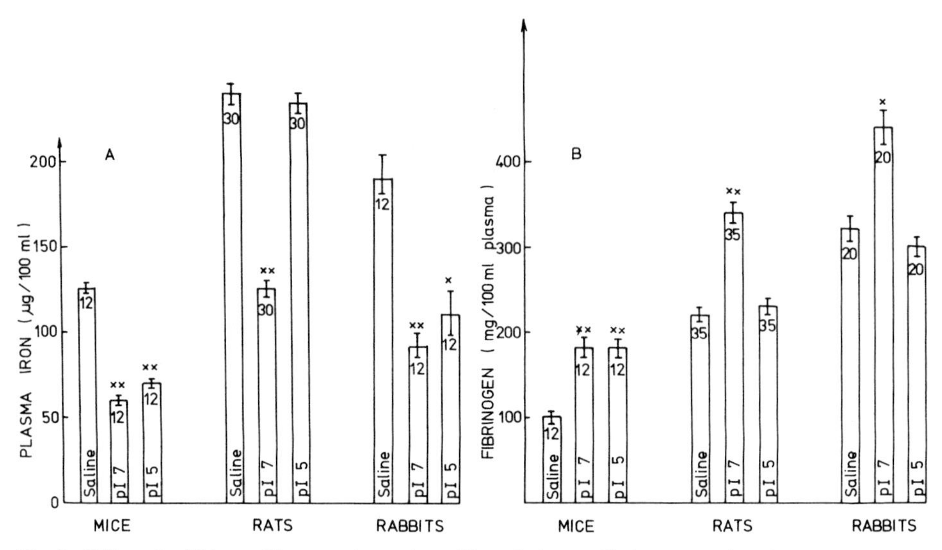

Fig. 1. Effect of rabbit cytokines on plasma iron (A) and plasma fibrinogen (B) in mice, rats and rabbits (after Kampschmidt et al., 1983). Two preparations of cytokines differing in isoelectric point (p*I* 5 and 7) obtained from rabbit peritoneal exudate cells were injected I.P. into mice and rats, and I.V. into rabbits. Brackets indicate the standard error while the number of animals in each group is given below. Significant difference from the saline control is shown as x (p <0.01) or xx (p<0.001).

the target cell. It is relevant that crude LEM is cleared from the circulation of the rat with a half-life of 6 – 10 min (Kampschmidt and Upchurch, 1980). An additional complication arises from the fact that crude preparations of rat LEM/IL-1 may contain an inhibitor which prevents stimulation of fibrinogen synthesis in rats but not in mice (R. Kampschmidt, personal communication). For these reasons experiments with cytokines in vivo are often difficult to interpret. However, some authors have been able to demonstrate a striking correlation between LAF activity determined in vitro and SAA or SAP increases in the blood of mice injected with purified murine and human IL-1 from stimulated monocytes (Sztein et al., 1981; Le et al., 1982) or epidermal cells (Sztein et al., 1982).

2. *Experiments with cytokines in hepatocyte cultures*

Further progress in understanding the action of cytokines has been related to a widespread use of primary hepatocyte cultures. Although Fouad et al. (1980) observed enhanced production of certain AP-proteins by rat hepatocytes after addition of rabbit LEM, a truly quantitative assay was introduced by Ritchie and Fuller who showed (by enzyme-linked immunosorbent assay) stimulation of fibrinogen synthesis and secretion by primary cultures of normal adult rat hepatocytes maintained in a medium enriched by addition of supernatant from rabbit peritoneal cells

(Ritchie and Fuller, 1981), or from LPS-stimulated human peripheral blood monocytes (Ritchie and Fuller, 1983). Almost simultaneously, McAdam and associates (Selinger et al., 1980; McAdam et al., 1982) and Sipe and co-workers (Tatsuta et al., 1983) demonstrated (by solid phase radioimmunoassay) an increased production of SAA and SAP by mouse hepatocytes when these were cultured with supernatants from mouse macrophages, or with preparations of partly purified human leucocytic pyrogen.

Baumann and co-workers who employed two-dimensional slab gel electrophoresis for analysis of the proteins secreted by cultured hepatocytes observed that dexamethasone caused a considerable increase in synthesis of rat α_1-AGP (Baumann et al., 1983a). Moreover, these authors found enhanced synthesis and secretion of α_1-AGP, haptoglobin, haemopexin and SAA by mouse liver cells cultured with the addition of conditioned medium from stimulated homologous monocytes (Baumann et al., 1983b).

All these results demonstrated that cultured rat and murine hepatocytes respond to monocyte-derived cytokines in a manner similar to the liver in vivo by increasing synthesis and secretion of typical AP-proteins. Some hormones, such as insulin and natural or synthetic glucocorticoids (dexamethasone) are essential for maintenance of the specialized functions (e.g. plasma protein synthesis) of hepatocytes when these have been cultured for several days. However, hepatocytes in culture show considerable differences compared with the intact organism:

(a) Non-stimulated hepatocytes produce certain AP-proteins which are barely detectable in the blood of healthy animals, i.e. SAA and SAP in cultures of mouse hepatocytes (Tatsuta et al., 1983);
(b) even the highest doses of cytokines cannot stimulate hepatocytes to the extent observed in vivo;
(c) the response pattern of plasma proteins synthesized in vivo and in vitro appears to be different.

In addition, whether stimulation of synthesis of all AP-proteins is due to a single factor is not yet known nor is the relationship between hepatocyte stimulating activity and the other biological activities which are attributed to IL-1. Unfortunately, in most of the studies mentioned above rather crude cytokine preparations were added to the tissue cultures and analysis was limited to one, or at most few, plasma proteins. In an attempt to close some of these gaps, experiments concerning the effect of human cytokines from two cellular sources on synthesis and secretion of several plasma proteins by cultured hepatic cells from man, rat and mouse were carried out (Baumann et al., 1984b; Koj et al., 1984a).

Crossed immunoelectrophoresis of hepatocyte medium demonstrated (Fig. 2) that

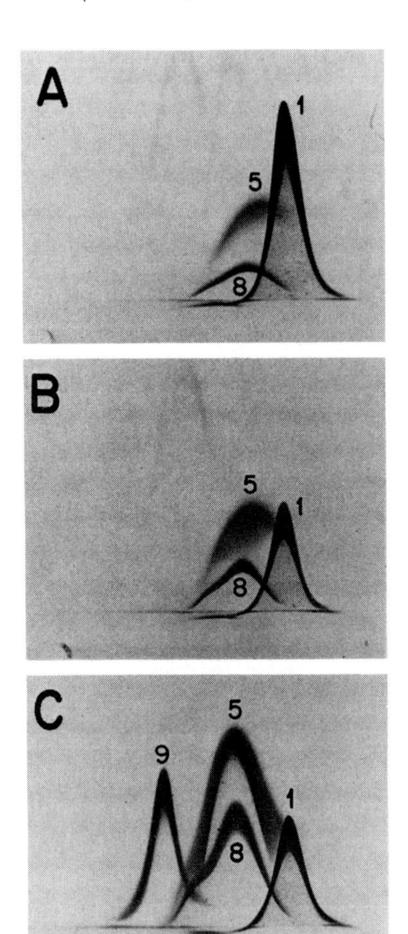

Fig. 2. Effect of crude cytokines from human monocytes on the secretion of some AP-reactants in hepatocyte culture (after Baumann et al., 1984b). Male rat hepatocytes were cultured for 48 h in control medium (A), in medium containing crude human cytokine preparation (supernatant from LPS-stimulated monocytes) (B), and in medium containing both human cytokine preparation and 1 μM dexamethasone (C). After labelling of the cells for 6 h with [^{35}S]methionine, 0.5 ml of the medium was concentrated to 30 μl and subjected to crossed immunoelectrophoresis. The second dimension gel contained a mixture of monospecific rabbit antibodies against the following rat plasma proteins: albumin (1), α_1-AGP (5), α_1-AP-globulin (8) and α_2-M (9). The autoradiograms were exposed for 24 h.

human monocyte cytokines stimulate synthesis and secretion of α_1-AGP, α_1-AP globulin and α_2-M, and decrease synthesis of albumin by rat hepatocytes. In addition, dexamethasone is required for enhanced synthesis of α_1-AGP and α_1-AP-globulin, and in the absence of this hormone α_2-M is not synthesized at all. The differences in synthesis of individual proteins following cytokine or dexamethasone treatment are directly related to the amount of translatable mRNA present (Fig. 3). Rat hepatocytes compared with murine and human liver cells are particularly sensitive to glucocorticoids and after 3 days in culture lacking any glucocorticoid, synthesis of all proteins is either considerably reduced (especially albumin, α_1-AGP) or totally arrested (α_2-M). However, subsequent addition of dexamethasone (0.05 – 1.0 μM) restores both plasma protein synthesis and response to cytokines (Koj et al., 1984a).

Quantitative determinations of individual plasma proteins secreted by hepatocytes into the incubation medium are easily performed by electroimmunoassay using

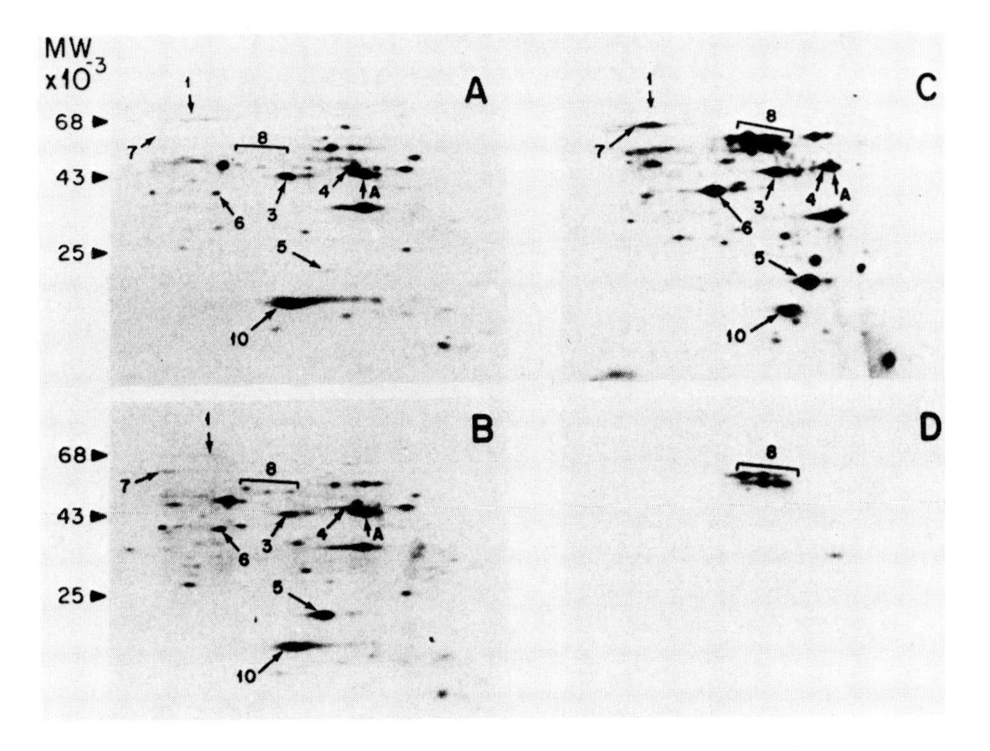

Fig. 3. Effect of cytokines from human COLO-16 cells on translatable mRNAs in cultured rat hepatocytes (after Baumann et al., 1984b). Male rat hepatocytes were cultured for 48 h with normal medium (A), or with medium containing 1 μM dexamethasone and crude cytokines from COLO-16 cells (B). The total cellular RNA was extracted by guanidine-HCl procedure and translated in a reconstituted cell-free system. The products were separated by two-dimensional gel electrophoresis. For comparison, the in vitro translation products from RNA of a liver 48 h after an in vivo inflammation were similarly separated (C). Identification of α_1-AP-globulin synthesized in the cell-free system was achieved by immunoprecipitation from the translation mixture (D). Spots indicated by numbers represent following proteins: 1, albumin; 3, α_1-PI; 4, α_1-ACh; 5, α_1-AGP; 6, haptoglobin; 7, haemopexin; 8, α_1-AP-globulin; 10, α_{2u}-globulin; A, actin.

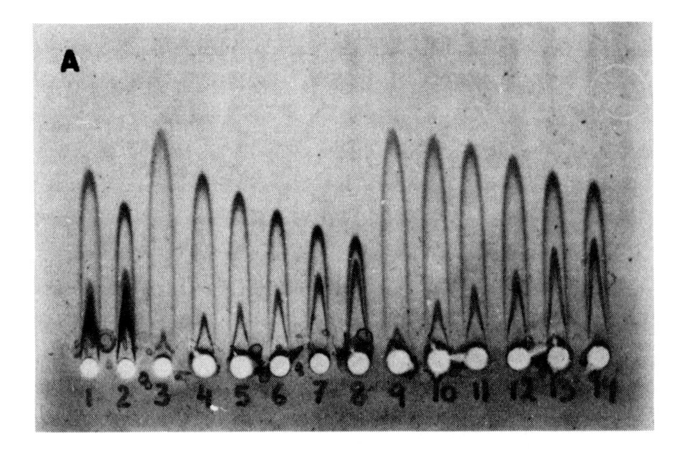

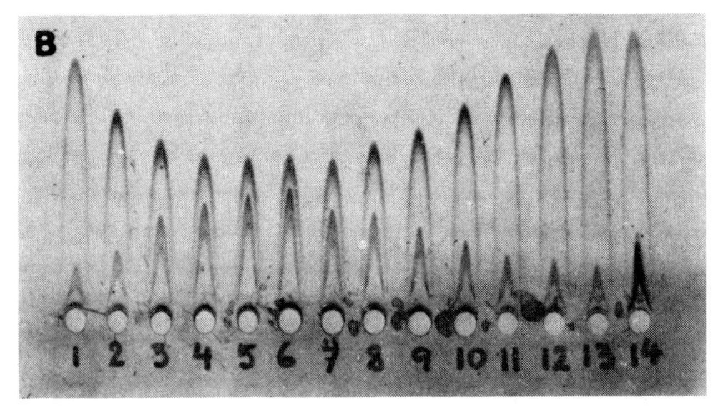

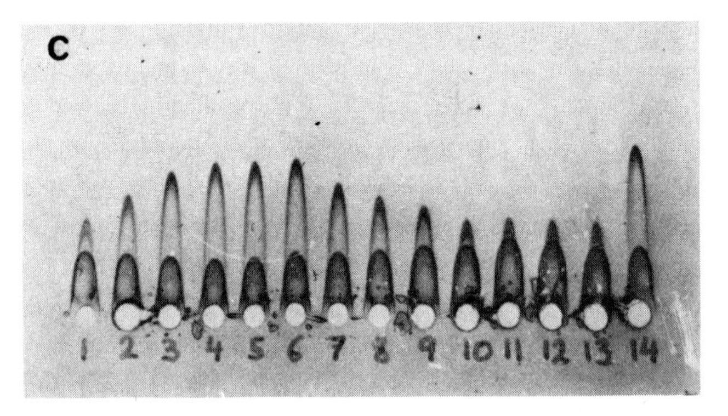

Fig. 4. Quantification of plasma proteins in the medium of cultured hepatocytes by rocket immunoelectrophoresis (after Koj et al., 1984a). (A) Effect of the supernatant from stimulated monocytes on plasma protein synthesis. The standards (wells 1 and 2) are: albumin (tall rockets), 84 and 63 ng; α_2-M (middle-sized, faint rockets), 75 and 100 ng; α_1-M (small, intense rockets), 360 and 480 ng. Wells 3 and 9, media from control cultures, wells 4 – 8, media from cells cultured in the presence of 5, 10, 20, 40 and 80 μl →

monospecific antisera and suitable protein standards. As shown in Figs. 4 and 5, fresh and dialyzed supernatants from LPS-stimulated human monocytes enhance synthesis of α_2-M and inhibit synthesis of albumin. In log-log plots the dose-dependent responses of these two proteins conformed to a straight line, though a

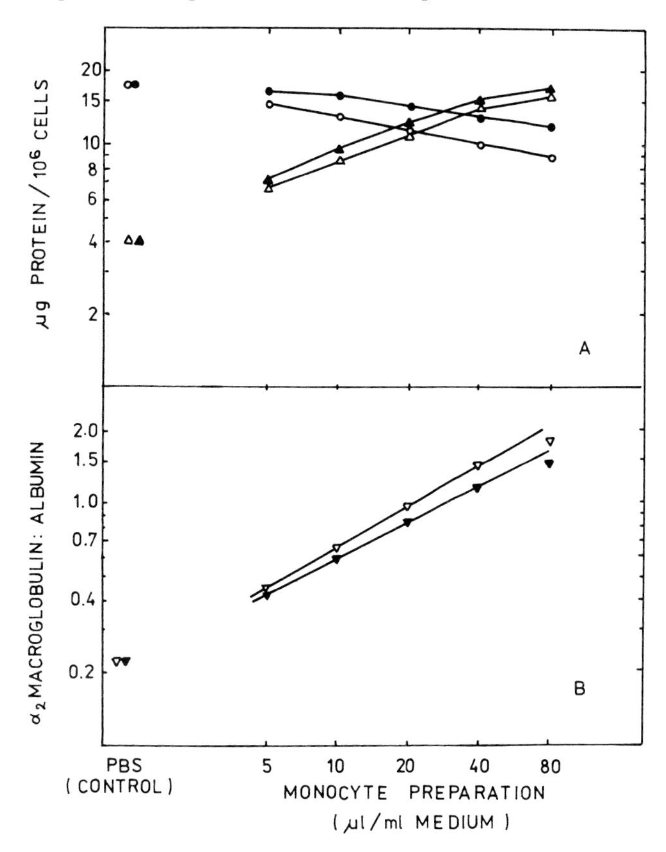

Fig. 5. Response of albumin (○,●) and α_2-M (△,▲) synthesis to different doses of cytokine (after Koj et al., 1984a). Data in panel A was calculated from plate (A) in Fig. 4. The source of cytokine was supernatant from stimulated human monocytes and it was either dialyzed (open symbols) or not dialyzed (full symbols). The α_2-M-to-albumin ratios shown in panel B were calculated from the data in panel A.

---

of dialyzed supernatant from stimulated monocytes; wells 10 – 14, the same as wells 4 – 8 except that the supernatant was not dialyzed. The arc formed by α_1-M is just above the rims of wells 3 – 14. For the dose-response curves drawn for this plate see Fig. 5. (B, C) The effect of column-purified cytokine from monocytes on plasma protein synthesis. The standards (well 14 in each plate) for B are: albumin (105 ng), α_1-M (280 ng) and α_2-M (60 ng), and for C: α_1-AP-globulin (38 ng, tall rocket) and transferrin (85 ng, small rocket). In both plates, the media analyzed in wells 1 – 13 were from hepatocyte cultures which had been incubated with 0.25 ml portions of every second fraction eluted from a Sephadex G-100 column in the molecular mass range between 50,000 and 4,000 daltons.

tendency to levelling off became apparent when cytokine concentrations were increased beyond a certain point; still higher concentrations of cytokine sometimes brought about inhibition of the synthesis of all the plasma proteins for which the tests were conducted. In view of the fact that increased synthesis of a positively reacting AP-protein was always accompanied by a negative effect on albumin synthesis, it seemed appropriate to express these two effects combined together, i.e. as the ratio of protein concentrations (α_2-M/Alb in Fig. 5). The advantage of these ratios, which can be obtained from a single electrophoretic plate, is that they not only increase sensitivity of the HSF assay but also reduce errors arising from variations in the number of hepatocytes present and in the sample volume loaded into the agarose wells.

A tacit assumption underlying the use of protein concentration ratios is that both the positive and negative effects of cytokines are exerted by the same component. Support for this assumption was gained in experiments when concentrated supernatants from LPS-stimulated human monocytes were chromatographed on Sephadex G-100 and individual fractions incubated with rat hepatocytes (Fig. 4). Within the experimental error synthesis of all tested proteins was affected by the same fractions (Koj et al., 1984a).

Analysis of human cytokine-induced changes in 11 plasma proteins produced by cultured rat hepatocytes (Table 1) permit four groups of proteins to be distinguished:

(1) Strongly positive reactants (α_2-M and fibrinogen) which regularly responded with maximum increases of 3 to 5-fold;
(2) weakly positive reactants, notably α_1-AP-globulin, α_1-AGP, haptoglobulin and α_1-PI. Of these α_1-AP-globulin gave the highest (maximum 2-fold) and most consistent increases. Only rarely did these proteins give dose-response relationships to cytokines;
(3) neutral proteins, α_1-M, AT III and prothrombin, which were practically unaffected by cytokines;
(4) negative AP-proteins, i.e. albumin and transferrin. Depressed albumin synthesis was the most consistent response to active cytokines in all cultures tested, while changes in transferrin synthesis were small and irregular.

Unlike in hepatocyte cultures, haptoglobin, α_1-AGP and α_1-AP-globulin in vivo are classified as strong reactants. The reason for this difference in responsiveness is not known but existence of additional stimulatory or regulatory factors operating in the whole organism but missing in the cytokine preparations, or in the hepatocyte culture, should be taken into consideration.

Another striking observation made during these studies concerned differences in

TABLE 1
Concentrations of some plasma proteins in rat hepatocyte culture and rat blood

Protein	Hepatocyte culture				Rat plasma			
	Control		Cytokine-treated		Control		Turpentine-injected	
	μg protein	% alb.	μg protein	% alb.	mg/ml	% alb.	mg/ml	% alb.
Albumin	19.66 ± 3.21	100.0	7.8	100	26.20 ± 0.59	100.0	17.3	100.0
Transferrin	10.41 ± 2.95	52.9	7.3	93	4.10 N.D.	15.6	3.6	20.8
α_2-M	3.30 ± 1.28	16.8	16.5	211	0.06 N.D.	0.2	1.3	7.5
Fibrinogen	9.04 ± 1.77	45.9	27.12	348	2.75 ± 0.28	10.5	7.7	44.9
α_1-AP-globulin	2.34 ± 1.09	11.9	4.68	60	0.94 ± 0.10	3.6	3.85	22.2
Haptoglobin	10.95 ± 4.06	55.7	15.33	196	1.58 ± 0.27	6.0	4.89	28.3
α_1-AGP	4.86 ± 1.33	24.7	6.80	87	0.09 ± 0.01	0.3	0.74	4.3
α_1-PI	10.38 ± 1.83	52.8	14.53	186	1.10 ± 0.09	4.2	1.43	8.3
α_1-M	1.52 ± 0.54	7.7	1.55	20	1.85 ± 0.25	7.1	2.22	12.8
AT III	1.38 ± 0.28	7.0	1.38	18	0.42 ± 0.05	1.6	0.52	3.0
Prothrombin	0.58 ± 0.19	2.9	0.60	7	0.21 ± 0.01	0.8	0.23	1.3

In hepatocyte culture the results are given as μg protein accumulated in the medium during the last 24 h after 3 days of culture of 10^6 cells in Williams E medium containing 1 μM insulin and dexamethasone and 5% foetal calf serum. The results represent mean values (± S.D.) of control cultures, or maximum (minimum) values observed in human-cytokine-treated cultures (based on the data of Koj et al., 1984a). In rat plasma the results are given as mg/ml (± S.E.) in control animals, or maximum (minimum) values observed 24 – 48 h after turpentine injection (based on the data of Koj et al., 1982, except prothrombin – Koj et al., 1984b, and transferrin – Schreiber et al., 1982). For comparison, all the values are additionally expressed as % of albumin contents in hepatocyte medium or rat plasma.

the proportions of individual rat plasma proteins in the culture media in comparison with the liver donor's blood. As shown in Table 1, the protein-to-albumin ratios in control hepatocyte culture are much higher than in the blood and in many cases resemble those found in the plasma from rats injected with turpentine (Koj et al., 1982). This may only partly be explained by catabolism or transcapillary losses occurring in the in vivo system (cf. Chapter 20). Furthermore, α_2-M, which occurs only in traces in the plasmas of healthy rats, was always elevated in the media from normal hepatocytes cultured in the presence of dexamethasone. Thus it appears that hepatocytes from normal rats synthesize plasma proteins in a fashion which is akin to the pattern observed in vivo during the AP-response. The reason for this is unknown, but several mechanisms may be considered:

(a) a delayed effect due to the tissue injury induced during excision of the liver and its perfusion with collagenase necessary for isolation of the hepatocytes;
(b) cytokine production by some endothelial and Kupffer cells remaining in the culture (Sanders and Fuller, 1983);
(c) the presence of unidentified stimulants in the culture medium;
(d) cessation of repressive mechanisms which may be operational in the liver in situ.

Whatever the precise answer may be, the fact that cultured control hepatocytes are already in a state of basal stimulation help us to understand why the changes affected by cytokines in vitro are less spectacular than those produced by turpentine in vivo (Koj et al., 1982; Schreiber et al., 1982). Time-dependent subsidence of this stimulated state, observed also by Ritchie and Fuller (1981), may explain our finding that hepatocytes were more responsive to cytokines after 2 – 3 days in culture (Koj et al., 1985). Hepatocytes cultured in suspension have also been found to secrete plasma proteins in proportions which are distinct from their proportions in plasma (Jeejeebhoy et al., 1980).

3. Relationship between HSF and other biological activities of IL-1

The results of numerous and well documented studies suggest that various biological functions of IL-1 are carried out either by a single molecule, or closely related molecules. As demonstrated by Sztein et al. (1981) and McAdam et al. (1982), partly purified murine and human IL-1 elicited fever in rabbits, synthesis of SAA in mice (or mouse hepatocyte culture) and proliferation of mouse thymocytes in vitro. Highly purified human IL-1 (Lachman et al., 1983) stimulated mouse thymocytes in culture and after injecting to rats caused granulocytosis and increased blood level of fibrinogen, while in rabbits it elicited fever. These biological effects are listed above in the order of decreasing sensitivity of IL-1 detection, LAF assay being the most, and fever the least sensitive (cf. p. 310).

Additional evidence supporting the concept of identity or close similarity of LAF/EP/HSF has been provided by experiments on chemical modification of IL-1 molecules. Incubation of murine IL-1 with phenylglyoxal, which is known to react with arginine residues in proteins, simultaneously abrogated thymocyte proliferation activity in vitro and SAA-inducing activity in vivo (Sztein et al., 1981). Experiments of Dinarello et al. (1982) suggest that the active site of human IL-1 requires not only arginine but also the gamma-carboxyl group of glutamic acid. Both pyrogenic and neutrophil-releasing properties were decreased as a result of modification of these two amino acid residues. Furthermore, treatment with phenylglyoxal of partly purified human cytokines obtained from blood monocytes, epidermal cells and COLO-16 cells reduced both LAF activity with mouse thymocytes and HSF activity with rat hepatocyte culture (Sauder et al., unpublished). Similarities between HSF and LAF are also emphasized by the fact that both activities disappear when purified human cytokine preparations are filter-sterilized without pre-saturation of the Millipore filter with inert proteins (Koj et al., 1985).

However, in many recent studies differences have been found between various IL-1 activities in subfractions from cytokine preparations. The experiments of Kampschmidt et al. (1983) reported in Fig. 1 suggest that the two LEM fractions obtained from acute or chronic exudate from rabbit peritoneal cells, although both active in inducing fever in vivo, were different in respect to iron removal from plasma, and also in stimulation of fibrinogen synthesis in rats. The existence of multiple forms of at least two biochemically and immunologically distinct endogenous pyrogens produced by rabbit macrophages has been reported by Murphy and co-workers (1983). Also, Luger et al. (1983) found that cytokines (ETAF) produced by human squamous carcinoma cell line SCC can be separated into three distinct fractions in respect of isoelectric point (p*I* 7.2, 6.0 and 4.8), all showing high LAF activity. However, ETAF p*I* 7.2 was superior in evoking the chemotactic response of blood polymorphonuclear leucocytes while mononuclear cells responded better to ETAF p*I* 4.8 (cf. p. 103). Also synthesis of SAA in mice was more effectively stimulated by ETAF p*I* 7.2 and 6.0 than by ETAF p*I* 4.8. The interesting finding by Damais and colleagues (1983) that muramyl-dipeptide derivative (MDP-Gln-OnBu) stimulates production of LAF but not EP by human monocytes and rabbit peritoneal exudate cells may throw further light on this complex problem (Chapter 7).

As some of the effects described above may be related to different sensitivities of target organs to IL-1 the clear resolution of HSF and LAF activities achieved by using gel filtration and hepatocyte culture is important. In several experiments with human monocyte-derived cytokines subjected to chromatography on Sephadex

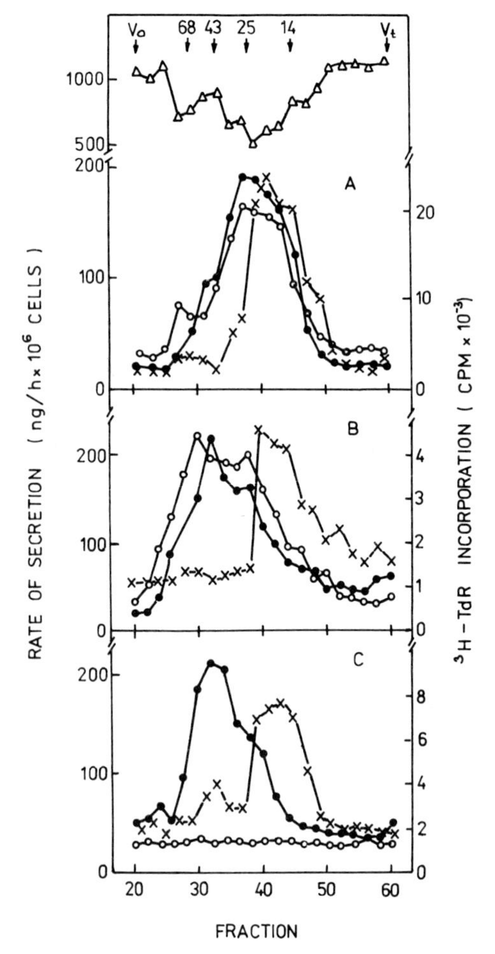

Fig. 6. Separation of hepatocyte and thymocyte stimulating activities by gel filtration (after Baumann et al., 1984b). Supernatants of tissue cultures from LPS-stimulated human peripheral blood monocytes (A), COLO-16 cells (B), and LPS-stimulated mouse peritoneal macrophages (C), each containing 10% foetal calf serum, were concentrated by ultrafiltration and separated on a Sephadex G-100 column (1.6 × 96 cm) in phosphate-buffered saline. To assay for hepatocyte stimulating activities, aliquots of filter-sterilized fractions were added to duplicate wells of Hep-G2 human cells or to rat hepatocytes in 24-well cluster plates (the medium for the latter cells contained 1 μM dexamethasone). After 48 h culture the amount of α_1-ACh (○) and albumin (△) secreted by Hep-G2 cells during 6 h, and α_2-M (●) by rat hepatocytes were determined by rocket immunoelectrophoresis. The mean values of the duplicate cultures are expressed as rate of protein secretion (ng/h × 10^6 cells). To assess thymocyte stimulating activity, aliquots of the diluted fractions were added to cultures of PHA-stimulated mouse thymocytes (1.5 × 10^6 cells). After 72 h, 0.5 μCi [^{3}H]thymidine was added to each culture and incubation was continued for an additional 6 h. The thymocyte stimulating activity is expressed as the amount of [^{3}H]thymidine incorporated into acid-insoluble cell components (x) (mean of triplicate cultures). Following separation of cytokines on a Sephadex column the elution positions of molecular mass standards were determined (marked in arrows in panel A, numbers indicate molecular mass in kilodaltons). Markers used were: Blue Dextran (exclusion volume, V_0), bovine serum albumin (68), ovalbumin (43), chymotrypsinogen (25), ribonuclease A (14), and free amino acids (total volume, V_t).

G-100, the HSF activity eluted as a molecule of approximately 30,000 daltons while LAF activity appeared in the region corresponding to 15,000 daltons, with an additional peak found occasionally at 45,000 daltons (Koj et al., 1984a). This chromatographic behaviour of HSF on Sephadex G-100 agrees well with the results of Fuller and co-workers (Ritchie and Fuller, 1983; Sanders and Fuller, 1983) who reported that the protein from human monocytes or rat liver macrophages, which stimulates fibrinogen synthesis by cultured rat hepatocytes, eluted from Sephadex G-75 with an apparent M_r of 25,000 – 30,000. Partial separation of HSF and LAF activities was also achieved using human cytokines from unstimulated squamous carcinoma cell line COLO-16 (Baumann et al., 1984b; Koj et al., 1985), and cytokines from LPS-stimulated mouse peritoneal macrophages (Baumann et al., 1984b). It should be emphasized that very similar results were obtained with rat hepatocyte culture and human hepatoma cells Hep-G2 (Fig. 6). The HSF activities in cytokine preparations from COLO-16 cells and mouse macrophages showed even higher M_r than HSF from human monocytes, while predominant LAF activity was in all cases located in the region of 15,000 daltons. Further distinction between LAF and HSF was observed during chromatofocusing: in a preliminary study (Koj, Gauldie, Sauder, Regoeczi and Sweeney, unpublished observations) Sephadex-purified cytokine from human monocytes was resolved into three distinct peaks with respect to LAF activity (p*I* 6.9, 5.5 and 5.1), whereas the hepatocyte stimulating activity eluted at pH 5.1 (Fig. 7). Rather similar conclusions may be drawn from the

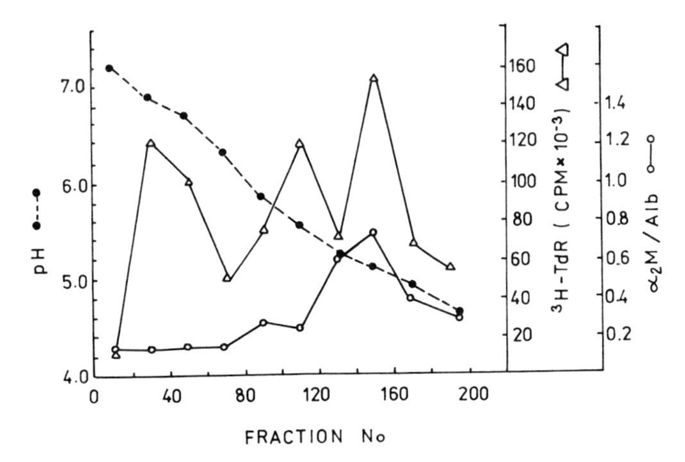

Fig. 7. Separation of hepatocyte and thymocyte stimulating activities during chromatofocusing. Human monocyte cytokine partly purified on Sephadex G-100 (molecular mass range 40,000 – 10,000 daltons) was loaded on a column (2 × 40 cm) of gel exchanger PBE 94 and eluted with polybuffer 74. After measuring pH the fractions were suitably pooled, concentrated 10-fold, dialyzed against PBS and used for LAF assay with mouse thymocytes (incorporation of [^{3}H]TdR) and for HSF assay with rat hepatocytes (α_2-M:Alb concentration ratio).

experiments with rat LEM preparations provided by Dr. R. Kampschmidt: crude material stimulated both proliferation of mouse thymocytes and synthesis of α_2-M and fibrinogen by rat hepatocytes while the material purified by electrofocusing and chromatography on Sephadex G-50 retained full LAF activity but showed only residual HSF activity (Koj, Kampschmidt, Gauldie and Sauder, unpublished observations). Although the final proof of existence of distinct molecular species of cytokines responsible for stimulation of thymocytes and hepatocytes can be provided only by complete purification of the components in question it is appropriate to designate them separately as LAF/IL-1 and HSF/IL-1 instead of using indiscriminately the term of interleukin 1.

At the present state of knowledge it appears that stimulated blood monocytes, peritoneal and lung macrophages, as well as cultured epidermal cells and certain established cell lines (such as COLO-16) produce a family of cytokines active at hormonal concentrations. These proteins share certain physicochemical properties, such as sensitivity to phenylglyoxal treatment or affinity to glass and hydrophobic surfaces, but occur in multiple molecular forms differing both in charge and molecular mass. The components of approximately 15,000 daltons exhibit a variety of biological effects and are usually identified as IL-1. As demonstrated by Mizel and co-workers and Dinarello and co-workers (cf. Chapter 3), both murine and human IL-1 are produced as higher molecular mass precursors (M_r of approximately 33,000). It is now being investigated whether this precursor is responsible for the bulk of HSF activity while the products of its partial proteolytic degradation exhibit mainly LAF and EP activities. Alternatively, individual cytokines are each the products of distinct but related genes, as in the case of interferon (Lawn et al., 1981). The latter hypothesis seems to be particularly attractive since individual members of the human interferon family were shown to exhibit different binding specificities toward target cells (Yonehara et al., 1983). Some of the experimental results with LEM/EP/LAF/HSF suggest that depending on a stimulus the competent cells may produce and secrete individual cytokines showing a characteristic profile of biological activities all useful in the adaptative response of the injured organism.

References, p. 232.

Gordon/Koj (eds.) The acute-phase response to injury and infection.

CHAPTER 18

Regulation of synthesis of acute-phase proteins

A. KOJ

1. *Hypothetical hepatocyte receptor to IL-1 and cellular second messengers*

In analogy to epidermal growth factor (Hunter and Cooper, 1981), to IL-2 (Leonard et al., 1983) or to interferon (Yonehara et al., 1983) it is to be expected that IL-1 will be recognized by a specific receptor on the plasma membrane of target cells. Until now such a receptor has not been identified although some indirect evidence suggests its presence, at least on cells of lymphocyte origin. Thus LAF/IL-1 can be absorbed from the medium by live and glutaraldehyde-fixed 1A5 cells (Gillis and Mizel, 1981). Even assuming the involvement of specific receptors for HSF/IL-1 direct membrane modification may lead to ion movements or protein phosphorylation; alternatively, the HSF-receptor complex may be internalized as happens with certain polypeptide hormones. As postulated by Dinarello (1984) local modification of the target cell membrane may result in nonspecific increases in intracellular levels of calcium, which activates membrane phospholipases. The subsequent release of arachidonic acid from membrane phospholipids provides substrates for the lipoxygenase pathway leading to stimulation of the sensitive cells of the hypothalamus, or cells of the immune system (cf. also Chapter 4). On the other hand, it is tempting to speculate that in order to affect protein synthesis in the hepatocyte the HSF/IL-1 molecule has to penetrate the cell. Such a double mechanism of action – immediate at the membrane level, and delayed after internalization of the active component – has been described for the epidermal-growth factor (EGF) (Hunter and Cooper,

1981). It is known that exposure of sensitive cells to high concentrations of EGF induces a loss of binding capacity for the hormone, presumably due to internalization followed by degradation or modification of the receptors (for references see Fernandez-Pol, 1981). This phenomenon has been termed receptor down regulation. A strikingly similar reduction in the magnitude of liver AP-response after repeated inflammatory insults has been reported by Macintyre et al. (1982) who observed progressively diminishing CRP synthesis in rabbits injected several times with turpentine or typhoid vaccine. As shown in Fig. 1, after the fifth and seventh injections of turpentine to rabbits, given at 3-day intervals, there was no additional CRP synthesis. These observations have been extended by Gahring et al. (1984) to a different model of injury and another experimental animal. The authors found that UV radiation enhances the release by keratinocytes, both in vitro and in vivo, of a cytokine similar to IL-1, epidermal cell thymocyte-activating factor (ETAF), which had been described earlier by Luger et al. (1981) and Sauder et al. (1982). As shown by Gahring et al. (1984), mice subjected to chronic UV radiation exhibited an initial elevation of fibrinogen, SAP and C3 complement that returned to normal concentrations within 7 days. The UV radiation-induced desensitization took place in spite of continuous production and presence in the blood of ETAF. In control animals the serum level of ETAF/IL-1 determined by the thymocyte proliferation assay was not measurable, whereas in the irradiated mice it was raised to 51 units/ml

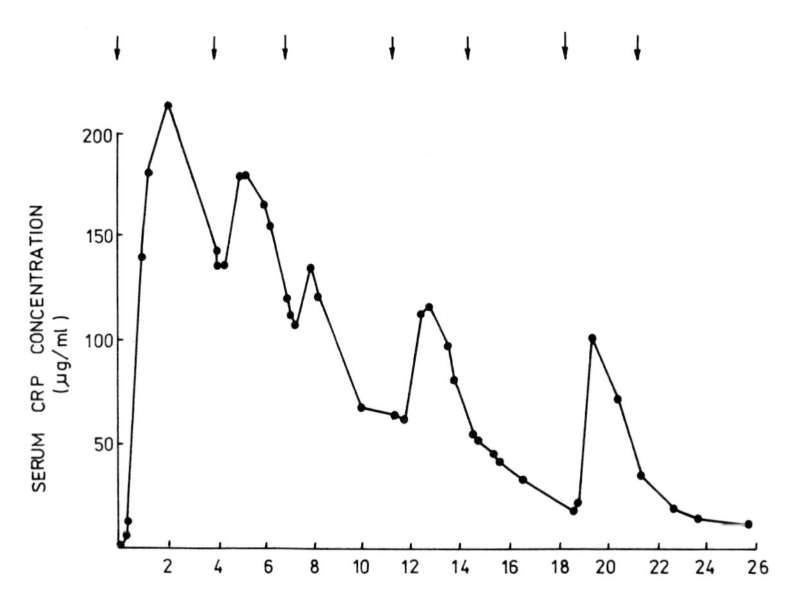

Fig. 1. Serum C-reactive protein response to repeated inflammatory stimuli (after Macintyre et al., 1982). A rabbit received biweekly intramuscular injections of 1 ml of turpentine as indicated by the arrows.

on day 3, 100 units/ml on day 28 and 63 units/ml on day 42. These experiments suggest that 'desensitization' may be taking place at the level of the target cell (hepatocyte) and may be the result of an impaired or modified sensitivity towards the endogenous signal (Gahring et al., 1984). Whether this is due to a down regulation of target cell surface receptor density, a depression of receptor affinity for ligand, or an elevation in specific inhibitors of ETAF/IL-1 is not yet known. Ritchie and Fuller (1983) speculate that expression of the hypothetical receptor on the hepatocyte membrane may require glucocorticoids known to be essential for the AP-response in rat liver.

So far search for cellular second messengers for IL-1 has been unsuccessful. Thus Wannemacher et al. (1975) demonstrated that LEM did not affect the hepatic level of cAMP, and reports of delayed elevation of cGMP in stimulated lymphocytes have also not been confirmed (Oppenheim and Gery, 1982). Like many polypeptide hormones, such as insulin (Kasuga et al., 1982), EGF (Hunter and Cooper, 1981) and platelet-derived growth factor (Heldin et al., 1983) which stimulate tyrosine-specific kinase activity when interacting with their receptors IL-1 may act in a similar way. However, evidence for this is not yet available. Experiments with calcium ionophore A23187 prompted Dinarello (1984) to conclude that rapid accumulation of intracellular calcium induces functional changes that are indistinguishable from those of IL-1 activity, at least in some cells. Evidently further studies are required to verify this hypothesis.

Despite some conflicting reports, prostaglandins do not seem to be involved in the hepatocyte AP-response. If this is so then the liver parenchymal cells differ from the hypothalamus, immune system, muscle or connective tissues (cf. Chapters 5, 6, 8 and 9, respectively). As shown by Schultz et al. (1982), indomethacin administered to rabbits in doses that inhibit prostaglandin synthesis failed to block the CRP response to intramuscular injection of turpentine. Also in primary cultures of rabbit hepatocytes neither the addition of arachidonic acid, the major precursor of prostaglandins, nor direct addition of PGE_1 or PGE_2 induced an increase in CRP synthesis. Similar observations on the lack of effect of indomethacin on synthesis of typical AP-proteins by primary cultures of rat hepatocytes both in the presence and absence of human cytokines have been made by Koj et al. (1984a). These findings suggest that indirect effects of prostaglandins in vivo are likely to have caused the stimulation of fibrinogen synthesis which occurred after infusion into rabbits of massive doses of PGE_1, PGE_2 or arachidonic acid (Carlson et al., 1977, 1978) or increases in CRP and SAA after administration of PGE_1 in man (Whicher et al., 1980, 1984). Pepys and Baltz (1983) suggest that prostaglandins may act through macrophages by enhancing release of IL-1. A similar explanation may apply to the observations of Voelkel et al. (1978) on the high levels of haptoglobin and ceruloplasmin found in rabbits bearing VX_2 carcinoma which is known to synthesize and secrete large amounts of PGE_2. On the other hand, certain direct effects of prostaglandins on protein synthesis in hepatocytes are available. Thus PGE_1 suppressed the induction of rat liver tyrosine aminotransferase by glucagon but not by hydrocortisone (Kajita and Hayashi, 1972) while PGE_1 and cyclic AMP derivatives induced another liver enzyme, ornithine decarboxylase, in primary cultures of foetal rat liver cells (Rupniak and Paul, 1978). Most studies have shown that prostaglandins of the E series suppress proliferative responses of various cells, presumably through their ability to alter intracellular levels of cyclic nucleotides (Goodwin et al., 1977; Dinarello et al., 1983). Also, participation of cyclic AMP derivatives in the specific induction of several liver enzymes is well documented (Ernest et al., 1977; Noda et al., 1983). It is interesting that although cyclic nucleotides are not involved in regulation of plasma protein synthesis in cultured adult hepatocytes, dibutyryl-cAMP considerably increased albumin mRNA level and albumin production by a permanent mouse hepatoma cell line, Hepa-2 (Brown and Papaconstantinou, 1979).

2. Hormonal control of AP-protein synthesis

Injury stimulates the pituitary-adrenal system, as well as other endocrine glands, and hormones represent additional humoral factors transported by blood and modifying the primary effect of HSF/IL-1 (cf. Chapter 11). As pointed out by Weimer and Coggshall (1967), adrenalectomy or administration of glucocorticoids may have multiple, and in some cases opposite, effects on the AP-response: regulatory, permissive, homeostatic (anti-inflammatory), catabolic, stimulatory (anabolic) or inhibitory. Rats injected with turpentine and low doses of cortisol (0.3 mg/100 g body weight) showed enhanced synthesis of fibrinogen and α_2-M in comparison with those injected with turpentine alone, but high, pharmacological doses (3 mg/100 g body weight) reduced the response, presumably by inhibiting IL-1 synthesis and release (anti-inflammatory effect of glucocorticoids). Thus variations in hormonal dosage and experimental design may account for some discrepancies in the results reported in the literature. Moreover, not all adrenal steroids are equally potent: the experiments with hepatocyte culture indicate that 11-hydroxyglucocorticoids, such as hydrocortisone and dexamethasone, are effective at nanomolar concentrations (Grieninger et al., 1978) whereas cortisol or corticosterone must be added to 20 – 60 μM concentration (Princen, 1983). Munck and co-workers (1984) postulate that the physiological function of stress-induced increases in glucocorticoid levels is to suppress certain defense mechanisms thus preventing them from overshooting which would further disturb homeostasis. This is primarily achieved by glucocorticoid-dependent inhibition of synthesis and release of various lymphokines, arachidonic acid metabolites and other mediators of inflammation (cf. also Chapter 11), but additionally by glucocorticoid-dependent synthesis of several AP-proteins showing anti-inflammatory and immunosuppressive properties (cf. Chapter 13).

The importance of several hormones in the regulation of synthesis of AP-proteins was demonstrated in the elegant studies of Miller and co-workers (John and Miller, 1969; Griffin and Miller, 1974). During 12 – 24 h perfusions of isolated rat livers, with medium enriched in insulin, cortisol, growth hormone, triiodothyronine and a mixture of amino acids, the synthesis of albumin slowly decreased while synthesis of fibrinogen, α_1-AGP and haptoglobin began to rise after 3 – 6 h. α_2-M was initially produced in trace amounts but after 6 h its output increased dramatically. The authors concluded that the hormones are directly responsible for the induced synthesis of AP-proteins. However, a more likely explanation is that liver macrophages stimulated during isolation of the organ began to produce HSF/IL-1, as was demonstrated with cultured Kupffer cells (Sanders and Fuller, 1983) or with adult rat hepatocytes co-cultured with another liver cell type (Guillouzo et al., 1984). In the experiments described above increases in production of all four AP-proteins were critically dependent on the presence of cortisol (John and Miller, 1969). On the other hand, triiodothyronine had no effect on fibrinogen synthesis but was beneficial for α_2-M, Hp and α_1-AGP (Griffin and Miller, 1974). When chicken embryo hepatocytes were used dexamethasone and triiodothyronine were the principal effectors of fibrinogen synthesis. In these experiments neither crude human, rabbit nor mouse IL-1 preparations had any effect (Grieninger et al., 1983).

Combination of glucocorticoids and adrenaline was shown to induce very high levels of α_2-M (α-macrofoetoprotein) in both adrenalectomized and normal rats, while haptoglobin and α_1-AP-globulin were much less affected (Van Gool et al., 1984). Differential responses of individual AP-proteins to hormones have been reported in vivo (Weimer and Coggshall, 1967; Thompson et al., 1976; Gordon and Limaos, 1979b) and with isolated hepatocytes (Jeejeebhoy et al., 1975, 1980; Crane and Miller, 1977; Hooper et al., 1981; Baumann et al., 1983b). In general, mixtures of several hormones have given the best results but haemopexin synthesis by suspensions of adult rat hepatocytes was found to be independent of hormonal supplementation (Jeejeebhoy et al., 1976) while haptoglobin formation was inhibited by glucagon in mouse hepatocyte cultures (Baumann et al., 1983b). Synthesis of C3 component was stimulated by glucocorticoids in some rat hepatoma cell lines but inhibited in other lines and in cultured mouse macrophages (Fey et al., 1983). The same authors observed that dexamethasone may affect the C3 mRNA and protein production in opposite ways.

Individual AP-proteins have been found to respond differently in males and females. The best known examples of such differences are α_1- and α_2-M of the rat (Bosanquet et al., 1976) and hamster female protein (Coe and Ross, 1983). In addition, several AP-proteins are known to be affected by oestrogens, androgens and anabolic steroids (Barbosa et al., 1971, Kueppers and Mills, 1983). Evidently in addition to the primary stimulation by HSF/IL-1, each AP-protein requires for its maximal synthesis optimal concentrations of particular hormones.

As shown in Table 1, injection of rabbit leucocyte factor, i.e. LEM/IL-1 into normal and sham-adrenalectomized rats, significantly increased synthesis of total

TABLE 1

Effect of LEM and cortisol on incorporation of orotate into liver RNA, and on serum levels of α_2-M and Hp in rats subjected to various treatment (after Thompson et al., 1976)

Animal treatment	[^{14}C]orotate incorporation (dpm/mg of DNA)	Concentration of	
		Serum α_2-M (units/ml)	Serum Hp (mg/100 ml)
Normal rats			
control	332 ± 18	0.4 ± 0.2	82 ± 4
LEM	449 ± 20[a]	37.7 ± 2.1[a]	155 ± 14[a]
cortisol	361 ± 19	0.6 ± 0.5	84 ± 7
cortisol + LEM	484 ± 23[a]	41.1 ± 2.8[a]	159 ± 7[a]
Hypophysectomized rats			
control	280 ± 30	1.4 ± 0.2	113 ± 12
LEM	240 ± 27	1.6 ± 0.3	153 ± 6[a]
cortisol	270 ± 20	2.0 ± 1.8	121 ± 9
cortisol + LEM	440 ± 18[a]	62.3 ± 3.4	162 ± 15[a]
Sham-adrenalectomized rats			
control	274 ± 20	25.8 ± 5.9	122 ± 6
LEM	411 ± 25[a]	59.8 ± 3.3[a]	204 ± 18[a]
cortisol	287 ± 10	25.5 ± 4.7	128 ± 9
cortisol + LEM	415 ± 20[a]	64.6 ± 4.9[a]	215 ± 20[a]
Adrenalectomized rats			
control	282 ± 26	1.3 ± 0.3	120 ± 9
LEM	246 ± 13	1.4 ± 0.3	199 ± 15[a]
cortisol	230 ± 14	27.4 ± 2.5[a]	125 ± 10
cortisol + LEM	381 ± 26[a]	60.5 ± 3.5	209 ± 19[a]

[a] $P<0.01$ when compared with heat-inactivated control.

Rats in each group were injected daily with either 0.5 mg cortisol or saline. After 7 days for normal or hypophysectomized rats and 3 days for adrenalectomized and sham-adrenalectomized groups, 12 rats from each group were injected i.p. with 1 ml of rabbit leucocyte-derived LEM, either fresh or heat inactivated (control). Then 4 h later six rats in each group were injected i.p. with 5 μCi of [^{14}C]orotate/100 g body weight. The rats were killed 4 h later and total hepatic DNA and RNA were extracted and analyzed for concentration and radioactivity. The remaining 6 rats in each group were killed 24 h after LEM, and serum α_2-M and Hp were determined. The results are means ± S.E.M. for 6 rats.

hepatic RNA, whereas injection of cortisol had no such effect. By contrast, adrenalectomy or hypophysectomy prevented LEM-induced increase of orotate incorporation, but this effect could be reversed by daily injections of cortisol. These observations constitute a classic example of glucocorticoids acting permissively, probably at the stage of transcription. In normal rats, serum α_2-M was increased almost 100-fold 24 h after LEM injection, and cortisol had no additional effect. At 3 days after surgery sham-adrenalectomized rats had significantly greater concentrations of serum α_2-M than control animals but LEM stimulated further increases of this protein. In LEM-injected rats adrenalectomy or hypophysectomy prevented the expected increase in serum α_2-M, but daily injections of cortisol restored the response. On the other hand, stimulation of haptoglobin synthesis by LEM was only slightly impaired in hypophysectomized and adrenalectomized rats and cortisol did not alter this response (Thompson et al., 1976). The results presented in Table 1 indicate that injury-induced synthesis of α_2-M and Hp in the rat is regulated differently. Later a similar conclusion was reached in experiments with rat hepatocyte cultures (Baumann et al., 1984b; Koj et al., 1984a).

The permissive effect of glucocorticoids on synthesis of AP-proteins in vivo is well documented but studies of Baumann et al. (1983a, 1984b) with rat hepatocyte culture indicate that dexamethasone by itself may stimulate transcription of mRNA for α_1-AGP, and enhance synthesis and secretion of other AP-proteins. If this occurs then synthesis of AP-proteins must in this respect be similar to synthesis of some inducible liver enzymes or other specific cellular proteins in cultured rat liver and hepatoma cells (Ivarie and O'Farrell, 1978; Baumann and Held, 1981; Marston et al., 1981; Augberger et al., 1983). However, Gross et al. (1984) observed a significant lag period of about 3 h before augmented synthesis of α_2-M or α_1-AGP could be detected following addition of dexamethasone (0.1 μM) to rat hepatocyte culture, whereas synthesis of tyrosine aminotransferase was induced almost immediately. A similar 3-h delay in the increase of fibrinogen mRNAs after addition of dexamethasone (0.48 μM) to freshly isolated rat hepatocytes has also been reported by Princen (1983). This lag phase for AP-proteins suggests that there may be intermediate steps between the binding of the steroid hormone-receptor complex to chromatin and the induction of specific mRNA. Unlike with rat hepatocytes dexamethasone has no effect on AP-protein synthesis in cultured hepatocytes from normal mice (Baumann et al., 1984b). On the other hand, increases of 30 to 80-fold in the synthesis and secretion of haptoglobin by a mouse hepatoma cell line did occur (Baumann and Jahreis, 1983).

3. Genetic factors involved in the AP-response and chromosomal localization of structural genes coding for AP-proteins

Surprisingly little attention has yet been paid to genetic control of the AP-response despite the well known fact that expression of all body proteins is genetically programmed. Early investigators noted considerable differences in the plasma level of ceruloplasmin (Meier and MacPike, 1968) and haptoglobin (Peacock et al., 1968) between various inbred strains of mice. Also Weimer et al. (1972) reported much higher synthesis of α_2-M in male Lewis rats injected with turpentine than in male Wistar or Sprague-Dawley rats, while Buffalo and Fisher strains occupied an intermediate position. Interestingly, the authors observed a characteristic 'genetic drift' in the AP-response in their Sprague-Dawley colony over 5 years of study.

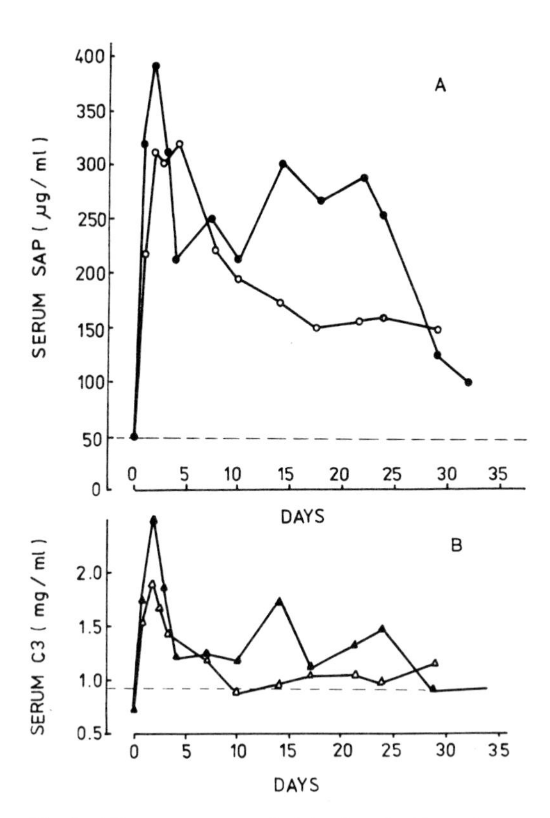

Fig. 2. AP-response to casein injections in CBA mice (●, ▲) and A/J mice (○, △) (modified from Baltz et al., 1980). Groups of five mice received 28 daily injections of casein. Serum SAP concentrations (A) and C3 concentrations (B) were estimated by electroimmunoassay. For the sake of clarity, standard deviation values are omitted. ---, The upper limit of pretreatment mean values of protein concentrations in all mice.

Wide individual variability of unknown origin and variability related to the season often occurs. Within the same strain of rats Gordon and Limaos (1979b) were able to distinguish both seasonal variability and regular and irregular responders to endotoxin injection.

The serum concentration of SAP in healthy mice varies considerably between strains (Pepys et al., 1979a), from 5 μg/ml in C57Bl/Cbi mice to 140 μg/ml in the DBA/2 strain, while CBA/Ca and A/J mice show intermediate values (30 – 60 μg/ml). In all strains there was initially a marked increase in response to casein injection but in A/J mice the level was not sustained despite continued daily injections, while in C57 and DBA mice it remained elevated for 25 days (Baltz et al., 1980). A similar but less pronounced phenomenon was observed with another AP-protein, the C3 component of complement (Fig. 2). Whether these strain-related differences reflect variations in the intensity of inflammation and production of mediators, or in capacity for synthesis, secretion and catabolism of each AP-protein is not yet known.

A considerable step forward in this respect has been made by Baumann and co-workers (1984a) who examined changes in amounts of translatable liver mRNAs during the AP-response in several inbred strains of mice. Unlike in rats, mice were found to have two forms of α_1-AGP differing in size and charge, termed AGP-1 and AGP-2. Polymorphism in the structure of AGP-1 among strains (Fig. 3) was used to map the structural gene, *Agp-1*, to chromosome 4 close to *Lps* locus. *Lps* was identified previously by Watson and co-workers on the basis of diminished response to bacterial LPS (endotoxin) in C3H/HeJ mice and appears to regulate all aspects of the AP-response, including lymphoid cell activation, hypothermia, and induction of both colony-stimulating factor and SAA levels in serum (Watson et al., 1978). Baumann et al. (1984a) noted the effect of the *Lps* locus on induction of mRNAs for all the AP-proteins detectable by in vitro translation assay. In a preliminary analysis of (C3H/HeJ × DE/Cv)-F_1 hybrids, the authors found that the effect of *Lps* on *Agp-1* expression is not *cis*-acting. Whatever the relationship between *Agp-1* and *Lps*, it is clear that not all AP-protein structural genes in the mouse are clustered on chromosome 4. Thus data on structural variants of SAA placed its gene, *Saa*, on chromosome 7, while recent experiments permitted location of the haptoglobin gene on chromosome 8 (H. Baumann, personal communication). Continuation of these studies may well lead to identification and characterization of these factors which modulate the AP-response corresponding to specific genes.

4. *The fine structure and expression of some AP-protein genes*

Independently of the molecular mechanism involved in HSF/IL-1 action on the liver

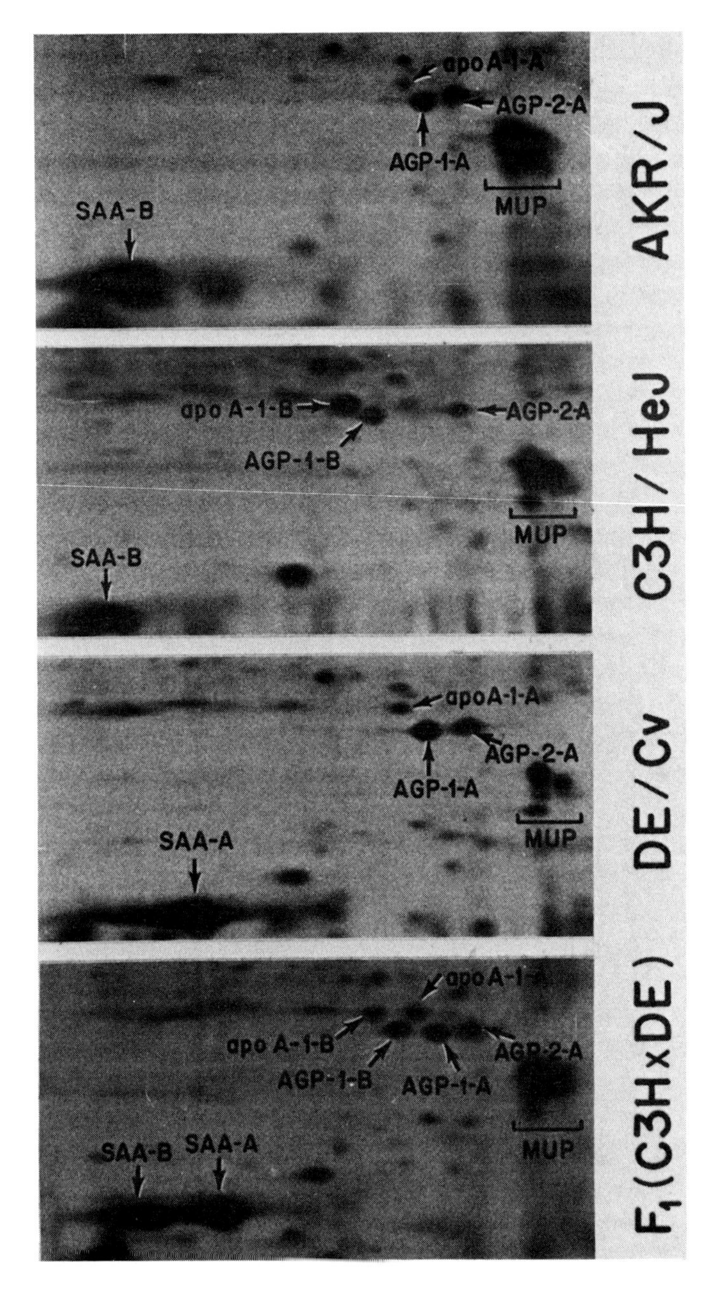

Fig. 3. Electrophoretic polymorphism of α_1-AGP and serum amyloid A protein (SAA) in inbred strains of mice (after Baumann et al., 1984a). Total liver RNA was extracted from males of each strain 24 h after turpentine injection and translated in a cell-free system with [^{35}S]methionine. The products in 5 μl of the translation mixtures were separated by two-dimensional polyacrylamide gel electrophoresis and exposed to a photographic plate for 24 h. Only the section of the fluorographic patterns containing SAA and AGP is reproduced. The positions of the major urinary proteins (MUP) and the polymorphic apolipoprotein A-1 (apo A-1A and apo A-1B) are shown as reference points.

cell, it is clear that ultimately the signal must affect the efficiency of transcription of the genes coding for individual AP-proteins. Sometimes gene activation may manifest itself with characteristic re-arrangement of chromatin structure; a well known example is the 'puffing pattern' of the salivary gland chromosomes of *Drosophila* and *Chironomus* in response to molting hormone, ecdysone (Ashburner, 1974). However, Schiaffonati et al. (1984) observed no comparable changes in rat liver chromatin structure during the AP-response.

How IL-1 activates genes is not yet known but possibly it may occur in a manner similar to induction of the human metallothionein-II_A gene by cadmium and by glucocorticoids; Karin et al. (1984) characterized DNA sequences involved in this induction. The regulatory region which they identified includes a characteristic tetranucleotide recognized by DNA polymerase II, the so-called TATA-box, and two other functional elements located further upstream of the DNA strand, i.e. toward the 5′ end. One of these is involved in heavy metal induction (and occurs in two copies), the other is important for hormonal induction and binds a specific protein, the glucocorticoid receptor. As indicated by the results of Durnam and Palmiter (1984) with mouse metallothionein-I the induction by heavy metals requires the presence of another protein, which argues against autoregulation of metallothionein genes. It is interesting that induction of mouse metallothionein-I mRNA by bacterial endotoxin is independent of metals and glucocorticoids (Durnam et al., 1984).

The molecular anatomy of the human metallothionein-II_A promoter shown in Fig. 4 encompasses certain features common to the majority of all eukaryotic promoters so far analyzed. These are relatively large in size and are composed of dif-

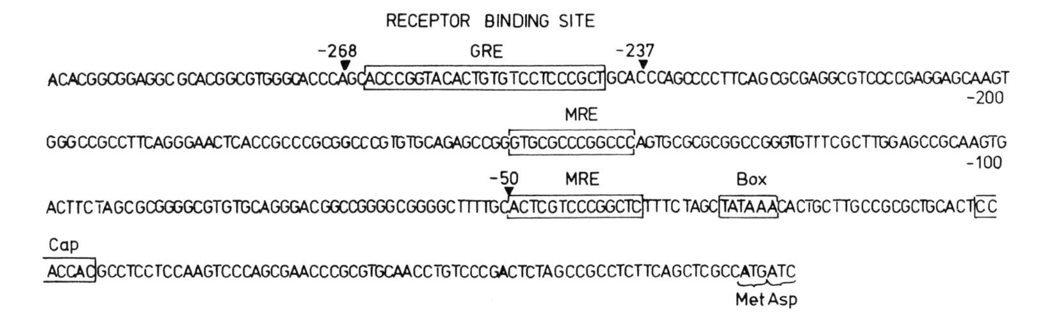

Fig. 4. Nucleotide sequence of the human metallothionein-II_A promoter (simplified from Karin et al., 1984). Sequence of the sense strand of DNA between positions −300 and +80 is shown including regions important for hormonal induction (GRE for glucocorticoid responsive element) and heavy metal induction (MRE for metal ion responsive element). Also indicated are TATA-box, cap region and the first two amino acids. Arrowheads and numbers above indicate the position of 5′ deletion end points important for the mapping of various regulatory elements.

ferent functional elements. Comparison of functionally related control elements has revealed conservation of nucleotide sequences which may serve as binding sites for various regulatory proteins (for references see Karin et al., 1984). By their ability to regulate RNA polymerase entry sites these elements can activate heterologous promoters when placed at various distances upstream. Alternatively, the interaction of regulatory proteins with the DNA at upstream sites may lead to changes in the conformation of the double helix and its local 'melting' at the TATA-box region, the site for RNA polymerase and initiation factors binding to DNA (Karin et al., 1984). It is tempting to speculate that promoters of all AP-proteins contain common regulatory sequences capable of recognizing HSF/IL-1, or its putative second messenger. Binding of HSF/IL-1 to these regions would then enhance transcription of the genes responsible for positive AP-proteins, or reduce expression of genes coding for negative AP-proteins. In addition, the promoters may also contain regulatory sequences recognized by hormones and other modulators. Such organization of the genes coding for AP-proteins would explain the permissive effects of glucocorticoids for the AP-response in the rat, or even the direct induction of these proteins by dexamethasone independently of IL-1. Moreover, by assuming that the regulatory elements can be reshuffled within the genome in the course of evolution the considerable species-related variability of response among AP-proteins becomes more comprehensible.

At present the fine structure and expression of several genes coding for AP-proteins is being analyzed, e.g. mouse and human C3 component of complement (Fey et al., 1983), rat fibrinogen (Fowlkes et al., 1984) and rat α_2-M (P.C. Heinrich and W. Northemann, personal communication). The mouse C3 gene occurs probably in a single copy per genome and is 24 kb long, i.e. four to five times longer than the mRNA coding sequences. This is due to the presence of intervening sequences (introns) typical for the majority of eukaryotic genes. The promoter region contains the TATA-box located 28 nucleotides upstream, while the ATG translation initiation triplet was found 56 nucleotides downstream from the cap site (Fey et al., 1983). No other regulatory elements have been identified in the C3 promoter region at the moment of writing of this chapter.

On the other hand, Fowlkes and co-workers (1984) discovered two unique regions of homology in the promoters of three genes coding for alpha, beta and gamma chains of rat fibrinogen. One region consists of 15 nucleotides that have a common nucleotide core, C-T-G-G-G-A, lying between −116 and −160 positions upstream of the cap. The other is approximately 100 nucleotides long and is located in the −165 to −472 region. As shown in Fig. 5, the beta and gamma fibrinogen genes are over 60% homologous in this region while alpha fibrinogen gene shows somewhat less homology. The authors speculate that these homologous regions serve as acceptors for the intracellular signals initiated by the HSF and their activation may lead to enhanced coordinated transcription of fibrinogen genes during the AP-response. Moreover, the beta fibrinogen gene was found to contain 22 nucleotides at position −480 that are homologous to sequences occurring in glucocorticoid-regulated genes. In the light of the swift progress of molecular biology, it is likely that the structures of the promoters of major AP-proteins and the mechanism of the increased transcription of corresponding mRNAs will be elucidated in the near future.

5. *Transcriptional versus translational control and other regulatory mechanisms of the AP-protein response*

There are many reports of considerable agreement between abundance of specific mRNAs in the liver cell and secretion rates of AP-proteins, both in vivo (Princen et al., 1981; Ricca et al., 1981; Northemann et al., 1983) and for hepatocyte cultures (Baumann et al., 1983a, b, 1984b). In rats injected with turpentine, Ricca et al. (1981) demonstrated a strict correlation in the increases of liver α_1-AGP-mRNA and plasma level of α_1-AGP (Fig. 6). Such correlations were observed both with synthesis-stimulated (positive) AP-proteins, such as fibrinogen, α_1-AGP or rat α_2-M, and in case of synthesis-inhibited albumin (cf. Chapter 15). Thus transcription appears to be a key regulatory point in expression of structural genes regulated by IL-1 or other injury-derived factors. In eukaryotes, however, mRNA levels depend not only on transcription rates and mRNA stability but also on post-transcriptional processing of the primary transcript, i.e. splicing, trimming, capping and poly-(A) addition (Revel and Groner, 1978; Innis and Miller, 1979). Vannice et al. (1984) demonstrated that glucocorticoids stimulate accumulation of α_1-AGP mRNA in HTC rat hepatoma cells by inducing an RNA processing factor that allows production of stable transcripts. There are some suggestions that injury-elicited changes in abundance of hepatocyte mRNA represent a more sensitive and specific indicator of the AP-response than amounts of secreted proteins. In vivo this is due to the fact that secreted proteins diffuse to the extravascular compartment

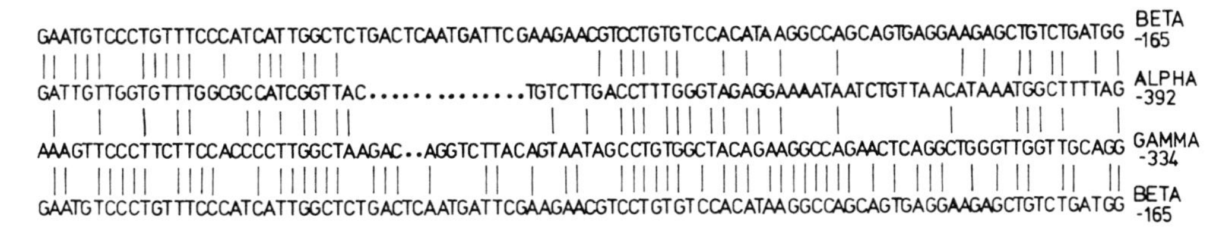

Fig. 5. The best alignment of the largest region of homology between the promoters of rat alpha, beta and gamma fibrinogen genes (after Fowlkes et al., 1984). The numbers at the right refer to position of nucleotides upstream of the cap site. For further details see text.

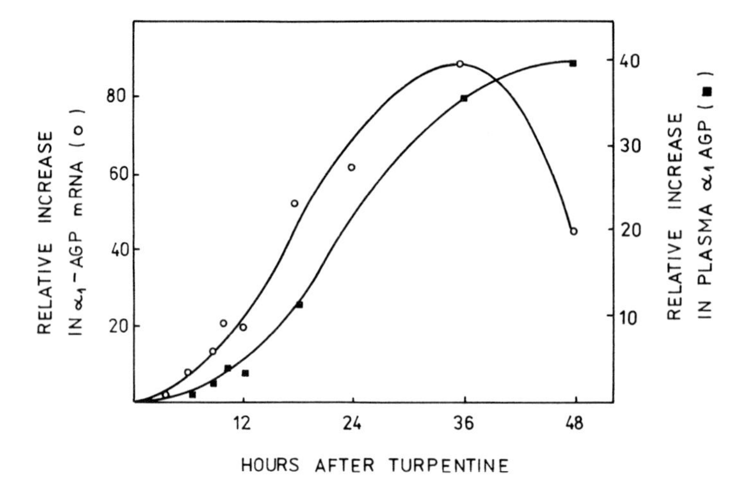

Fig. 6. Time course of the changes in plasma level of α_1-AGP and α_1-AGP-mRNA in the liver following turpentine-induced inflammation in the rat (after Ricca et al., 1981). The relative change in α_1-AGP-mRNA levels was determined from the hybridization of total liver RNA to ^{32}P-labelled α_1-AGP cDNA. Plasma levels of α_1-AGP were determined by rocket immunoelectrophoresis.

and are catabolized (cf. Chapter 20). In tissue cultures proteins may not be efficiently secreted from the cells as reported by Fey et al. (1983) for C3 component of complement in a mouse macrophage cell line stimulated by dexamethasone, and by Baumann et al. (1983b) for haptoglobin in mouse hepatocytes stimulated with monocyte cytokines and dexamethasone. However, with presently available techniques, quantitative determinations of AP-proteins represent a more convenient method of assessing the magnitude of the AP-response than measurements of the levels of functional mRNAs.

The transcriptional control of AP-protein synthesis is in agreement with the general regulatory mechanisms of expression of other genes coding for secretory proteins, such AFP. As demonstrated among others by Sala-Trepat et al. (1979), Schwartz et al. (1982) and Tilghman and Belayew (1982) changes in synthesis rates of AFP and albumin observed during development and carcinogenesis are directly proportional to the transcription rates of respective genes. In mice the two genes are closely linked in tandem on chromosome 5 and they are probably activated by *cis*-acting mechanisms (Tilghman and Belayew, 1982). Similarity in regulation of the two genes is reflected in their response to injury-derived factors: injection of turpentine to neonatal rats simultaneously decreased plasma AFP and albumin concentrations (Savu et al., 1983), and crude human IL-1 preparations similarly suppressed albumin and AFP synthesis by human Hep-G2 cells (Baumann et al., 1984b). On the other hand, AFP synthesis occurs only in dividing hepatocytes or hepatoma cells, and more specifically in the G_1 phase, whereas albumin is produced also in other phases of the cell cycle (Abelev, 1976; Guillouzo et al., 1979; Tilghman and Belayew, 1982). These observations may well illustrate complexity of regulation of those two closely related eukaryotic genes.

In addition to variable transcription rates, the rate of mRNA translation may control the overall protein production as has been shown with alpha and beta globin chain synthesis in reticulocytes, and with some other specialized proteins in animal and plant cells (for references see Innis and Miller, 1979; Godefroy-Colburn and Thach, 1981). Polypeptide chain initiation is generally regarded as a critical stage in regulating translation (Revel and Groner, 1978). It appears that in liver cell stimulated by injury, mRNA is abundant and many mRNA species compete fiercely for the translating machinery; hence the kinetic model of regulation of protein synthesis proposed by Godefroy-Colburn and Thach (1981) may be applied. The model assumes the existence of a discriminating factor, independent of the ribosome, which selects against those messages for which its affinity is the lowest, thereby

TABLE 2
Stages of the AP-response and factors affecting regulation of synthesis, secretion and breakdown of AP-proteins

Phenomenon	Modifying factor
1. Injury and IL-1 formation	Age and sex of the organism Anti-inflammatory drugs Modulators of immune response
2. Hepatocyte stimulation	Some hormones Hepatotoxic drugs Hepatitis, liver regeneration
3. Increased transcription and mRNA processing in the liver	Glucocorticoids, insulin and other hormones Actinomycin D, ethionine, galactosamine
4. Enhanced protein synthesis in the liver	Nutritional status of the injured organism Some hormones Cycloheximide
5. Intracellular protein transport and secretion from the liver	Glycosylation inhibitors Proteinase inhibitors Colchicine Hormones, prostaglandins
6. Increased metabolism of AP-proteins in blood and tissues	Pathological states leading to intravascular consumption or tissue deposition of SAA, SAP, CRP, proteinase inhibitors, fibrinogen, haptoglobin, α_1-AGP

reducing their initiation efficiency. It is interesting that AP-proteins show variable degrees of sensitivity to translational inhibitors: Tatsuta et al. (1983) reported that SAA synthesis in cultured mouse hepatocytes was strikingly resistant to cycloheximide inhibition as compared with SAP and other proteins. In experiments in vivo, Chih et al. (1977) observed selective stimulation of fibrinogen synthesis and imbalance in the fibrinogen chain ratio (Procyk et al., 1979) in rats recovering from intoxication with cycloheximide. With embryonic chick hepatocytes Plant et al. (1983) observed preferential resistance of fibrinogen synthesis to inhibition of polypeptide chain initiation.

Considerable variations in the sensitivity to hormones, in the amplitude of response and asynchrony in the kinetics of synthesis and secretion of several AP-proteins observed both in vivo and in tissue culture imply differences in their regulatory mechanisms. Even assuming that the reaction is started in all cases by the same triggering factor (presumably HSF/IL-1), hormones and other mediators then become involved. Most probably, the rise in AP-protein concentration in the blood following injury will prove to be regulated at many levels: transcription, translation, processing of protein precursors and secretion from the cell. The list of potential effectors at each level shown in Table 2 is far from complete, but their existence must always be borne in mind when attempts are made to explain the magnitude of the AP-response.

References, p. 232.

Gordon/Koj (eds.) The acute-phase response to injury and infection.

CHAPTER 19

Extrahepatic synthesis of acute-phase proteins

A. KOJ

1. Synthesis of AP-proteins by hepatoma cells

Formation and secretion of AP-proteins represent a highly specialized function of the hepatocyte and an important question is whether this ability is preserved during carcinogenesis. The best model for such studies is represented by spontaneous or chemically induced transplantable and well differentiated hepatomas, such as some Morris hepatomas. The experiments of Schreiber and co-workers (1978) demonstrated that in vivo Morris hepatomas synthesize plasma albumin but fail to secrete it to the blood stream. The mechanism of the defect is not clear, since cell suspensions from these hepatomas (Schreiber et al., 1978), or slices of the tumour (Redman et al., 1979), are capable of processing intracellular albumin precursor and secreting albumin into the incubation medium. However, albumin represents only a small fraction of the hepatoma proteins destined for export, being replaced in this respect by AFP. Phenotypic expression of AFP occurs during liver regeneration following partial hepatectomy and after chemically induced necrosis of hepatocytes or exposure to carcinogens (for references see Smith and Kelleher, 1980). Sell et al. (1979) and Sala-Trepat and associates (1979) found no evidence for amplification, deletion or rearrangement of AFP and albumin genes during rat liver carcinogenesis, the level of expression of these genes being regulated by the relative abundance of functional mRNAs (cf. also Chapter 18). They confirmed earlier observations of Sell and Morris on the great variability of synthesis and secretion of AFP by different lines of hepatomas.

Much less is known about the synthesis of AP-proteins by hepatomas. Fibrinogen production not followed by secretion was demonstrated in Zajdela rat ascites hepatoma in vitro (Sarcione and Smalley, 1976). On the other hand, fibrinogen and α_2-M synthesis and secretion by slices of two lines of Morris hepatoma has been reported (Koj, 1980). Since some AP-proteins are produced by hepatomas, the question arises as to whether these tumours can respond to a tissue injury inflicted upon the host. In order to address this question local inflammation was induced by subcutaneous injection of turpentine into control and hepatoma-bearing Buffalo rats and 20 h later tissue slices from the liver and tumour were prepared. The relative synthesis and secretion rates of several proteins are shown in Table 1. It is evident that liver from injured rats synthesizes considerably more fibrinogen, α_1-AP-globulin, α_1-AGP or haptoglobin, and relatively less albumin, than does that from control rats. Furthermore, larger amounts of labelled proteins were released into the medium in liver slices from turpentine-stimulated donors (Koj, 1980).

In the case of Morris hepatoma 7777, almost half the labelled protein released into the medium during 2 h incubation was AFP. Synthesis of albumin was reduced

TABLE 1

Synthesis and secretion of some plasma proteins by slices of rat liver and Morris hepatoma 7777 from control and turpentine-injected donors (modified from Koj et al., 1983)

Protein		Control liver	Turpentine liver	Control hepatoma	Turpentine hepatoma
AFP	Medium Tissue	–	–	40.08 4.09 (0.98)	39.15 4.08 (0.96)
Albumin	Medium Tissue	38.24 8.85 (1.85)	21.46 4.85 (2.08)	4.88 0.77 (0.63)	4.95 0.81 (0.61)
Fibrinogen	Medium Tissue	2.95 2.48 (0.51)	7.86 6.27 (0.59)	4.32 1.14 (0.38)	4.62 1.16 (0.41)
α_1-AP-globulin	Medium Tissue	1.84 1.27 (0.62)	6.14 4.51 (0.64)	2.28 0.51 (0.45)	2.33 0.58 (0.40)
Haptoglobin	Medium Tissue	2.62 1.90 (0.59)	5.75 4.10 (0.66)	0.21 0.23 (0.09)	0.19 0.17 (0.11)
α_1-AGP	Medium Tissue	0.51 0.29 (0.75)	1.96 1.10 (0.84)	0.92 0.21 (0.43)	0.83 0.18 (0.46)

In the groups with local inflammation livers or tumours were isolated 20 h after s.c. injection of 0.2 ml turpentine per 100 g body weight. The results are means of 4 – 6 experiments and are given as percentages of total protein radioactivities in the medium or tissue as recovered in the immunoprecipitates of individual proteins after 2 h incubation of tissue slices with [^{14}C]leucine. The medium to tissue radioactivity ratios reported in parentheses were calculated directly from protein counts (Koj, 1980).

to the rate of fibrinogen formation. As shown in Table 1, neither of these proteins represented more than 5% of the total labelled protein in the incubation medium. In relative terms, the production of fibrinogen, α_1-AP-globulin and α_1-AGP by hepatoma is comparable to that in the liver, while synthesis of albumin and haptoglobin is considerably diminished. Secretory rates of all plasma proteins in hepatoma preparations, as reflected by medium-to-tissue radioactivity ratios, are lower than in liver slices, although the difference is striking only with regard to albumin and haptoglobin. In the latter case this is not a secondary phenomenon resulting from the uptake of haemoglobin-haptoglobin complexes by hepatoma cells (Kasperczyk, Magielska and Koj, unpublished observations).

Impairment of secretion of plasma proteins synthesized by Morris hepatoma cells in vivo may be even more pronounced, as witnessed by the levels of plasma AFP in tumour-bearing animals. Even at the terminal stage when the tumour weight exceeds that of the liver 4 to 5-fold, the concentration of AFP in the blood only rarely reaches 5 mg/ml (Koj et al., 1982), despite the fact that AFP is the major protein product of Morris hepatoma 7777. However, this 'secretory inefficiency' is not a special feature of the neoplastic cells themselves but may be due to profound differences in the histological organization of the liver and Morris hepatoma 7777.

The results shown in Table 1 do not indicate increased synthesis of AP-proteins by hepatoma slices from turpentine-stimulated donors. Unlike in the liver, the percentage of radioactivity incorporated into the proteins synthesized by hepatoma, at least in the case of the proteins actually examined, is similar in the control and injured groups. It should be noted here that the relative synthesis rates of fibrinogen, α_1-AP-globulin and α_1-AGP are higher in hepatoma than in the control liver, suggesting that their synthesis might already have been 'derepressed' in the tumour. This is supported by the fact that neither ethionine nor galactosamine administered to rats bearing Morris hepatoma 7777 had any influence on the formation of any of the tested plasma proteins by slices of the tumour. In contrast the same treatment considerably reduced the synthesis of fibrinogen, α_1-AP-globulin and haptoglobin in slices from livers from turpentine-stimulated rats (Kasperczyk and Koj, 1983). It is possible that the ability to regulate AP-protein synthesis in response to injury-derived factors has been lost during liver carcinogenesis. However, it cannot be concluded that Morris hepatoma is insensitive to HSF/IL-1 since in turpentine-injected animals characteristic changes in the cell ultrastructure and stimulation of the Golgi complex were observed both in hepatocytes and in hepatoma cells (Koj et al., 1983).

Baumann and co-workers (1984b) studied the effect of human cell-derived cytokines on the synthesis and secretion of AP-proteins in the human cell line, Hep-G2. This line was shown to produce at least 20 human plasma proteins, including typical AP-reactants such as fibrinogen, haptoglobin, α_1-AGP, α_1-AT, α_1-ACh or

C3 complement (Knowles et al., 1980; Morris et al., 1982). However, after supplementation of the culture medium with cytokines containing HSF, the only reproducible change in plasma protein synthesis was increased formation of α_1-ACh and reduced formation of albumin, transferrin and AFP (Baumann et al., 1984b). Thus it appears that hepatoma cells may respond to HSF/IL-1 but synthesis rates of only certain AP-proteins are affected. This finding may have relevance to future understanding of the regulatory mechanisms of AP-protein synthesis in normal hepatocytes.

Considerable differences between cultured hepatocytes and transformed liver cells in responsiveness to regulatory factors and hormones other than HSF have been described by several authors (cf. also Chapter 18). Thus Selten and co-workers (1981) observed that rat hepatoma line SY/1/80 during growth in vitro requires homologous serum for expression of AFP and albumin genes. Glucocorticoids exert variable effects on synthesis of several secretory proteins, not only in two distinct rat hepatoma lines, HTC cells and Reuber H-35 cells (Baumann et al., 1980), but also in low and high passage variants of a single mouse hepatoma cell line, Hepa-2 (Morales and Papaconstantinou, 1982). Baumann and Jahreis (1983) found that a cloned line of mouse hepatoma cells, Hepa-1, responded to treatment with dexamethasone by a 30 to 80-fold increase in synthesis and secretion of functional haptoglobin. No such effect of dexamethasone was observed with prenatal or adult mouse hepatocytes. However, when several different clones of hybrid cells formed from adult mouse hepatocytes and rat hepatoma cells were treated with dexamethasone the synthesis of mouse haptoglobin was always elevated. It appears that haptoglobin expression in mouse liver cells is potentially sensitive to glucocorticoids, but this is manifested only in transformed cells and their derivatives.

2. *Synthesis of AP-proteins by cells of non-liver origin*

The literature abounds in papers claiming extrahepatic synthesis of various plasma proteins but only a few reports stand up to close scrutiny. Demonstration of a protein in any cell by immunological or other methods is clearly not enough since its presence may be secondary due to adsorption or endocytosis. A good example is provided by Anderson (1983) who showed the presence of prealbumin, albumin, α_1-AGP, α_1-AT, haptoglobin and transferrin in human lymphocytes, monocytes and granulocytes. By using crossed immunoelectrophoresis these six proteins were consistently detected in leucocytes, although sometimes as slightly different forms than those present in the plasma. However, after incubation of cells with [^{14}C]leucine only the α_1-AT peak was found to be labelled in the lysates of monocytes and granulocytes. Synthesis of human α_1-AT by human monocytes was independently demonstrated by Wilson et al. (1980) and Ikuta et al. (1982), while Rogers et al. (1983) even detected antitrypsin mRNA in human leucocytes. Thus it is safe to conclude that human leucocytes certainly can produce some α_1-PI although a direct comparison of their synthetic potential with liver cells is lacking.

On the other hand, Andersen (1983) could not substantiate the production of α_1-AGP by human lymphoblasts as reported by Gahmberg and Andersson (1978). The discrepancy may be due to the use of cells at different stages of maturity and variable expression of the genetic potential of cells maintained in vitro. Lemonnier et al. (1979) demonstrated incorporation of [^{14}C]leucine into human albumin by human fibroblasts, but this only occurred in some of the cultures. It is doubtful whether this synthesis of traces of albumin by fibroblasts has any significance in the organism.

Substantial evidence of extrahepatic synthesis of complement proteins has been provided by Colten and co-workers. They showed that the small proportions of the components of both the classical and alternative complement pathways, at least up to C5, are synthesized in the cells of the monocyte/macrophage origin (Einstein et al., 1976; Colten et al., 1979, cf. Chapter 11, Section 6). The importance of this fact needs to be emphasized as an aid towards better understanding of the inflammatory reaction. Thus local production of complement proteins by the macrophage may provide the initial response to tissue injury or microbial invasion (Colten, 1982). This will still be true even if the monocyte contribution to the overall production of complement proteins is small as suggested by the studies of Alper and co-workers (cited from Colten, 1982) who found that after liver transplantation at least 90% of the serum C3, C6, C8 and factor B are synthesized in the liver.

Although freshly isolated peripheral blood monocytes synthesize C2 only after a lag of 3 days in culture (Einstein et al., 1976; Whaley, 1980), macrophages isolated from breast milk or broncho-alveolar lavage produce C2 immediately and at rates several times above that in blood monocytes (Cole, 1980; Colten, 1982). Synthesis and secretion of C2 by macrophages is also affected by activation of these cells as occurs when IL-1 is elicited. Although a cytokine responsible for stimulation of C2 synthesis and secretion has not yet been identified, it is possible that the same factor enhances production of complement proteins in hepatocytes and macrophages. Stimulation of α_1-AT synthesis in cultures of human lymphocytes by conditioned media from human monocytes was reported by Ikuta et al. (1982).

The data presented above indicate that at least two types of human AP-proteins: α_1-PI and some complement components are synthesized by blood leucocytes as well as in the liver. Moreover, their synthesis appears to be enhanced in the same conditions as those which lead to stimulation of AP-protein synthesis in liver cells. Although in general liver is the main source of plasma AP-proteins the situation may be different during response to injury, especially in local areas infiltrated by macrophages.

References, p. 232.

Gordon/Koj (eds.) The acute-phase response to injury and infection.

CHAPTER 20

Catabolism and turnover of acute-phase proteins

A. KOJ

1. Plasma protein turnover in health and during the AP-response

Due to numerous studies with radioiodine-labelled plasma proteins carried out mainly in the years 1950 – 1970, a relatively clear picture of the turnover of these plasma components has emerged (for references see McFarlane, 1964; Schultze and Heremans, 1966; Rothschild and Waldmann, 1970). The body pool of fibrinogen, which is one of the most thoroughly investigated proteins, is maintained in a steady state by constant fractional catabolic rate and variable absolute synthetic rate with no apparent coordination between production and elimination (Regoeczi, 1970, 1974). It may be recalled here that fractional catabolic rate (FCR) is defined as the fraction of the intravascular mass of a protein which is catabolized per day whereas the absolute rate is the total mass destroyed (or synthesized) during such a period of time. Although Freeman showed that FCR of albumin and IgG increases with their plasma concentration while transferrin and haptoglobin appear to be catabolized faster at reduced plasma levels (cf. Schultze and Heremans, 1966), the rule of constant FCR is valid for the majority of proteins. The most remarkable example has been provided by Kushner and co-workers who showed that the fractional catabolic rate of CRP in rabbits was unaffected by changes of the plasma concentration of this protein ranging from 2 μg/ml in healthy animals to 100 μg/ml in rabbits which had been injected with turpentine or endotoxin (Chelladurai et al., 1983). At the same time, the absolute amounts of CRP catabolized in stimulated rabbits were of course increased in proportion to plasma level of the CRP.

The phenomenon of constant FCR suggests that protein molecules are removed from plasma by endocytosis rather than specific adsorption and then degraded at rates independent of the time elapsed since synthesis. Degradation of plasma proteins occurs presumably in lysosomes after their fusion with endocytic vesicles. The catabolic sites are still disputable (McFarlane and Koj, 1970) but they remain in a prompt equilibrium with the intravascular space; in fact a certain proportion of the protein may be degraded in transit from plasma to the extravascular compartment, although liver, kidney, intestine and cells of the reticuloendothelial system have also been implicated (McFarlane, 1964; Schultze and Heremans, 1966; Regoeczi, 1974).

The constant FCR value and existence of the extravascular compartment are together responsible for the fact that the apparent increases in plasma concentration of positive AP-reactants, e.g. fibrinogen, are always much less pronounced than the changes in the true synthetic rates in inflammatory states (Koj, 1968, 1984). In addi-

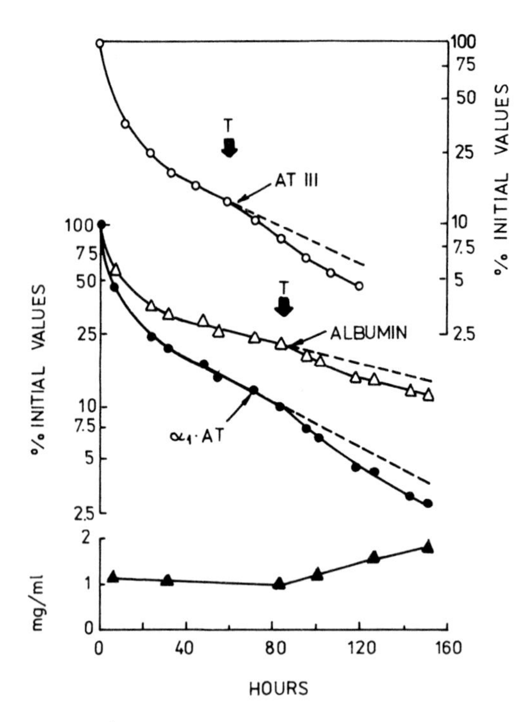

Fig. 1. Effect of turpentine (T) on the plasma slopes of homologous [^{131}I]AT III, [^{125}I]albumin and [^{131}I]α_1-AT in the rabbit (After Koj and Regoeczi, 1978). Albumin and α_1-AT were studied simultaneously in the same recipient, AT III in another rabbit. Plots represent protein-bound radioactivities per millilitre of plasma; ---, the expected slopes. α_1-AT concentrations in plasma from the dual-label experiment are shown in the lower diagram (▲).

tion to the effects of catabolism the plasma concentration of some proteins may decline due to deposition at the inflammatory site. Already in 1969 Robinson noted that a biosynthetically labelled rat plasma glycoprotein was removed from the intravascular pool more quickly in turpentine-injected than in normal animals. Later Koj and Regoeczi (1978) observed that injection of turpentine into rabbits resulted in a transient acceleration of the plasma slopes of iodine-labelled albumin, α_1-AT and AT III (Fig. 1). The plasma curves of all three proteins were similarly affected despite the fact that during the inflammatory reaction the concentration of albumin (a synthesis-inhibited AP-protein) was slightly reduced and α_1-PI significantly increased (a synthesis-stimulated AP-protein). Loss of these proteins into the inflammatory exudate may explain this phenomenon since increased concentration of iodine-labelled plasma albumin and α_1-AGP (Shibata et al., 1978) and albumin and α_1-AT (Ishibashi et al., 1978) were found in a granuloma induced by subcutaneous implantation of polyurethane foam, compared with the amounts in the surrounding tissues. Schreiber et al. (1982) observed a transient retardation of the total body slope of ^{125}I-labelled transferrin in rats which had been injected with turpentine. The change represented almost a mirror image of the experiments depicted in Fig. 1, since it was based on the total body counts instead of plasma radioactivities. Taken together, these results indicate that redistribution of several plasma proteins occurs in acute inflammatory states with the shift toward the extravascular space. In experiments based solely on measurement of plasma protein radioactivities this effect may be wrongly interpreted as a sudden burst of catabolism. Due to this phenomenon the increased liver output of positive AP-reactants will not be fully reflected by their concentrations in plasma. Analysis of the plasma protein radioactivity curves enables calculation of not only the fractional catabolic rates but also the intravascular and extravascular pool ratios and other turnover parameters as described by numerous authors (McFarlane, 1964; Nosslin, 1973; Regoeczi, 1974; Jones et al., 1978). Unfortunately, changes in protein pool size during the AP-response so complicate interpretation that few comprehensive studies on the turnover of AP-proteins are yet available.

The fractional catabolic rates of each plasma protein in different species vary considerably but are roughly proportional to the body weight to the power of 0.68 as calculated by Munro (1969) for albumin and ceruloplasmin, and confirmed recently by Regoeczi and Hatton (1980) for transferrin. Interestingly albumin in all species so far examined has a significantly longer half-life in the blood than other liver-produced plasma proteins while transferrin occupies an intermediate position. As shown in Table 1, turnover of synthesis-stimulated AP-proteins in man, rabbit and rat is 2 – 3 times faster than that of albumin, and CRP in the rabbit is catabolized even 30 times faster than albumin (Chelladurai et al., 1983). Although comparison of these results should be treated with caution since they were obtained by different

TABLE 1
Half-lives of some iodine-labelled plasma proteins as determined from the terminal slopes of plasma curves

	Plasma half-life of the protein (days)		
Protein	Man	Rabbit	Rat
Albumin	14.8[a]	8.2[b]	2.45[d]
Transferrin	8.7[b]	4.58[l]	2.06[d]
Fibrinogen	3.2[a]	2.58[c]	0.96[k]
Haptoglobin	3.5[a]	3.5[h]	0.79[i]
α_1-AGP	5.2[a]	?	1.3[d]
α_1-PI	4.6[g]	1.25[f]	1.4[j]
CRP	<2[m]	0.24[e]	?

[a] Mean value from the literature, after Koj (1974).
[b] Mean value from the literature, after Schultze and Heremans (1966).
[c] Regoeczi et al. (1963).
[d] Schreiber et al. (1982).
[e] Chelladurai et al. (1983).
[f] Regoeczi et al. (1980).
[g] Jones et al. (1978).
[h] Borel et al. (1963).
[i] Krauss and Sarcione (1966).
[j] Computed from the data of Ishibashi et al. (1978).
[k] Nieuwenhuizen et al. (1980).
[l] Regoeczi et al. (1974).
[m] Estimated from the non-labelled CRP plasma curve after myocardial infarction (Kushner et al., 1978).
[?] Unknown to the author.

investigators it appears that positive AP-reactants belong to the group of plasma proteins highly sensitive to proteinases (Dice and Goldberg, 1976) and therefore may be more easily degraded on transit from plasma to the extravascular space.

2. *Intravascular consumption or tissue deposition of some AP-proteins in pathological states*

The observed rapid turnover of AP-proteins does not necessarily mean their increased catabolism by indirect additional routes related to their biological functions. However, as briefly discussed previously (Koj, 1974) in certain pathological states such additional pathways do become operational. For example, increased loss from the plasma of fibrinogen occurs during the intravascular disseminated coagulation, in burned patients and in fibrinolytic disorders (Reeve et al., 1966; Kukral et al., 1969; Müller-Berghaus, 1969). An interesting feed-back mechanism has been demonstrated in vivo by Ritchie et al. (1982) in which fibrinogen degradation products indirectly stimulate synthesis of fibrinogen and other AP-proteins. This was found to occur due to enhanced formation of HSF/IL-1 by human peripheral blood leucocytes.

Haptoglobin is another protein which disappears rapidly in some diseases. According to Noyes and Garby (1967) a significant proportion of haptoglobin is always removed from plasma by an indirect route after binding to haemoglobin. This pathway may increase considerably in haemolytic conditions resulting in transient ahaptoglobinaemia (Herman, 1961). The Hp-Hb complex is quickly removed from the blood in mammals (cf. Koj, 1974 and Chapter 13, Section 6) but not in birds in which haptoglobin differs in structure and in haemoglobin-binding properties (Dobryszycka et al., 1979). The increased catabolism of α_1-AGP observed after severe injuries and in sustained inflammation may be due to the loss of this protein in the urine (Zeineh et al., 1972; Zeineh and Kukral, 1970).

Increased consumption of plasma inhibitors of proteinases in injury and infection has been described by several authors (Ohlsson, 1974; Heimburger, 1975; Fritz, 1980; Chapter 13, Section 1). In various febrile and inflammatory diseases the decrease in plasma concentration of inter-α-trypsin inhibitor (ITI) is accompanied by a rise of its degradation products. This is understandable since ITI has been shown to be a strong inhibitor of granulocyte elastase which is released into the inflammatory exudate (Jochum and Bittner, 1983). Although ITI is regarded as a negative AP-protein (cf. Chapter 12, Table 1), its reduced synthesis has never been proved. Indeed, it would not be surprising if ITI is produced at enhanced rates but very quickly degraded.

Proteins of the pentraxin family: CRP and SAP appear to have exceptionally short half-lives of 5 – 8 h in rabbits (Skinner et al., 1982; Chelladurai et al., 1983). Furthermore, CRP shows a high affinity for the phosphoryl choline present in damaged plasma membranes (Pepys and Baltz, 1983; Chapter 14, Section 3). Deposited in this form, CRP has been demonstrated in necrotic tissues (Kushner and Kaplan, 1961). Despite this fact the fractional catabolic rate of CRP does not increase in experimental inflammation (Chelladurai et al., 1983) suggesting that its tissue sequestration represents a negligible proportion of the total CRP removal from the circulation. Only in inflammation induced by $\varkappa$-carrageenan has increased removal of CRP been clearly demonstrated (Preston et al., 1983). SAP is known to bind to isolated amyloid fibrils of both primary and secondary types (Pepys et al., 1979b; Pepys and Baltz, 1983). However, iodine-labelled SAP was detected in similar amounts in normal and amyloidotic tissues hence the association of SAP to the fibril may involve more than a simple binding mechanism (Skinner et al., 1982;

Cohen et al., 1983). The main fibril protein of secondary amyloidosis is a polymer of amyloid A protein deriving by proteolytic degradation of its plasma precursor, SAA (Pepys and Baltz, 1983). SAA is a spectacular AP-reactant in mouse and man occurring in the plasma as a complex with high density lipoprotein. It appears that an amyloid enhancing factor produced by repeated episodes of inflammation has an essential role in the formation of amyloid fibrils from SAA (Cohen et al., 1983). These facts illustrate the complexity of the processes which lead to removal of plasma proteins such as SAA from the circulation but also emphasize the need for further investigation of the underlying mechanisms.

In particular, the data presented above indicate that increased removal from the circulation of AP-proteins due to tissue deposition constitutes a serious difficulty in any attempt to assess the real magnitude of their enhanced production. In experimental animals this difficulty may be overcome by direct measurement of rates of synthesis using labelled amino acids but in humans the problem is still not resolved.

References (Chapters 12 – 20)

Abd-el-Fattah, M., Scherer, R., Fouad, F.M. and Ruhehstroth-Bauer, G. (1981) Cancer Res. *41*, 2548 – 2555.

Abelev, G.I. (1976) in Oncodevelopmental Gene Expression (Fishman, W.H. and Sell, S., eds.) pp. 191 – 202, Academic Press, NY and London.

Abramson, F.P., Jenkins, J. and Ostchega, Y. (1982) Clin. Pharmacol. Ther. *32*, 659 – 663.

Ades, E.W., Hinson, A., Chapuis-Cellier, C. and Arnaud, P. (1982) Scand. J. Immunol. *15*, 109 – 113.

Aisen, P. and Listowsky, I. (1980) Ann. Rev. Biochem. *49*, 357 – 393.

Albrecht, G.J., Hochstrasser, K. and Salier, J.P. (1983) Hoppe Seyler's Z. Physiol. Chem. *364*, 1703 – 1708.

Algranati, I.D. and Sabatini, D.D. (1979) Biochem. Biophys. Res. Commun. *90*, 220 – 226.

Allen, P.C., Hill, E.A. and Stokes, A.M. (1977) Plasma Proteins. Analytical and Preparative Techniques, Blackwell Sci. Publ., Oxford-London-Edinburgh-Melbourne.

Alper, C.A. (1974) in Structure and Function of Plasma Proteins (Allison, A.C. ed.) Vol. 1, pp. 195 – 222, Plenum Press, London and New York.

Amrani, D.L., Plant, P.W., Pindyck, J., Mosesson, M.W. and Grieninger, G. (1983) Biochim. Biophys. Acta *743*, 394 – 400.

Andersen, M.M. (1983) Scand. J. Clin. Lab. Invest. *43*, 49 – 59.

Andersen, P. and Eika, C. (1979) Thromb. Haemost. *42*, 299.

Andersen, P., Kierulf, P., Elde, A.T. and Godal, H.C. (1980) Thromb. Res. *19*, 401 – 408.

Andersen, P., Kierulf, P. and Godal, H.C. (1981) Thromb. Res. *22*, 593 – 602.

Andus, T., Gross, V., Tran-Thi, T.A., Schreiber, G., Nagashima, M. and Heinrich, P.C. (1983a) Eur. J. Biochem. *133*, 561 – 571.

Andus, T., Gross, V., Tran-Thi, T.A. and Heinrich, P.C. (1983b) Eur. J. Biochem. *136*, 253 – 257.

Aoki, N., Saito, H., Kamiya, T., Koie, K., Sokata, Y. and Kobakura, M. (1979) J. Clin. Invest. *63*, 877 – 884.

Apffel, C.A. and Peters, J.H. (1969) Progr. Exp. Tumor Res. *12*, 1 – 54.
Aronsen, K.F., Ecklund, G., Kindmark, C.O. and Laurell, C.B. (1972) Scand. J. Lab. Invest. *29* (Suppl. 124), 127 – 136.
Ashburner, M. (1974) Dev. Biol. *39*, 141 – 157.
Augberger, P., Samson, M. and LeCam, A. (1983) Biochem. J. *214*, 679 – 685.
Baldo, B.A. and Fletcher, T.C. (1973) Nature (London) *246*, 145 – 146.
Balegno, H.F. and Neuhaus, O.W. (1970) Life Sci. *9* (II), 1039 – 1044.
Baltz, M.L., Gomer, K., Davies, A.J.S., Evans, D.J., Klaus, G.G.B. and Pepys, M.B. (1980) Clin. Exp. Immunol. *39*, 355 – 360.
Baltz, M.L., de Beer, F.C., Feinstein, A., Munn, E.A., Milstein, C.P., Fletcher, T.C., March, J.F., Taylor, J., Bruton, C., Clamp, J.R., Davies, A.J.S. and Pepys, M.B. (1982) Ann. N.Y. Acad. Sci. *389*, 49 – 75.
Banks, B.E.C., Vernon, C.A. and Warner, J.A. (1984) Neurosci. Lett. *47*, 41 – 45.
Baracos, V., Rodemann, H.P., Dinarello, C.A. and Goldberg, A.L. (1983) N. Engl. J. Med. *308*, 553 – 558.
Barbosa, J., Seal, U.S. and Doe, R.P. (1971) J. Clin. Endocrinol. Metab. *33*, 388 – 398.
Barna, B.P., Deodhar, S.D., Grautam, S., Yen-Lieberman, B. and Roberts, B. (1984) Cancer Res. *44*, 305 – 310.
Barrett, A.J., Brown, M.A. and Sayers, C.A. (1979) Biochem. J. *181*, 401 – 418.
Baskies, A.M., Chretien, P.B., Weiss, J.F., Makush, R.W., Beveridge, R.A., Catalona, W.J. and Spiegel, W.E. (1980) Cancer *45*, 3050.
Bata, J. and Revillard, J. (1981) Agents Actions, *11*, 614 – 616.
Baumann, H. and Held, W.A. (1981) J. Biol. Chem. *256*, 10145 – 10155.
Baumann, H. and Jahreis, G.P. (1983) J. Cell Biol. *97*, 728 – 736.
Baumann, H., Gelehrter, T.D. and Doyle, D. (1980) J. Cell Biol. *85*, 1 – 8.
Baumann, H., Firestone, G.L., Burgess, T.L., Gross, K.W., Yamamoto, K.R. and Held, W.A. (1983a) J. Biol. Chem. *258*, 563 – 570.
Baumann, H., Jahreis, G.P. and Gaines, K. (1983b) J. Cell Biol. *97*, 866 – 876.
Baumann, H., Held, W.A. and Berger, C. (1984a) J. Biol. Chem. *259*, 566 – 573.
Baumann, H., Jahreis, G.P., Sauder, D.N. and Koj, A. (1984b) J. Biol. Chem. *259*, 7331 – 7342.
Bayard, B., Kerckaert, J.P., Laine, A. and Hayem, A. (1982) Eur. J. Biochem. *124*, 371 – 376.
Bayer, B.M., Lo, T.N. and Beaven, M.A. (1980) J. Biol. Chem. *255*, 8784 – 8790.
Beisel, W.R. (1975) Ann. Rev. Med. *26*, 9 – 20.
Benditt, E.P., Hoffman, J.S., Eriksen, N., Parmelee, D.C. and Walsh, K.A. (1982) Ann. N.Y. Acad. Sci. *389*, 183 – 189.
Bennett, M. and Schmid, K. (1980) Proc. Natl. Acad. Sci. USA *77*, 6109 – 6113.
Berger, E.G., Buddecke, E., Kamerling, J.P., Kobata, A., Paulson, J.C. and Vliegenthart, J.F.G. (1982) Experientia *38*, 1129 – 1162.
Bernheim, H.A. and Kluger, M.J. (1977) J. Physiol. (London) *267*, 659 – 666.
Bernuau, D., Rogier, E. and Feldmann, G. (1983) Hepatology *3*, 29 – 33.
Bieth, J. (1980) Bull. Eur. Physiopathol. Respir. *16* (Suppl.) 183 – 195.
Billingham, M.E.J., Gordon, A.H. and Robinson, B.V. (1971) Nature (London) *231*, 26 – 29.
Blatteis, C.M., Hunter, W.S., Llanos, Q.J., Ahokas, R.A. and Mashburn, T.A. (1983) Brain Res. Bull. *11*, 519 – 526.
Bøg-Hansen, T., Bjerrum, O.J. and Ramlau, J. (1975) Scand. J. Immunol. (Suppl. 2), 141 – 147.
Borel, J., Dobryszycka, W., Moretti, J. and Jayle, M.F. (1963) Bull. Soc. Chim. Biol. *45*, 203 – 210.
Borges, D.R. and Gordon, A.H. (1976) J. Pharm. Pharmacol. *28*, 44 – 48.
Bornstein, D.L. and Walsh, E.C. (1978) J. Lab. Clin. Med. *91*, 236 – 245.

Bosanquet, A.G., Chandler, A.M. and Gordon, A.H. (1976) Experientia, 1348 – 1349.
Bowen, M., Raynes, J.G. and Cooper, E.H. (1982) in Lectins – Biology, Biochemistry, Clinical Biochemistry (Bøg-Hansen, T.C., ed.) Vol. 2, pp. 403 – 411, Walter de Gruyter and Co., Berlin and New York.
Boyer, R.F. and Schori, B.E. (1983) Biochem. Biophys. Res. Commun. *116*, 244 – 250.
Brady, P.O. (1982) Trends Biochem. Sci. *7*, 143 – 146.
Brown, P.C. and Papaconstantinou, J. (1979) J. Biol. Chem. *254*, 9379 – 9384.
Carlson, J. and Stenflo, J. (1982) J. Biol. Chem. *257*, 12987 – 12994.
Carlson, T.H., Fradl, D.C., Leonard, B.D., Wentland, S.H. and Reeve, E.B. (1977) Am. J. Physiol. *233*, H1 – H9.
Carlson, T.H., Wentland, S.H., Leonard, B.D., Ruder, M.A. and Reeve, E.B. (1978) Am. J. Physiol. *235*, H223 – H230.
Carp, H. and Janoff, A. (1980) J. Clin. Invest. *66*, 987 – 995.
Carrell, R.W., Boswell, D.R., Brennan, S.O. and Owen, M.C. (1980) Biochem. Biophys. Res. Commun. *93*, 399 – 402.
Carrell, R.W., Jeppson, J.O., Laurell, C.B., Brennan, S.O., Owen, M.C., Vaughan, L. and Boswell, D.R. (1982) Nature (London) *298*, 329 – 334.
Chandler, A.M. and Neuhaus, O.W. (1968) Biochim. Biophys. Acta *166*, 186 – 194.
Charlwood, P.A., Hatton, M.W.C. and Regoeczi, E. (1976) Biochim. Biophys. Acta *453*, 81 – 92.
Chelladurai, M., Macintyre, S.S. and Kushner, I. (1983) J. Clin. Invest. *71*, 604 – 610.
Cheresh, D.A., Haynes, D.H. and Distasio, J.A. (1984) Immunology *51*, 541 – 548.
Chih, J.J., Procyk, R. and Devlin, T.M. (1977) Biochem. J. *162*, 501 – 507.
Chiu, K.M., Mortensen, R.F., Osmand, A.P. and Gewurz, H. (1977) Immunology *32*, 997.
Chow, V., Murray, R.K., Dixon, J.D. and Kurosky, A. (1983) FEBS Lett. *153*, 275 – 279.
Clowes, G.H.A., George, B.C., Villee, Jr., C.A. and Saravis, C.A. (1983) N. Engl. J. Med. *308*, 545 – 552.
Coe, J.E. and Ross, M.J. (1983) J. Exp. Med. *157*, 1421 – 1433.
Cohen, A.S., Shirahama, T., Sipe, J.D. and Skinner, M. (1983) Lab. Invest. *48*, 1 – 4.
Cole, F.S., Matthews, Jr., W.J., Marino, J.T., Gash, D.J. and Colten, H.R. (1980) J. Immunol. *125*, 1120 – 1124.
Collen, D. and Wiman, B. (1978) Blood *51*, 563 – 569.
Coller, B.S. (1980) Thromb. Res. *18*, 579 – 584.
Colten, H.R. (1982) Molec. Immunol. *19*, 1279 – 1285.
Colten, H.R., Ooi, Y.M. and Edelson, P.J. (1979) Ann. N.Y. Acad. Sci. *332*, 482 – 490.
Cordier, G. and Revillard, J.P. (1980) Experientia *36*, 603 – 604.
Costello, M., Fiedel, B.A. and Gewurz, H. (1979) Nature (London) *281*, 677 – 678.
Courtoy, P.J., Lombart, C., Feldmann, G., Moguilevsky, N. and Rogier, E. (1981) Lab. Invest. *44*, 105 – 115.
Courtoy, P.J., Moguilevsky, N., Retegui, L.A., Castranne, C.E. and Masson, P.L. (1984) Lab. Invest. *50*, 329 – 334.
Crane, L.J. and Miller, D.L. (1977) J. Cell Biol. *72*, 11 – 25.
Damais, C., Riveau, G., Parant, M. and Chedid, L. (1983) in Interleukins, Lymphokines and Cytokines (Oppenheim, J.J., Cohen, S. and Landy, M., eds.) pp. 465 – 472, Academic Press, New York and London.
Daniels, J.C., Larson, D.I., Abston, S. and Ritzmann, S.E. (1974) J. Trauma *14*, 153 – 162.
Davies, P. and Allison, A.C. (1976) in Lysosomes in Biology and Pathology (Dingle, J.T. and Dean, R.T., eds.) Vol. 5, pp. 61 – 98, Elsevier/North-Holland, Amsterdam – Oxford – New York.
DeBeer, F.C., Baltz, M.L., Munn, E.A., Feinstein, A., Taylor, J., Bruton, C., Clamp, J.R. and Pepys,

M.B. (1982) Immunology *45*, 55 – 70.
Dendo, C.W. (1979) Agents Actions *9*, 333 – 336.
Dice, J.F. and Goldberg, A.L. (1976) Nature (London) *262*, 514 – 516.
Dinarello, C.A. (1984) Rev. Inf. Dis. *6*, 51 – 95.
Dinarello, C.A., Bendtzen, K. and Wolff, S.M. (1982) Inflammation *6*, 63 – 78.
Dinarello, C.A., Mornoy, S.O. and Rosenwasser, L.J. (1983) J. Immunol. *130*, 890 – 895.
Dinarello, C.A., Clowes, Jr., G.H.A., Gordon, A.H., Saravis, C.A. and Wolff, S.M. (1984) J. Immunol., *133*, 1332 – 1338.
Dobryszycka, W. and Bec-Katnik, I. (1975) Acta Biochim. Polon. *22*, 143 – 153.
Dobryszycka, W. and Krawczyk, E. (1979) Comp. Biochem. Physiol. *62B*, 111 – 113.
Dobryszycka, W., Woźniak, M., Krawczyk, E. and Furmaniak-Kazimierczak, E. (1979) Int. J. Biochem. *13*, 739 – 743.
Doolittle, R.F. (1975) in The Plasma Proteins (Putman, W., ed.) Vol. 1, pp. 110 – 161, Academic Press, New York – San Francisco – London.
Doolittle, R.F. (1983) Ann. N.Y. Acad. Sci. *408*, 13 – 27.
Duff, G.W. and Durum, S.K. (1983) Nature (London) *304*, 449 – 451.
Durnam, D.M. and Palmiter, R.D. (1984) Molec. Cell Biol. *4*, 481 – 491.
Durnam, D.M., Hoffman, J.S., Quaife, C.J., Benditt, E.P., Chen, H.Y., Brinster, R.L. and Palmiter, R.D. (1984) Proc. Natl. Acad. Sci. USA *81*, 1053 – 1056.
Duthie, E.S. (1955) J. Gen. Microbiol. *13*, 383 – 393.
Eaton, J.W., Brandt, P., Mahoney, J.R. and Lee, Jr., J.T. (1982) Science *215*, 691 – 693.
Eddington, C.L., Upchurch, H.F. and Kampschmidt, R.F. (1971) Proc. Soc. Exp. Biol. Med. *136*, 159 – 164.
Edwards, K., Nagashima, M., Dryburgh, H., Wykes, A. and Schreiber, G. (1979) FEBS Lett. *100*, 269 – 272.
Einstein, L.P., Schneeberger, E.E. and Colten, H.R. (1976) J. Exp. Med. *143*, 114 – 126.
Eisen, V. (1977) in Inflammation: Mechanisms and the Impact on Therapy (Bonta, I.L., Thompson, J. and Brune, K., eds.), pp. 9 – 16, Birkhauser-Verlag, Basel and Stuttgart.
Elbein, A.D. (1983) Methods Enzymol. *98*, 135 – 154.
Ernest, M.J., Chen, C.L. and Feigelson, P. (1977) J. Biol. Chem. *252*, 6783 – 6791.
Esnard, F. and Gauthier, F. (1983) J. Biol. Chem. *258*, 12443 – 12447.
Fantone, J.C. and Ward, P.A. (1982) Am. J. Pathol. *107*, 397 – 418.
Fernandez-Pol, J.A. (1981) Biochemistry *20*, 3907 – 3912.
Fey, G. and Colten, H.R. (1981) Fed. Proc. *40*, 2099 – 2104.
Fey, G., Domdey, H., Wiebauer, K., Whitehead, A.S. and Odink, K. (1983) Springer Semin. Immunopathol. *6*, 119 – 147.
Fiedel, B.A., Simpson, R.N. and Gewurz, H. (1982) Ann. N.Y. Acad. Sci. *389*, 263 – 271.
Fleck, A. (1976) in Metabolism and the Response to Injury (Wilkinson, A.W. and Cuthbertson, D., eds.), pp. 229 – 236, Pitman Medical, London.
Fouad, F.M., Scherer, R., Abd-el-Fattah, M. and Ruhenstroth-Bauer, G. (1980) Eur. J. Cell Biol. *21*, 175 – 179.
Fournet, B., Montreuil, J., Strecker, G., Dorland, L., Haverkamp, J., Vliegenthart, J.F.G., Binette, J.P. and Schmid, K. (1978) Biochemistry *17*, 5206 – 5214.
Fowlkes, D.M., Mullis, N.T., Comeau, C.M. and Crabtree, G.R. (1984) Proc. Natl. Acad. Sci. USA *81*, 2313 – 2316.
Franck, C. and Pedersen, J.Z. (1983) Scand. J. Clin. Lab. Invest. *43*, 151 – 155.
Fridovich, I. (1975) Ann. Rev. Biochem. *44*, 147 – 159.
Frieden, E. (1973) Nutrit. Rev. *31*, 41 – 42.

Frieden, E. (1979) Ciba Foundation Symp. *74*, 93 – 124.
Frieden, E. and Hsieh, H.S. (1976) Adv. Enzymol. *44*, 187 – 236.
Friesen, A.D. and Jamieson, J.C. (1980) Can. J. Biochem. *58*, 1101 – 1111.
Fritz, H. (1980) Ciba Foundation Symp. *75*, 351 – 379.
Gahmberg, C.G. and Andersson, L.C. (1978) J. Exp. Med. *148*, 507 – 521.
Gahring, L., Baltz, M.L., Pepys, M.B. and Daynes, R. (1984) Proc. Natl. Acad. Sci. U.S.A. *81*, 1198 – 1202.
Ganguly, M. and Westphal, U. (1968) J. Biol. Chem. *243*, 6130 – 6139.
Ganrot, P.O. and Schersten, B. (1967) Clin. Chim. Acta *15*, 113 – 120.
Gauthier, F. and Ohlsson, K. (1978) Hoppe-Seyler's Z. Physiol. Chem. *359*, 987 – 992.
Geiger, T., Northemann, W., Schmelzer, E., Gross, V., Gauthier, F. and Heinrich, P.C. (1982) Eur. J. Biochem. *126*, 189 – 195.
Geisow, M.J. and Gordon, A.H. (1978) Trends Biochem. Sci. *3*, 169 – 171.
Gejyo, F. and Schmid, K. (1981) Biochim. Biophys. Acta *671*, 78 – 84.
Gerdin, B., Saldeen, T., Roszkowski, W., Szmigielski, S., Stachurska, J. and Kopeć, M. (1980) Thromb. Res. *18*, 461 – 468.
Giblett, E. (1974) in Structure and Function of Plasma Proteins (Allison, A.C., ed.) Vol. 1, pp. 55 – 72, Plenum Press, London and New York.
Gillis, S. and Mizel, S.B. (1981) Proc. Natl. Acad. Sci., U.S.A. *78*, 1133 – 1137.
Godefroy-Colburn, T. and Thach, R.E. (1981) J. Biol. Chem. *256*, 11762 – 11773.
Goldstein, I.M., Kaplan, H.B., Edelson, H.S. and Weissmann, G. (1979) J. Biol. Chem. *254*, 4040 – 4045.
Goldstein, I.M., Kaplan, H.B., Edelson, H.S. and Weissmann, G. (1982) Ann. N.Y. Acad. Sci. *389*, 368 – 379.
Goodwin, J., Bankhurst, A. and Messner, R. (1977) J. Exp. Med. *146*, 1719.
Gordon, A.H. (1976a) in Plasma Protein Turnover (Bianchi, R., Mariani, G. and McFarlane, A.S., eds.), pp. 381 – 394, University Park Press, Baltimore.
Gordon, A.H. (1976b) Biochem. J. *159*, 643 – 650.
Gordon, A.H. and Limaos, E.A. (1979a) Br. J. Exp. Pathol. *60*, 441 – 446.
Gordon, A.H. and Limaos, E.A. (1979b) Br. J. Exp. Pathol. *60*, 434 – 440.
Goutner, A., Summler, M.C., Tapon, J. and Rosenfeld, C. (1976) Differentiation, *5*, 171 – 173.
Gravagna, P., Giamazza, E., Arnaud, P., Neels, M. and Ades, E.W. (1982) J. Reticuloendothel. Soc. *32*, 125 – 130.
Greenwald, R.A. and Moy, W.M., (1979) Arthritis Rheum. *22*, 251 – 259.
Grieninger, G., Hertzberg, K.M. and Pindyck, J. (1978) Proc. Natl. Acad. Sci., U.S.A. *75*, 5506 – 5510.
Grieninger, G., Plant, P.W., Liang, T.J., Kalb, R.G., Amrani, D., Mosesson, M.W., Hertzberg, K.M. and Pindyck, J. (1983) Ann. N.Y. Acad. Sci. *408*, 469 – 489.
Griffin, E.E. and Miller, L.L. (1974) J. Biol. Chem. *249*, 5062 – 5069.
Gross, V., Tran-Thi, T.A., Vosbeck, K. and Heinrich, P.C. (1983) J. Biol. Chem. *258*, 4032 – 4036.
Gross, V., Andus, T., Tran-Thi, T.A., Bauer, J., Decker, K. and Heinrich, P.C. (1984) Exp. Cell Res. *151*, 46 – 54.
Guillouzo, A., Boisnard-Rissel, M., Belanger, L. and Bourel, M. (1979) Biochem. Biophys. Res. Commun. *91*, 327 – 331.
Guillouzo, A., Delers, F., Clement, B., Bernard, N. and Engler, R. (1984) Biochem. Biophys. Res. Commun. *120*, 311 – 317.
Gutteridge, J.M.C. (1983) FEBS Lett. *157*, 37 – 40.
Hanley, J.M., Haugen, T.H. and Heath, E.C. (1983) J. Biol. Chem. *258*, 7858 – 7869.
Hanson, D.F., Murphy, P.A., Silicano, R. and Shin, H.S. (1983) J. Immunol. *130*, 216 – 221.

Hatton, M.W.C., März, L. and Regoeczi, E. (1983) Trends Biochem. Sci. *8*, 287 – 291.
Haugen, T.H., Hanley, J.M. and Heath, E.C. (1981) J. Biol. Chem. *256*, 1055 – 1057.
Havemann, K. and Janoff, A. (eds.) (1978) Neutral Proteases of Human Polymorphonuclear Leukocytes, Urban and Schwarzenberg. Baltimore and Munich.
Hawiger, J., Timmons, S., Strong, D.D., Cottrell, B.A., Riley, M. and Doolittle, R.F. (1982) Biochemistry *21*, 1407 – 1413.
Hawiger, J., Kloczewiak, M., Timmons, S., Strong, D. and Doolittle, R.F. (1983) Ann. N.Y. Acad. Sci. *408*, 521 – 535.
Hedner, U., Martinsson, G. and Berquist, D. (1983) Haemostasis *13*, 219 – 226.
Heimburger, N. (1975) in Protease and Biological Control, Cold Spring Harbor Conferences on Cell Proliferation (Reich, E., Rifkin, D.B. and Shaw, E., eds.), Vol. 2, pp. 367 – 386, Cold Spring Harbor Laboratory.
Heldin, C.H., Westermark, B., Mellstrom, K., Johnsson, A., Ek, B., Nister, M., Bertholtz, C., Rönnstrand, L. and Wateson, A. (1983) Surv. Synth. Pathol. Res. *1*, 153 – 164.
Henschen, A., Lottspeich, F., Kehl, M. and Southan, C. (1983) Ann. N.Y. Acad. Sci. *408*, 28 – 43.
Herman, Jr., E.C. (1961) J. Lab. Clin. Med. *57*, 834.
Hirano, K., Okumura, Y., Hayakawa, S., Adachi, T. and Sugiura, M. (1984) Hoppe-Seyler's Z. Physiol. Chem. *365*, 27 – 32.
Hoffmann, M., Feldmann, S.R. and Pizzo, S.V. (1983) Biochim. Biophys. Acta *760*, 421 – 423.
Högstrop, H., Jakobson, H. and Carlin, G. (1981) Thromb. Res. *21*, 247 – 253.
Hooper, D.C., Steer, C.J., Dinarello, C.A. and Peacock, A.C. (1981) Biochim. Biophys. Acta *653*, 118 – 129.
Hornung, M. and Fritschi, S. (1971) Nature (London) *230*, 84 – 85.
Hsieh, H.S. and Frieden, E. (1975) Biochem. Biophys. Res. Commun. *67*, 1326 – 1331.
Huang, J.S., Huang, S.S. and Deuel, T.F. (1984) Proc. Natl. Acad. Sci. U.S.A. *81*, 342 – 346.
Hudig, D., Haverty, T., Fulcher, C., Redelman, D. and Mendelsohn, J. (1981) J. Immunol. *126*, 1569 – 1574.
Humes, J.L., Bonney, R.J., Pelus, L., Dahlgren, M.E., Sadowski, S.J., Kuehl, Jr., F.A. and Davies, P. (1977) Nature (London) *269*, 149 – 151.
Hunt, L.T. and Dayhoff, M.O. (1980) Biochem. Biophys. Res. Commun. *95*, 864 – 871.
Hunter, T. and Cooper, J.A. (1981) Cell *24*, 741 – 752.
Ikuta, T., Okubo, H., Kudo, J., Ishibashi, H. and Inoue, T. (1982) Biochem. Biophys. Res. Commun. *104*, 1509 – 1516.
Innis, M.A. and Miller, D.L. (1979) J. Biol. Chem. *254*, 9148 – 9154.
Ishibashi, H., Shibata, K., Okubo, H., Tsuda-Kawamura, K. and Yanase, T. (1978) J. Lab. Clin. Med. *91*, 575 – 583.
Ivarie, R.D. and O'Farrell, P.H. (1978) Cell *13*, 41 – 55.
Jackson, C.M. and Nemerson, Y. (1980) Ann. Rev. Biochem. *49*, 765 – 811.
James, K. (1980) Trends Biochem. Sci. *5*, 43 – 47.
James, K., Baum, L.L., Vetter, M.L. and Gewurz, H. (1982) Ann. N.Y. Acid. Sci. *389*, 274 – 285.
Jamieson, J.C. (1983) in Plasma Protein Secretion by the Liver (Glaumann, H., Peters, T. and Redman, C., eds.), pp. 257 – 284, Academic Press, London and New York.
Jamieson, J.C. and Ashton, F.E. (1973) Can. J. Biochem. *51*, 1034 – 1045.
Jamieson, J.C., Kaplan, H.A., Woloski, B.M.R.N.J., Hellman, N. and Ham, K. (1983) Can. J. Biochem. Cell Biol. *61*, 1041 – 1048.
Javaid, J.I., Hendricks, H. and Davis, J.M. (1983) Biochem. Pharmacol. *32*, 1149 – 1153.
Jayle, M.F. (1951) Bull. Soc. Chim. Biol. *33*, 876 – 880.
Jeejeebhoy, K.N., Ho, J., Greenberg, G.R., Phillips, M.J., Bruce-Robertson, A. and Sodtke, U. (1975) Biochem. J. *146*, 141 – 155.

Jeejeebhoy, K.N., Bruce-Robertson, A., Ho, J., Kida, S. and Muller-Eberhard, U. (1976) Can. J. Biochem. *54*, 74 – 78.
Jeejeebhoy, K.N., Phillips, M.J., Ho, J. and Bruce-Robertson, A. (1980) Ann. N.Y. Acad. Sci. *349*, 18 – 27.
Jesty, J. (1978) Arch. Biochem. Biophys. *185*, 165 – 173.
Jochum, M. and Bittner, A. (1983) Hoppe-Seyler's Z. Physiol. Chem. *364*, 1709 – 1715.
John, D.W. and Miller, L.L. (1969) J. Biol. Chem. *244*, 6134 – 6142.
Johnson, W.J., Pizzo, S.V., Imber, M.J. and Adams, D.O. (1982) Science *218*, 574 – 576.
Jones, E.A., Vergalla, J., Steer, C.J., Bradley-Moore, P.R. and Vierling, J.M. (1978) Clin. Sci. Molec. Med. *55*, 139 – 148.
Jue, D.M., Shim, B.S. and Kang, Y.S. (1983) Molec. Cell Biochem. *51*, 141 – 148.
Kajita, Y. and Hayashi, O. (1972) Biochim. Biophys. Acta *261*, 281 – 283.
Kalsheker, N.A., Bradwell, A.R. and Burnett, D. (1981) Experientia *37*, 447 – 448.
Kampschmidt, R.F. (1978) J. Reticuloendothel. Soc. *23*, 287 – 297.
Kampschmidt, R.F. (1981) in Infection: The Physiologic and Metabolic Responses of the Host (Powanda, M.C. and Canonico, P.G., eds.), pp. 56 – 74, Elsevier/North-Holland Biomedical Press, Amsterdam.
Kampschmidt, R.F. and Pulliam, L.A. (1975) J. Reticuloendothel. Soc. *17*, 162 – 169.
Kampschmidt, R.F. and Pulliam, L.A. (1978) Proc. Soc. Exp. Biol. Med. *158*, 32 – 35.
Kampschmidt, R.F. and Upchurch, H.F. (1969a) Proc. Soc. Exp. Biol. Med. *131*, 864 – 867.
Kampschmidt, R.F. and Upchurch, H.F. (1969b) Am. J. Physiol. *216*, 1287 – 1291.
Kampschmidt, R.F. and Upchurch, H.F. (1974) Proc. Soc. Exp. Biol. Med. *146*, 904 – 907.
Kampschmidt, R.F. and Upchurch, H.F. (1980) Proc. Soc. Exp. Biol. Med. *164*, 537 – 539.
Kampschmidt, R.F., Upchurch, H.F. and Worthington, III, M.L. (1983) Infect. Immun. *41*, 6 – 10.
Kaplan, A.P., Silverberg, M., Dunn, J.T. and Ghebrehiwet, B. (1982) Ann. N.Y. Acad. Sci. *389*, 25 – 38.
Kaplan, H.A. and Jamieson, J.C. (1977) Life Sci. *21*, 1311 – 1316.
Karin, M., Haslinger, A., Holtgreve, H., Richards, I., Krauter, P., Westphal, H. and Beato, M. (1984) Nature (London) *308*, 513 – 519.
Kasperczyk, H., and Koj, A. (1983) Br. J. Exp. Pathol. *64*, 277 – 285.
Kasuga, M., Zick, Y., Blithe, D.L., Crettaz, M. and Kahn, C.R. (1982) Nature (London) *298*, 667 – 669.
Kazim, A.L. and Atasn, M.Z. (1981) Biochem. J. *197*, 507 – 510.
Kino, K., Tsonoo, H., Higa, Y., Takami, M., Hamaguchi, H. and Nakajima, H. (1980) J. Biol. Chem. *255*, 9616 – 9620.
Kisilevsky, R., Benson, M.D., Axelrad, M.A. and Boudreau, L. (1979) Lab. Invest. *41*, 206 – 210.
Klempner, M.S., Dinarello, C.A. and Gallin, J.I. (1978) J. Clin. Invest. *61*, 1330 – 1336.
Kluft, C., Vellenga, E. and Brommer, E.J.P. (1979) Lancet *ii*, 206 – 207.
Kluger, M.J. (1979) Fed. Proc. *38*, 30 – 34.
Kluger, M.J. (1981) in Infection: The Physiologic and Metabolic Responses of the Host (Powanda, M.C. and Canonico, P.G., eds.), pp. 75 – 95, Elsevier/North-Holland Biomedical Press Amsterdam.
Kluger, M.J. and Rothenburg, B.A. (1979) Science *203*, 374 – 376.
Knowles, B.B., Howe, C.C. and Aden, D.P. (1980) Science *209*, 497 – 499.
Koide, T., Odani, S., Takahashi, K., Ono, T. and Sakuragawa, N. (1984) Proc. Natl. Acad. Sci. U.S.A. *81*, 289 – 293.
Koj, A. (1968) Biochim. Biophys. Acta *165*, 97 – 107.
Koj, A. (1970) Folia Biol. (Krakow) *18*, 275 – 286.
Koj, A. (1972) Abstracts of 8th FEBS Meeting, Amsterdam, No. 916.
Koj, A. (1974) in Structure and Function of Plasma Proteins, Vol. 1, (Allison, A.C., ed.), pp. 73 – 131, Plenum Press, London and New York.

Koj, A. (1980) Br. J. Exp. Pathol. *61*, 332 – 338.

Koj, A. (1984) in Pathophysiology of Plasma Protein Metabolism (Mariani, G., ed.), pp. 221 – 248, Macmillan, London.

Koj, A. and Dubin, A. (1976) Br. J. Exp. Pathol. *57*, 733 – 741.

Koj, A. and McFarlane, A.S. (1968) Biochem. J. *108*, 137 – 146.

Koj, A. and Regoeczi, E. (1978) Br. J. Exp. Pathol. *59*, 473 – 481.

Koj, A. and Regoeczi, E. (1981) Int. J. Peptide Protein Res. *17*, 519 – 526.

Koj, A., Hatton, M.W.C., Wong, K.L. and Regoeczi, E. (1978a) Biochem. J. *169*, 589 – 596.

Koj, A., Regoeczi, E., Toews, C.J., Leveille, R. and Gauldie, J. (1978b) Biochim. Biophys. Acta *539*, 496 – 504.

Koj, A., Dubin, A., Kasperczyk, H., Bereta, J. and Gordon, A.H. (1982) Biochem. J. *206*, 545 – 553.

Koj, A., Dubin, A., Kasperczyk, H., Kaczmarski, F. and Stankiewicz, D. (1983) Folia Histochem. Cytochem. *21*, 211 – 218.

Koj, A., Gauldie, J., Regoeczi, E., Sauder, D.N. and Sweeney, G.D. (1984a) Biochem. J., *224*, 505 – 514.

Koj, A., Regoeczi, E., Chindemi, P.A. and Gauldie, J. (1984b) Br. J. Exp. Pathol. *65*, 691 – 700.

Koj, A., Gauldie, J., Sweeney, G.D., Regoeczi, E. and Sauder, D.N. (1985) J. Immunol. Methods *76*, 317 – 328.

Konijn, A.M., Carmel, N., Levy, R. and Herschko, C. (1981) Br. J. Haematol. *49*, 361 – 370.

Koo, P.H. (1982) Cancer Res. *42*, 1788 – 1797.

Kopeć, M., Roszkowski, B., Gerdin, B. and Saldeen, T. (1982) in Fibrinogen – Recent Biochemical and Medical Aspects (Henschen, A., Graeff, H. and Lottspeich, F., eds.), pp. 355 – 360, Walter de Gruyter and Co., Berlin and New York.

Krauss, S. and Sarcione, E.J. (1966) Proc. Soc. Exp. Biol. Med. *122*, 1019 – 1022.

Ku, C.S.L. and Fiedel, B.A. (1983) J. Exp. Med. *158*, 767 – 780.

Kudryk, B., Okada, M., Redman, C.M. and Blomback, B. (1982) Eur. J. Biochem. *125*, 673 – 682.

Kueppers, F. and Black, L.F. (1974) Am. Rev. Respir. Dis. *110*, 176 – 194.

Kueppers, F. and Mills, J. (1983) Science *219*, 182 – 184.

Kukral, J.C., Zeineh, R., Dobryszycka, W., Pollitt, J. and Stone, N. (1969) Clin. Sci. *36*, 221 – 230.

Kurdowska, A., Koj, A. and Jaśkowska, A. (1982) Acta Biochim. Polon. *29*, 95 – 103.

Kurosky, A., Barnett, D.R., Lee, T., Touchstone, B., Hay, R.E., Arnott, M.S., Bowmann, B.H. and Fitch, W.M. (1980) Proc. Natl. Acad. Sci. U.S.A. *77*, 3388 – 3392.

Kushner, I. (1982) Ann. N.Y. Acad. Sci. *389*, 39 – 48.

Kushner, I. and Feldmann, G. (1978) J. Exp. Med. *148*, 466 – 477.

Kushner, I. and Kaplan, M.H. (1961) J. Exp. Med. *114*, 961.

Kushner, I., Edgington, T.S., Trimble, C., Siem, H.H. and Müller-Eberhard, U. (1972) J. Lab. Clin. Med. *80*, 18 – 25.

Kushner, I., Broder, M.L. and Karp, D. (1978) J. Clin. Invest. *61*, 235 – 242.

Kushner, I., Gewurz, H. and Benson, M.D. (1981) J. Lab. Clin. Med. *97*, 739 – 749.

Kwan, S.W. and Fuller, G.M. (1977) Biochim. Biophys. Acta *475*, 659 – 668.

Lachman, L.B., Atkins, E. and Kampschmidt, R.F. (1983) in Interleukins, Lymphokines and Cytokines (Oppenheim, J.J., Cohen, S. and Landy, M., eds.), pp. 441 – 446, Academic Press, New York and London.

Langstaff, J.M. and Burton, D.N. (1982) Can. J. Biochem. *60*, 712 – 770.

Lawn, R.M., Adelman, J., Dull, T.J., Gross, M., Goeddel, D.V. and Ullrich, A. (1981) Science *212*, 1159 – 1162.

Le, P.T., Muller, M.T. and Mortensen, R.F. (1982) J. Immunol. *129*, 665 – 672.

Lebreton de Vonne, T., Gutman, N. and Mouray, H. (1970) Clin. Chim. Acta *30*, 603 – 607.

Lebreton, J.P., Joisel, F., Raoult, J.P., Lannuzel, B., Rogez, J.P. and Humbert, G. (1979) J. Clin. In-

vest. *64*, 1118 – 1129.
Leicht, M., Long, G.L., Chandra, T., Kurachi, K., Kidd, V.J., Mace, Jr., M., Davie, E.W. and Woo, S.L.C. (1982) Nature (London) *297*, 655 – 659.
Lemonnier, F., Gautier, M. and Nguyen-Dingh, F. (1979) Biochimie *61*, 483 – 486.
Leonard, W.J., Depper, J.M., Robb, R.J., Waldmann, T.A. and Greene, W.C. (1983) Proc. Natl. Acad. Sci. U.S.A. *80*, 6957 – 6961.
Levin, J. and Bang, F.B. (1968) Thromb. Diath. Haemorrh. *19*, 186 – 197.
Lewis, D.A. (1977) Biochem. Pharmacol. *26*, 693 – 698.
Li, J.J., McAdam, K.P.W.J. and Bausserman, L.L. (1982) Ann. N.Y. Acad. Sci. *389*, 456.
Limaos, E.A., Borges, D.R., Souza-Pinto, J.C., Gordon, A.H. and Prado, J.L. (1981) Br. J. Exp. Pathol. *62*, 591 – 594.
Linder, M.C. and Moor, J.R. (1977) Biochim. Biophys. Acta *499*, 329 – 336.
Lipinski, B., Hawiger, J. and Jeljaszewicz, J. (1967) J. Exp. Med. *126*, 979 – 998.
Liu, A.Y. and Neuhaus, O.W. (1968) Biochim. Biophys. Acta *166*, 195 – 204.
Liu, T.Y., Robey, F.A. and Wang, C.M. (1982) Ann. N.Y. Acad. Sci. *389*, 151 – 162.
Lombart, C., Nebut, M., Ollier, M.P., Jayle, M.F. and Hartmann, L. (1968) Rev. Franc. Etud. Clin. Biol. *13*, 258.
Lorand, L. (ed.) (1981) Methods Enzymol. *80*, 919.
Loustad, R.A. (1981) Int. J. Biochem. *13*, 221 – 224.
Lowe, M.A. and Ashwell, G.A. (1982) Arch. Biochem. Biophys. *216*, 704 – 710.
Luger, T.A., Stadler, B.M., Katz, S.I. and Oppenheim, J.J. (1981) J. Immunol. *127*, 1493 – 1498.
Luger, T.A., Sztein, M.B., Charon, J.A. and Oppenheim, J.J. (1983) in Interleukins, Lymphokines and Cytokines (Oppenheim, J.J., Cohen, S. and Landy, M., eds.), pp. 447 – 454, Academic Press, New York and London.
Lustbader, J.W., Arcoleo, J.P., Birken, S. and Greer, J. (1983) J. Biol. Chem. *258*, 1227 – 1234.
Lynch, R.E. and Fridovich, I. (1978) J. Biol. Chem. *253*, 1838 – 1845.
Macartney, H.W. and Tschesche, H. (1983) Eur. J. Biochem. *130*, 85 – 92.
Macintyre, S.S., Schultz, D. and Kushner, I. (1982) Ann. N.Y. Acad. Sci. *389*, 76 – 87,
Macintyre, S.S., Kushner, I. and Samols, D. (1985) J. Biol. Chem. *260*, in press.
Mannick, J.A., Constantin, M., Pardrige, D., Saporoschetz, I. and Badger, A. (1977) Cancer Res. *37*, 3066.
Marciniak, E., Farley, C.H. and de Simone, P.A. (1974) Blood *43*, 219 – 224.
Marston, F.A.O., Dickson, A.J. and Pogson, C.I. (1981) Mol. Cell Biochem. *34*, 59 – 64.
Marx, J., Hilbig, R. and Rahmann, H. (1984) Compr. Biochem. Physiol. *77*, 483 – 488.
Matheson, N.R., Janoff, A. and Travis, J. (1982) Moll. Cell Biol. *45*, 65 – 71.
Matsuda, M., Wakabayashi, K., Aoki, N. and Marioka, Y. (1980) Thromb. Res. *17*, 527 – 532.
Matsumoto, M., Tsuda, M., Kusumi, T., Takada, S., Shimamura, T. and Katsunuma, T. (1981) Biochem. Biophys. Res. Commun. *100*, 478 – 482.
Matsumoto, M., Yamamura, M., Tsuda, M., Takada, S. and Katsunuma, T. (1982) J. Biochem. (Tokyo) *92*, 1979 – 1983.
Mbikay, M. and Garrick, M.D. (1981) Can. J. Biochem. *59*, 321 – 327.
McAdam, K.P.W.J., Elin, R.J., Sipe, J.D. and Wolff, S.M. (1978) J. Clin. Invest. *61*, 390 – 394.
McAdam, K.P.W.J., Li, J., Knowles, J., Foss, N.T., Dinarello, C.A., Rosenwasser, L.J., Sclinger, M.J., Kaplan, M.M. and Goodman, R. (1982) Ann. N.Y. Acad. Sci. *389*, 126 – 136.
McDuffie, F.G., Giffin, C., Niedringhaus, R., Mann, K.G., Owen, Jr., C.A., Bowie, E.J.W., Peterson, J., Clark, G. and Hunder, G.C. (1979) Thromb. Res. *16*, 759 – 773.
McFarlane, A.S. (1964) in Mammalian Protein Metabolism (Munro, H.N. and Allison, J.B., eds.), pp. 298 – 342, Vol. 1, Academic Press, New York.
McFarlane, A.S. and Koj, A. (1970) J. Clin. Invest. *49*, 1903 – 1911.

Meier, H. and MacPike, A.D. (1968) Proc. Soc. Exp. Biol. Med. *128*, 1185 – 1190.
Merriman, C.R., Upchurch, H.F. and Kampschmidt, R.F. (1978) Proc. Soc. Exp. Biol. Med. *157*, 669 – 671.
Misumi, Y., Tanaka, Y. and Ikehara, Y. (1981) Biochem. Biophys. Res. Commun. *114*, 729 – 736.
Moguilevsky, N., Retegui, L.A., Courtoy, P.J., Castracane, C.E. and Masson, P.L. (1984) Lab. Invest. *50*, 323 – 328.
Mold, C., Du Clos, T.W., Nakayama, S., Edwards, K. and Gewurz, H. (1982) Ann. N.Y. Acad. Sci. *389*, 251 – 262.
Montreuil, J. (1980) Adv. Carbohydr. Chem. Biochem. *37*, 157 – 223.
Morales, M.H. and Papaconstantinou, J. (1982) Arch. Biochem. Biophys. *218*, 592 – 602.
Morii, M. and Travis, J. (1983) Biochem. Biophys. Res. Commun. *111*, 438 – 443.
Moroi, M. and Aoki, N. (1976) J. Biol. Chem. *251*, 5956 – 5965.
Moroi, M. and Aoki, N. (1977) J. Biochem. (Tokyo) *82*, 969 – 972.
Morris, K.M., Aden, D.P., Knowles, B.B. and Colten, H.R. (1982) J. Clin. Invest. *70*, 906 – 913.
Mortensen, R.F., Osmand, A.P. and Gewurz, H. (1975) J. Exp. Med. *141*, 821 – 839.
Mortensen, S.B., Sottrup-Jensen, L., Hansen, H.F., Rider, D., Petersen, T.E. and Magnusson, S. (1981) FEBS Lett. *129*, 314 – 317.
Mosesson, M.W. and Doolittle, R.F. (eds.) (1983) Ann. N.Y. Acad. Sci., 408.
Movat, H.Z. (1985) The Inflammatory Reaction, Elsevier Science Publishers (Biomedical Division), in press.
Müller, W.E., Stillbauer, A.E. and El-Gamal, S. (1983) J. Pharm. Pharmacol. *35*, 684 – 686.
Müller-Berghaus, G. (1969) Thromb. Diath. Haemorrh. Suppl. *36*, 45 – 62.
Müllertz, S. and Clemensen, I. (1976) Biochem. J. *159*, 545 – 553.
Munck, A., Guyre, P.M. and Holbrook, N.J. (1984) Endocrine Rev. *5*, 25 – 44.
Munro, H.N. (1969) in Mammalian Protein Metabolism (Munro, H.N., ed.), Vol. 3, pp. 133 – 182, Academic Press, New York.
Murphy, P.A., Cebula, T.A., Levin, J. and Windle, B.E. (1981) Infect. Immun. *34*, 177 – 183.
Nagashima, M., Urban, J. and Schreiber, G. (1980) J. Biol. Chem. *255*, 4951 – 4956.
Nakagawa, H., Watanabe, K., Shuto, K. and Tsurufuji, S. (1983) Biochem. Pharmacol. *32*, 1191 – 1196.
Nakagawa, H., Watanabe, K. and Tsurufuji, S. (1984) Biochem. Pharmacol. *33*, 1181 – 1186.
Nakamura, S. and Levin, J. (1982) Biochem. Biophys. Res. Commun. *108*, 1619 – 1623.
Nakamura, S., Takagi, T., Iwanaga, S., Niwa, M. and Takahashi, K. (1976) J. Biochem. (Tokyo) *80*, 649 – 652.
Nelles, L.P. and Schnebli, H.P. (1982) Hoppe-Seyler's Z. Physiol. Chem. *363*, 677 – 682.
Nicollet, I., Lebreton, J.P., Fontaine, M. and Hiron, M. (1981) Biochim. Biophys. Acta *668*, 235 – 245.
Nicollet, I., Lebreton, J.P., Fontaine, M. and Hiron, M. (1982) in Lectins – Biology, Biochemistry, Clinical Chemistry (Bøg-Hansen, T.C., ed.),Vol. 2, pp. 413 – 421, Walter de Gruyter and Co., Berlin and New York.
Nieuwenhuizen, W., Emeis, J.J., Vermond, A., Kurver, P. and van der Heide, D. (1980) Biochem. Biophys. Res. Commun. *97*, 49 – 55.
Nilsson, T. and Wiman, B. (1982) Biochim. Biophys. Acta *705*, 271 – 276.
Noda, C., Shinjyo, F., Nakamura, T. and Ichihara, A. (1983) J. Biochem. (Tokyo) *93*, 1677 – 1684.
Northemann, W., Andus, T., Gross, V., Nagashima, M., Schreiber, G. and Heinrich, P.C. (1983) FEBS Lett. *161*, 319 – 322.
Nosslin, B. (1973) Ciba Foundation Symp. *9*, 113 – 127.
Noyes, W.D. and Garby, L. (1967) Scand, J. Clin. Lab. Invest. *20*, 33 – 38.
Ohlsson, K. (1971) Biochim. Biophys. Acta *236*, 84 – 91.
Ohlsson, K. (1974) in Proteinase Inhibitors (Fritz, H., Tschesche, V., Greene, L.J. and Truscheit, E.,

eds.), pp. 96 – 105, Bayer Symposium V, Springer-Verlag, New York.
Ohlsson, K. (1975) Ann. N.Y. Acad. Sci. *256*, 409 – 419.
Ohlsson, K. (1978) in Neutral Proteases of Human Polymorphonuclear Leucocytes (Havemann, K. and Janoff, A., eds.), pp. 167 – 177, Urban and Schwarzenberg, Baltimore and Munich.
Ohlsson, K. and Akesson, U. (1976) Clin. Chim. Acta *73*, 285 – 292.
Okubo, H. and Chandler, M. (1974) Proc. Soc. Exp. Biol. Med. *146*, 1159 – 1162.
Olden, K., Bernard, B.A., White, S.L. and Parent, J.B. (1982) J. Cell Biochem. *18*, 313 – 335.
Oliveira, E.B., Gotschlich, E.C. and Liu, T. (1979) J. Biol. Chem. *254*, 489 – 502.
Olson, J.P., Miller, L.L. and Troup, S.B. (1966) J. Clin. Invest. *45*, 690 – 701.
Oppenheim, J.J. and Gery, I. (1982) Immunology Today *3*, 113 – 119.
Osmand, A.P., Friedenson, B., Gewurz, H., Painter, R.H., Hofmann, T. and Shelton, E. (1977) Proc. Natl. Acad. Sci. U.S.A. *74*, 739 – 743.
Owen, M.C., Brennan, S.O., Lewis, J.H. and Carrell, R.W. (1983) N. Engl. J. Med. *309*, 694 – 698.
Owen, M.R. and Miller, L.L (1980) Biochim. Biophys. Acta *627*, 30 – 39.
Pagano, M., Nicola, M.A. and Engler, R. (1982) Can. J. Biochem. *60*, 631 – 637.
Peacock, A.C., Gelderman, A.H., Ragland, R.H. and Hoffman, H.A. (1967) Science *158*, 1703 – 1704.
Pekala, P.H., Kawakami, M., Angus, C.W., Lane, M.D. and Cerami, A. (1983) Proc. Natl. Acad. Sci. U.S.A. *80*, 2743 – 2747.
Pekarek, R.S., Wannemacher, R.W. and Beisel, W.R. (1972a) Proc. Soc. Exp. Biol. Med. *140*, 685 – 688.
Pekarek, R.S., Powanda, M.C. and Wannemacher, Jr., R.W. (1972b) Proc. Soc. Exp. Biol. Med. *141*, 1029 – 1031.
Pekarek, R.S., Wannemacher, R., Powanda, M., Abeles, F., Mosher, D., Dinterman, R. and Beisel, W. (1974) Life Sci. *14*, 1765 – 1776.
Pellegrini, A. and von Fellenberg, R. (1980) Biochim. Biophys. Acta *616*, 351 – 361.
Pepys, M.B. (1979) in The Science and Practice of Clinical Medicine: Rheumatology and Immunology (Cohen, A.S., ed.), pp. 85 – 88, Grune and Stratton, New York.
Pepys, M.B. (1981a) Lancet *i*, 653 – 657.
Pepys, M.B. (1981b) Clin. Immun. Allergy *1*, 77 – 101.
Pepys, M.B. and Baltz, M.L. (1983) Adv. Immunol. *34*, 141 – 242.
Pepys, M.B., Dash, A.C., Munn, E.A., Feinstein, A., Skinner, M., Cohen, A.S., Gewurz, H., Osmand, A.P. and Painter, R.H. (1977) Lancet *i*, 1029 – 1031.
Pepys, M.B., Dash, A.C., Fletcher, T.C., Richardson, N., Munn, E.A. and Feinstein, A. (1978) Nature (London) *273*, 168 – 170.
Pepys, M.B., Baltz, M.L., Gomer, K., Davies, A.J.S. and Doenhoff, M. (1979a) Nature (London) *278*, 259 – 261.
Pepys, M.B., Dyck, R.F., de Beer, F.C., Skinner, M. and Cohen, A.S. (1979b) Clin. Exp. Immunol. *38*, 284 – 293.
Pepys, M.B., Becker, G.J., Dyck, R.F., McCraw, A., Hilgard, P., Merten, R.F. and Thomas, D.P. (1980) Clin. Chim. Acta *105*, 83 – 91.
Piafsky, K.M., Borga, O., Odar-Cederlof, I., Johansson, C. and Sjoquist, F. (1978) N. Engl. J. Med. *299*, 1435 – 1439.
Piccoletti, R., Aletta, M.G., Cajone, F. and Bernelli-Zazzera, A. (1984) Br. J. Exp. Pathol. *65*, 419 – 430.
Pindyck, J., Beuving, G., Hertzberg, K.M., Liang, T.J., Amrani, D. and Grieninger, G. (1983) Ann. N.Y. Acad. Sci. *408*, 660 – 661.
Plant, P., Martini, O., Koch, G. and Grieninger, G. (1983) Ann. N.Y. Acad. Sci. *408*, 662 – 663.
Plonka, A. and Metodieva, D. (1979) Biochem. Biophys. Res. Commun. *95*, 978 – 984.
Plow, E.F., Edgington, T.S. and Cierniewski, C.S. (1983) Ann. N.Y. Acad. Sci. *408*, 45 – 59.

Pontet, M., D'Asniers, M., Gache, D., Escaig, J. and Engler, R. (1981) Biochim. Biophys. Acta *671*, 202–210.
Poole, S., Gordon, A.H., Baltz, M. and Stenning, B.M. (1984) Br. J. Exp. Pathol. *65*, 431–440.
Potempa, L.A., Siegel, J.N. and Gewurz, H. (1982) Ann. N.Y. Acad. Sci. *389*, 461–462.
Powanda, M.C., Cockerell, G.L. and Pekarek, R.S. (1973) Am. J. Physiol. *225*, 399–401.
Preston, J., Kushner, I., Schultz, D., Mahajan, D. and Macintyre, S. (1983) Fed. Proc. *42*, 421 (Abstract 753).
Princen, J.M.G. (1983) Synthesis of fibrinogen and albumin under physiological circumstances and after tissue injury. Thesis, University of Nijmegen.
Princen, J.M.G., Nieuwenhuizen, W., Mol-Backx, G.P.B.M. and Yap, S.H. (1981) Biochem. Biophys. Res. Commun. *102*, 717–723.
Procyk, R., Devlin, T.M. and Chih, J.J. (1979) Biochem. J. *178*, 501–504.
Putnam, F.W. (ed.) (1975) The Plasma Proteins, Vol. 1, Academic Press, New York – San Francisco – London.
Quigley, J.P. and Armstrong, P.B. (1983) J. Biol. Chem. *258*, 7903–7906.
Rangei, G., Bensi, G., Colantuoni, V., Romano, V., Santoro, C., Constanzo, F. and Cortese, R. (1983) Nucleic Acid. Res. *11*, 5811–5819.
Redman, C.M., Yu, S., Bannerjee, D. and Morris, H.P. (1979) Cancer Res. *39*, 101–111.
Reeve, E.B., Takeda, Y. and Atencio, A.C. (1966) Prot. Biol. Fluids *14*, 283–294.
Regoeczi, E. (1970) Clin. Sci. *38*, 111–121.
Regoeczi, E. (1974) in Structure and Function of Plasma Proteins (Allison, A.C., ed.), Vol. 1, pp. 133–167, Plenum Press, London and New York.
Regoeczi, E. and Hatton, M.W.C. (1980) Am. J. Physiol. *238*, R306–R310.
Regoeczi, E., Henley, G.E., Holloway, R.C. and McFarlane, A.S. (1963) Br. J. Exp. Pathol. *44*, 397–403.
Regoeczi, E., Hatton, M.W.C. and Wong, K.L. (1974) Can. J. Biochem. *52*, 155–161.
Regoeczi, E., Koj, A. and Lam, L.S.L. (1980) Biochem. J. *192*, 929–934.
Revel, M. and Groner, Y. (1978) Ann. Rev. Biochem. *47*, 1079–1126.
Reynolds, W.W. and Casterlin, M.E. (1982) in Pyretics and Antipyretics (Milton, A.S., ed.), pp. 649–668, Springer-Verlag, Berlin.
Ricca, G.A., Hamilton, R.W., McLean, J.W., Conn, A., Kalinyak, J.E. and Taylor, J.M. (1981) J. Biol. Chem. *256*, 10362–10368.
Ritchie, D.G. and Fuller, G.M. (1981) Inflammation *5*, 275–287.
Ritchie, D.G. and Fuller, G.M. (1983) Ann. N.Y. Acad. Sci. *408*, 491–502.
Ritchie, D.G., Levy, B.A., Adams, M.A. and Fuller, G.M. (1982) Proc. Natl. Acad. Sci. U.S.A. *79*, 1530–1534.
Roberts, Jr., N.J. (1979) Microbiol. Rev. *43*, 241–259.
Roberts, Jr., N.J. and Steigbigel, R.T. (1977) Infect. Immun. *18*, 673–679.
Robey, R.A. and Liu, T.Y. (1981) J. Biol. Chem. *256*, 969–975.
Robey, F.A., Tanaka, T. and Liu, T.Y. (1983) J. Biol. Chem. *258*, 3889–3894.
Robey, F.A., Jones, K.D., Tanaka, T. and Liu, T.Y. (1984) J. Biol. Chem. *259*, 7311–7316.
Robinson, G.B. (1969) Biochem. J. *114*, 635–640.
Rogers, J., Kalsheker, N., Wallis, S., Speer, A., Coutelle, C.H., Woods, D. and Humphries, S.E. (1983) Biochem. Biophys. Res. Commun. *116*, 375–382.
Roszkowski, W., Stachurska, J., Gerdin, B., Saldeen, T. and Kopeć, M. (1981) Eur. J. Cancer *17*, 889–892.
Rothschild, M.A. and Waldmann, T. (eds.) (1970) Plasma Protein Metabolism, Academic Press, New York and London.
Ruhenstroth-Bauer, G., Scherer, R., Hornberger, M. and Tongendorf, G. (1981) Inflammation *5*, 343–351.

Rupniak, H.T. and Paul, D. (1978) Biochim. Biophys. Acta *543*, 10 – 15.
Ryley, H.C. (1979) Biochem. Biophys. Res. Commun. *89*, 871 – 878.
Sacks, T., Moldow, C.F., Craddock, P.R., Bowers, T.K. and Jacob, H.S. (1978) J. Clin. Invest. *62*, 1161 – 1167.
Saeed, S.A., McDonald-Gibson, W.J., Cuthbert, J., Copas, J.L., Schneider, C., Gardiner, P.J., Butt, N.M. and Collier, H.J. (1977) Nature (London) *270*, 32 – 36.
Sala-Trepat, J.M., Dever, J., Sargent, T.D., Thomas, K., Sell, S. and Bonner, J. (1979) Biochemistry *18*, 2167 – 2178.
Saldeen, T. (1983) Ann. N.Y. Acad. Sci. *408*, 424 – 437.
Salvesen, G.S. and Barrett, A.J. (1980) Biochem. J. *187*, 501 – 695.
Salvesen, G.S., Sayers, C.A. and Barrett, A.J. (1981) Biochem. J. *195*, 453 – 461.
Samak, R., Edelstein, R. and Israel, L. (1982) Cancer Immunol. Immunother. *13*, 38 – 43.
Sanders, K.D. and Fuller, G.M. (1983) Thromb. Res. *32*, 133 – 145.
Sarcione, E.J. and Smalley, J.R. (1976) Cancer Res. *36*, 3203 – 3206.
Sas, G., Blaskó, G., Bánhegyi, D, Jákó, J. and Pálos, L.V. (1974) Thromb. Diath. Haemorrh. *32*, 105 – 111.
Sauder, D.N., Carter, C.S., Katz, S.I. and Oppenheim, J.J. (1982) J. Invest. Dermatol. *79*, 34 – 39.
Sauder, D.N., Mounessa, N.L., Katz, S.I., Dinarello, C.A. and Gallin, J.I. (1984) J. Immunol. *132*, 828 – 832.
Savu, L., Zouaghi, H. and Nunez, E.A. (1983) Biochem. Biophys. Res. Commun. *110*, 796 – 803.
Sayers, C.A. and Barrett, A.J. (1980) Biochem. J. *189*, 255 – 261.
Schapira, M., Silver, L.D., Scott, C.F. and Colman, R.W. (1982) Blood *59*, 719 – 724.
Scherer, R. and Ruhenstroth-Bauer, G. (1978) Blut *36*, 327 – 330.
Scherer, R., Abd-el-Fattah, M. and Ruhenstroth-Bauer, G. (1977) in Perspectives in Inflammation (Willoughby, D.A., Giroud, J.P. and Velo, G.P., eds.), pp. 437 – 444, MTP.
Schiaffonati, L., Bardella, L., Cairo, G., Giancotti, V. and Bernelli-Zazzera, A. (1984) Biochem. J. *219*, 165 – 171.
Schmidt, J.A., Mizel, S.B., Cohen, D. and Green, I. (1982) J. Immunol. *128*, 2177 – 2180.
Schreiber, G., Urban, J., Dryburgh, H. and Bradley, T.R. (1978) in Morris Hepatomas, Mechanism of Regulation (Morris, H.P. and Criss, W.E., eds.), pp. 565 – 582, Plenum Press, New York and London.
Schreiber, G., Dryburgh, D., Millership, A., Matsuda, Y., Inglis, A., Phillips, J., Edwards, K. and Maggs, J. (1979) J. Biol. Chem. *254*, 12013 – 12019.
Schreiber, G., Howlett, G., Nagashima, G., Millersip, A., Martin, H., Urban, J. and Kotler, L. (1982) J. Biol. Chem. *257*, 10271 – 10277.
Schultz, D., Macintyre, S., Chelladurai, M. and Kushner, I. (1982) Ann. N.Y. Acad. Sci. *389*, 465 – 466.
Schultze, H.E. and Heremans, J.F. (1966) Molecular Biology of Human Proteins with Special Reference to Plasma Proteins, Vol. 1, Elsevier Publ. Co., Amsterdam.
Schwartz, C.E., Gabryelak, T., Smith, C.J., Taylor, J.M. and Chiu, J.F. Biochem. Biophys. Res. Commun. *107*, 235 – 245.
Scott, C.F., Schapira, M., James, H.L. and Cohen, A.B. (1982) J. Clin. Invest. *69*, 844 – 852.
Sebag, J., Reed, W.P. and Williams, Jr., R.C. (1977) Infect. Immun. *16*, 947 – 954.
Selinger, M.J., McAdam, K.P.W.J., Kaplan, M.M., Sipe, J.D., Vogel, S.N. and Rosenstreich, D.L. (1980) Nature (London) 285, 498.
Sell, S., Thomas, K., Michalson, M., Sala-Trepat, J. and Bonner, J. (1979) Biochim. Biophys. Acta *564*, 173 – 178.
Selten, G.C.M., Selten-Versteegen, A.M.E. and Yap, S.H. (1981) Biochem. Biophys. Res. Commun. *103*, 278 – 284.
Shapira, E., Martin, C.L. and Nadler, H.L. (1977) J. Biol. Chem. *252*, 7923 – 7929.

Shibata, Z., Balegno, H.F. and Neuhaus, O.W. (1970) Biochem. Biophys. Res. Commun. *38*, 692 – 696.
Shim, B.S. (1976) Korean J. Biochem. *8*, 7 – 11.
Shutler, G.G., Langstaff, J.M., Jamieson, J.C. and Burton, D.N. (1977) Arch. Biochem. Biophys. *183*, 710 – 717.
Sim, R.B. and Sim, E. (1981) Biochem. J. *193*, 129 – 141.
Simon, P.L., Willoughby, J.B.W. and Willoughby, W.F. (1983) in Interleukins, Lymphokines and Cytokines (Oppenheim, J.J., Cohen, S. and Landy, M., eds.), pp. 487 – 494, Academic Press, New York.
Sipe, J.D. (1978) Br. J. Exp. Pathol. *59*, 305 – 310.
Sipe, J.D., Vogel, S.N., Ryan, J.L., McAdam, K.P.W.J. and Rosenstreich, D.L. (1979) J. Exp. Med. *150*, 597 – 606.
Sipe, J.D., Vogel, S.N., Sztein, M.B., Skinner, M. and Cohen, A.S. (1982) Ann. N.Y. Acad. Sci. *389*, 137 – 149.
Skinner, M., Sipe, J.D., Yood, R.H., Shirahama, T. and Cohen, A.S. (1982) Ann. N.Y. Acad. Sci. *389*, 180 – 198.
Smith, C.J.P. and Kelleher, P.C. (1980) Biochim. Biophys. Acta *605*, 1 – 32.
Smith, J.B., Knowlton, R.P. and Agarval, S.S. (1978) J. Immunol. *121*, 691 – 694.
Smith, S.J., Bos, G., Essevald, M.R., van Eijk, H.G. and Gerbrandy, J. (1977) Clin. Chim. Acta *81*, 75 – 86.
Snellman, O. and Sylven, B. (1974) Experientia *30*, 1114 – 1155.
Snyder, S. and Coodley, E.L. (1976) Arch. Int. Med. *136*, 778 – 781.
Sobocinski, P.Z. and Canterbury, Jr., W.J. (1982) Ann. N.Y. Acad. Sci. *389*, 354 – 367.
Solum, N.O. (1973) Thromb. Res. *2*, 55 – 70.
Sottrup-Jensen, L., Petersen, T.E. and Magnusson, S. (1981) FEBS Lett. *128*, 127 – 132.
Starkey, P.M. and Barrett, A.J. (1977) in Proteinases in Mammalian Cells and Tissues (Barrett, A.J., ed.), pp. 663 – 696, Elsevier/North-Holland Publ. Co., Amsterdam.
Starkey, P.M. and Barrett, A.J. (1982a) Biochem. J. *205*, 91 – 95.
Starkey, P.M. and Barrett, A.J. (1982b) Biochem. J. *205*, 105 – 115.
Starkey, P.M., Fletcher, T.C. and Barrett, A.J. (1982) Biochem. J. *205*, 97 – 104.
Staruch, M.J. and Wood, D.D. (1983) J. Immunol. *130*, 2191 – 2194.
Stead, N., Kaplan, A.P. and Rosenberg, R.D. (1976) J. Biol. Chem. *251*, 6481 – 6488.
Steinbuch, M. and Audran, R. (1974) in Proteinase Inhibitors, Bayer Symposium V (Fritz, H., Tschesche, H., Green, L.J. and Truscheit, E., eds.), pp. 78 – 95, Springer-Verlag, Berlin – Heidelberg – New York.
Struck, D.K., Siuta, P.B., Lane, M.D. and Lennarz, W.J. (1978) J. Biol. Chem. *253*, 5332 – 5337.
Sundsmo, J.S. and Fair, D.S. (1983) Clin. Physiol. Biochem. *1*, 225 – 284.
Suttie, J.W. and Jackson, C.M. (1977) Physiol. Rev. *57*, 1 – 70.
Sztein, M.B., Vogel, S.N., Sipe, J.D., Murphy, P.A., Mizel, S.B., Oppenheim, J.J. and Rosenstreich, D.L. (1981) Cell Immunol. *63*, 164 – 176.
Sztein, M.B., Luger, T.A. and Oppenheim, J.J. (1982) J. Immunol. *129*, 87 – 90.
Tai, J.Y. and Liu, T.Y. (1977) J. Biol. Chem. *252*, 2178 – 2181.
Tatsuta, E., Sipe, J.D., Shirahama, T., Skinner, M. and Cohen, A.S. (1983) J. Biol. Chem. *258*, 5414 – 5418.
Thompson, W.L., Abeles, F.B., Beall, F.A., Dinterman, R.E. and Wannemacher, Jr., R.W. (1976) Biochem. J. *156*, 25 – 32.
Tilghman, S.M. and Belayew, A. (1982) Proc. Natl. Acad. Sci. U.S.A. *79*, 5254 – 5257.
Tobias, P.S., McAdam, K.P.W.J. and Ulevitch, R.J. (1982) J. Immunol. *128*, 1420.
Travis, J. and Salvesen, G.S. (1983) Ann. Rev. Biochem. *52*, 665 – 709.
Travis, J., Bowen, J. and Baugh, R. (1978) Biochemistry *17*, 5651 – 5656.

Tsukada, K., Oura, H., Nakashima, S. and Hayasaki, N. (1968) Biochim. Biophys. Acta *165*, 218 – 224.
Tunstal, A.M., Merriman, J.M., Milne, I. and James, K. (1975) J. Clin. Pathol. *28*, 133 – 139.
Turchik, J.B. and Bornstein, D.L. (1980) Infect. Immun. *30*, 439 – 444.
Uhlenbruck, G., Sölter, J. and Janssen, E. (1981) J. Clin. Chem. Clin. Biochem. *19*, 1201 – 1208.
Uhlenbruck, G., Sölter, J., Janssen, E. and Haupt, H. (1982) Ann. N.Y. Acad. Sci. *389*, 476 – 479.
Ulevitch, R.J., Johnston, A.R. and Weinstein, D.B. (1981) J. Clin. Invest. *67*, 827.
Van Gool, J., Schreuder, J. and Ladiges, N.C.J.J. (1974) J. Pathol. *112*, 245 – 262.
Van Gool, J., Ladiges, N.C.J. and Boers, W. (1982) Inflammation *6*, 127 – 136.
Van Gool, J., Boers, W., Sala, M. and Ladiges, N.C.J.J. (1984) Biochem. J. *220*, 125 – 131.
Van Halbeek, H., Dorland, L., Vliegenthart, J.F.G., Montreuil, J., Fournet, B. and Schmid, K. (1981) J. Biol. Chem. *256*, 5588 – 5590.
Vannice, J.L., Taylor, J.M. and Ringold, G.M. (1984) Proc. Natl. Acad. Sci. U.S.A. *81*, 4241 – 4245.
Van Oss, C.J., Gillman, C.F., Bronson, P.M. and Border, J.R. (1974) Immunol. Commun. *3*, 321 – 328.
Van Snick, J.L., Masson, P.L. and Heremans, J.F. (1974) J. Exp. Med. *140*, 1068 – 1084.
Vaughan, L., Lorier, M.A. and Carrell, R.W. (1982) Biochim. Biophys. Acta *701*, 339 – 345.
Voelkel, E.F., Levine, L., Alper, C.A. and Tashjian, A.H. (1978) J. Exp. Med. *147*, 1078 – 1088.
Volanakis, J.E. (1982) Ann. N.Y. Acad. Sci. *389*, 235 – 250.
Wannemacher, Jr., R.W., Pekarek, R.S. and Beisel, W.R. (1972) Proc. Soc. Exp. Biol. Med. *139*, 128 – 132.
Wannemacher, Jr., R.W., Pekarek, R.S., Thompson, W.L., Curnow, R.T., Beall, F.A., Zenser, T.V., de Rubertis, F.R. and Beisel, W.R. (1975) Endocrinology *96*, 651 – 661.
Watson, J., Largen, M. and McAdam, K.P.W.J. (1978) J. Exp. Med. *147*, 39 – 49.
Weimer, H.E. and Coggshall, V. (1967) Can. J. Physiol. Pharmacol. *45*, 767 – 775.
Weimer, H.E., Roberts, D.M. and Comb, J.C. (1972) Br. J. Exp. Pathol. *53*, 253 – 257.
Weinberg, E.D. (1974) Science *184*, 952.
Weissmann, G. (ed.) (1980) The Cell Biology of Inflammation, Elsevier/North-Holland Biomedical Press, Amsterdam.
Weström, B.R., Karlsson, B.W. and Ohlsson, K. (1983) Hoppe-Seyler's Z. Physiol. Chem. *364*, 375 – 381.
Whaley, K. (1980) J. Exp. Med. *151*, 501 – 516.
Whicher, J.T., Martin, M.F.R. and Dieppe, P.A. (1980) Lancet *ii*, 1187.
Whicher, J.T., Bell, A.M., Martin, M.F.R., Marshall, L.A. and Dieppe, P.A. (1984) Clin. Sci. *66*, 165 – 171.
White, A., Fletcher, T.C., Pepys, M.B. and Baldo, B.A. (1981) Compr. Biochem. Physiol. *69C*, 325 – 329.
Williams, D.M., Lee, G.R. and Cartwright, G.E. (1974) Am. J. Physiol. *227*, 1094 – 1097.
Williams, R.C., Kilpatrick, K.A., Kassaby, M. and Abdin, Z.H. (1978) J. Clin. Invest. *59*, 1384 – 1393.
Wilson, G.B., Walker, J.H., Watkins, J.H. and Wolgroch, D. (1980) Proc. Soc. Exp. Biol. Med. *164*, 105 – 114.
Wolley, D.E., Roberts, D.R. and Evanson, J.M. (1976) Nature (London) *261*, 325 – 327.
Woloski, B.M.R.N.J., Kaplan, H.A. and Jamieson, J.C. (1983a) Compr. Biochem. Physiol. *74A*, 813 – 816.
Woloski, B.M.R.N.J., Kaplan, H.A., Gospodarek, E. and Jamieson, J.C. (1983b) Biochem. Biophys. Res. Commun. *112*, 14 – 19.
Yonehara, S., Yonehara-Takahashi, M., Ishii, A. and Nagata, S. (1983) J. Biol. Chem. *258*, 9046 – 9049.
Yother, J., Volanakis, J.E. and Briks, D.E. (1982) J. Immunol. *128*, 2374 – 2376.
Yu, S., Sher, B., Kudryk, B. and Redman, C.M. (1983) J. Biol. Chem. *258*, 13407 – 13410.
Zeineh, R.A. and Kukral, J.C. (1970) J. Trauma *10*, 493 – 498.
Zeineh, R.A., Barrett, B., Niemirowski, L. and Fiorella, B.J. (1972) Am. J. Physiol. *222*, 1326 – 1332.
Ziegler, I., Maier, K. and Fink, M. (1982) Cancer Res. *42*, 1567 – 1573.

PART IV

Clinical aspects of the acute-phase response

Gordon/Koj (eds.) The acute-phase response to injury and infection.

CHAPTER 21

Diagnostic and prognostic significance of the acute-phase proteins

A. FLECK and M.A. MYERS

1. Nature of the AP-response in man

The acute-phase response (AP-response) occurs with minor differences in a large variety of animal species and increasing attention is being given to the possible clinical applications of measurements of the changes in concentration of various plasma proteins which occur during it. Our present knowledge and understanding of the AP-response is an extension of earlier attempts to recognise changes in the electrophoresis patterns of serum proteins as being specific to certain diseases (Sunderman, 1964). One example is the 'pattern of acute infection' in which the main changes are reduced albumin and increased α_1-globulin fractions (Owen, 1967). Although it had been recognised for many years that the fractions obtained on simple electrophoresis of serum were composed of many proteins, it was only with the application of the convenient quantitative immunological methods of determining individual proteins that it has become possible to resolve further patterns of changes in individual proteins which occur during the AP-response in man. The phylogenetic aspects of the response have been reviewed by Pepys and Baltz (1983) and our aim in this chapter is to discuss the AP-response in man, with emphasis on possible clinical applications.

Since the discovery of C-reactive protein (CRP) in the serum of patients with pneumonia by Tillet and Francis in 1930, interest in the phenomenon of the AP-response gradually increased. The aim of some studies was to determine whether the changes of the AP-response might provide clinically useful information. An example is the study by Yocum and Doerner(1957) of 729 patients from which they concluded that CRP was a very good indicator of disease and recommended its use in preference to the Erythrocyte Sedimentation Rate (ESR). They found that CRP was always elevated in bacterial infections, acute rheumatic fever, acute myocardial infarction and widespread malignant disease. They also found that CRP was commonly but not consistently raised in rheumatoid arthritis, viral infections and active tuberculosis, but in contrast rarely elevated in limited primary carcinoma with no clinical evidence of metastases. In a subsequent series of clinical studies Hedlund (1961) found that CRP was elevated in every instance where elevations of temperature were associated with inflammation. Two years later, Belfrage (1963) expanded on these by measuring several other proteins in various diseases. From this study of more than 900 cases of acute infection of various types in which he found an increase in most instances of CRP, fibrinogen, haptoglobin, ESR and the α-globulin region of a protein electrophoresis strip, he concluded that CRP was the best indicator of infection, and that virus infections are the only common causes of fever which are not usually associated with an increase in CRP in the serum. There are now descriptions of the AP-response during a variety of infections. Examples include: typhoid fever (Bostian et al., 1976) in which the changes were stated to be similar to those following surgery (described below); fungal septicaemia in which CRP is measured (Kostiala, 1984); and the acute stage of malaria which is accompanied by increases in the α-globulins, α_1-antitrypsin (α_1-AT) and haptoglobin (Klainer et al., 1968; Murphy et al., 1971). In malaria there is an interesting difference from the usual pattern of the AP-response because the concentration of haptoglobin falls once haemolysis begins due to the haptoglobin-haemoglobin complex being cleared by the kidneys (Allison and Rees, 1957; Klainer et al., 1968). There are numerous descriptions of the AP-response after surgery and trauma. Werner and Odenthal (1967) and Werner (1969) measured several plasma proteins after operation and reported rises in the concentrations of CRP, α_1-AT, α_1-acid glycoprotein (α_1-AGP), haptoglobin and ceruloplasmin; decreases in the concentration of transferrin and albumin; and no change in that of α_2-macroglobulin (α_2-M).

It is possible to compare the changes in plasma proteins after cholecystectomy, reported by Aronsen et al. (1972) with those following myocardial infarction (Johannsson et al. 1972). The general patterns in both studies were similar and in agreement with those of Werner and Odenthal (1967), and Werner (1969) following surgical operation. In addition, Aronsen et al. (1972) observed that CRP and α_1-antichymotrypsin (α_1-ACh) both peaked early (2 days after injury) and best reflected the magnitude of injury. The maximum concentration of fibrinogen occurred at 3 days, that of haptoglobin and α_1-AGP at 5 days and of α_1-AT at $3-5$ days. Aronsen et al. (1972) also found that hemopexin, β_1-C-globulin, prothrombin, ceruloplasmin, GC-globulin, and α_1-easily-precipitable-glycoprotein, plasminogen and cold insoluble globulin were not very good indicators of the AP-response because the increases in concentration of these proteins were only moderate and were prolonged.

Many attempts have been made to confirm the impression of a 'dose-response relation' in the AP-response, i.e. that there is a direct relationship between degree of trauma or extent of tissue damage and the magnitude and duration of the changes in the plasma proteins.

The common minor inflammatory conditions such as gingivitis or minor cuts and bruises may cause minor elevations in CRP with peak concentration of about 10 mg/l (Morley and Kushner, 1982), which contrast with rises of over 200 mg/l commonly seen in more severe infections. In general, the more extensive surgical operations such as cholecystectomy seem to lead to greater increases in the AP-proteins than less extensive ones such as mastectomy (Aronsen et al., 1972). A similar conclusion was reached by Dominioni et al. (1980) who followed changes in CRP and α_1-AT after a variety of surgical operations. Studies of the magnitude of the AP-response after myocardial infarction do not present such a clear trend. Although generally higher levels of CRP were reported by Kushner et al. (1978) after extensive infarcts than after smaller ones, variation was considerable: one patient with a mild infarct having a peak CRP of 93 mg/1 while one with a severe infarct had a maximum CRP of 84 mg/1. It is therefore not surprising that others (Johannsson et al., 1972; Laurent, 1982) have concluded that the magnitude of the AP-response was not related to the magnitude of the myocardial infarction. In our studies of the AP-response after the relatively minor operation of herniorrhapy (Colley et al., 1983) we found wide differences in the peak concentration of CRP, the range being 38 – 196 mg/1.

At this stage we can summarise some facts about the AP-response in man:

(1) It occurs promptly (i.e. within 2 days) after trauma or cell damage, prior to the specific immune response;
(2) that it is non-specific; a wide variety of types of tissue damage and different infections promote a similar response;
(3) the many different AP-proteins in man (see Chapter 12) do not behave in exactly the same way and certain proteins (e.g. CRP, SAA) are much more responsive to tissue damage than others (i.e. α_1-AGP, caeruloplasmin).

2. *The sequence of events of the AP-response*

The pattern of the AP-response after elective surgery not preceded by infection is shown in Figs. 1 and 2. The earliest change is a rise in cortisol which peaks at 6 h, followed by a rise in the white blood cells, which peak at 10 h. There appears to be a delay of about 6 h before any rise in the AP-proteins occurs. CRP then rises very rapidly, from normal levels of less than 10 mg/l to levels above 100 mg/l, and peaks at about 48 h. The initial rise in fibrinogen and α_1-AGP also appears to be at about 6 h but the rate of increase is not as great as shown by CRP and the peak concentrations are at 96 and 120 h, respectively.

This pattern, with minor deviations, has been reported frequently (Aronsen et al., 1972; Johannsson et al., 1972; Fischer et al., 1976). It is therefore apparent that CRP is a good marker for injury as it rises to levels far above baseline and returns to normal rapidly, usually within 8 days. Fibrinogen and α_1-AGP do not return to normal levels until several days or even over a week after CRP has returned to baseline levels. The rise in the AP-proteins α_1-AGP and fibrinogen after surgery is preceded by a slight fall in concentration almost immediately after operation (Fig. 3).

Albumin and transferrin differ from many other plasma proteins, the concentration of which is unaffected by injury (Chapter 12, Table 2), in that their concentrations in the plasma decrease almost immediately after operation and remain depressed for several days after surgery (Fig. 4).

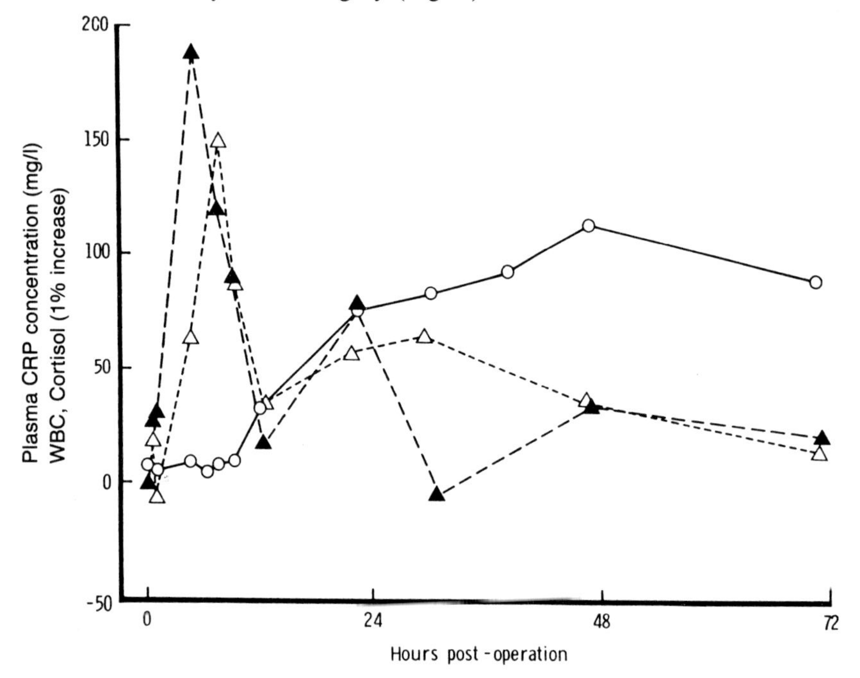

Fig. 1. Changes in cortisol white blood cell count (WBC) and CRP after herniorrhaphy. Figures are given as % increase from baseline levels at zero time for WBC and cortisol and in mg/l for CRP. ▲, Cortisol; △, WBC, ○, C-reactive protein.

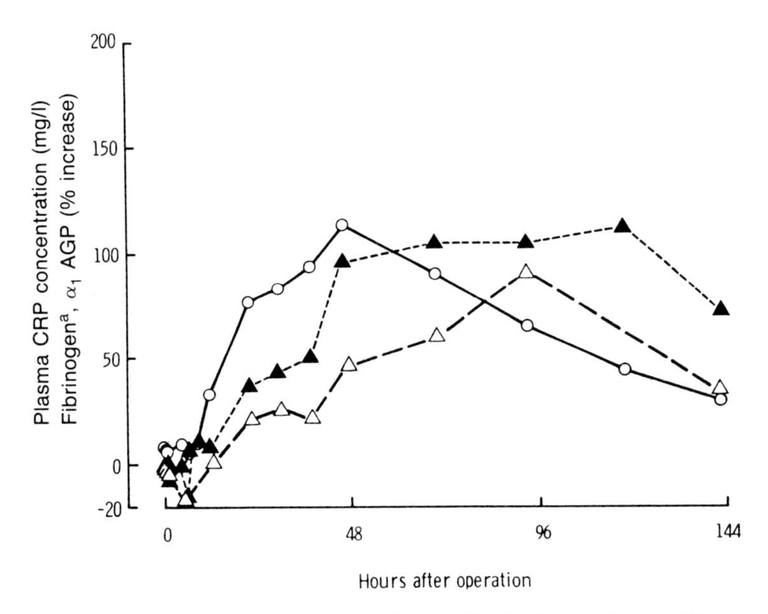

Fig. 2. Changes in CRP, fibrinogen and α_1-AGP after herniorrhaphy. Figures are given as % increase from baseline levels at zero time for fibrinogen and α_1-AGP and in mg/l for CRP. ○, C-reactive protein; ▲, α_1-AGP; △, plasma fibrinogen[a].

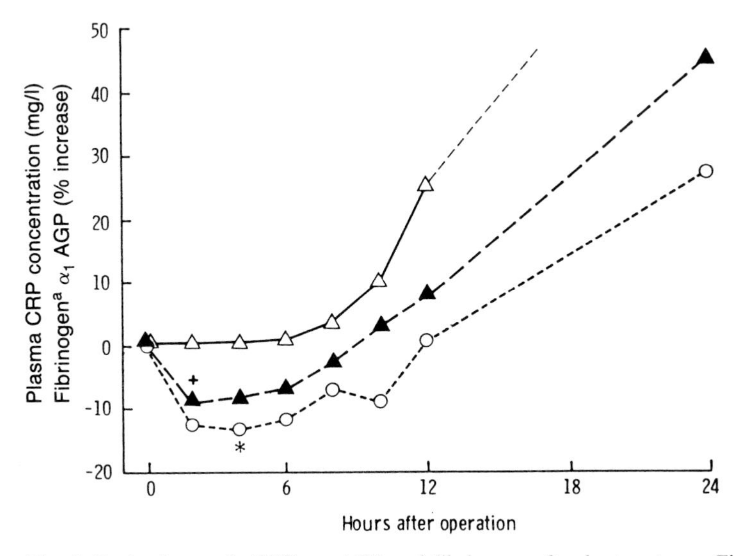

Fig. 3. Early changes in CRP, α_1-AGP and fibrinogen after hysterectomy. Figures are given as % increase from baseline levels at zero time, for α_1-AGP and fibrinogen and in mg/l for CRP. The concentration of CRP increased to 85 mg/l at 24 h. See text for explanation. △, C-reactive protein; ▲, α_1-AGP; ○, fibrinogen. The earliest significant fall in α_1-AGP (+ $p<0.001$) and plasma fibrinogen[a] (*$p<0.02$) is shown.

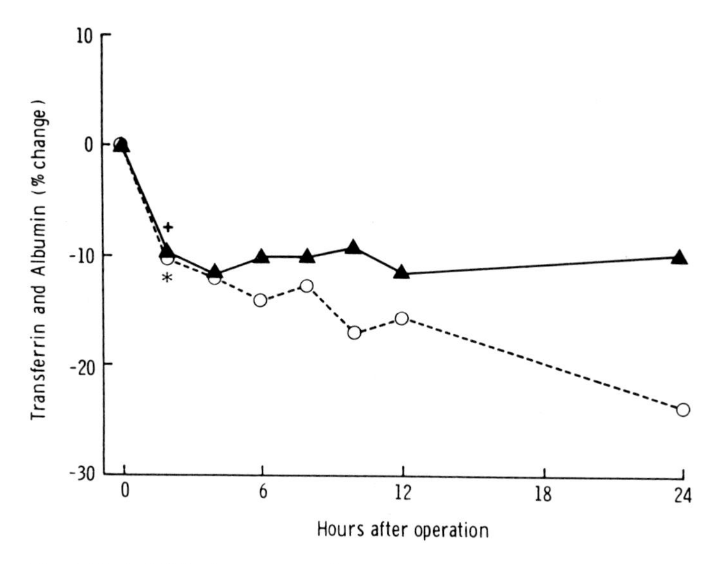

Fig. 4. Early fall in the concentrations of albumin and transferrin after hysterectomy. Figures are given as % decrease from baseline levels at zero time. ▲, Albumin; ○, transferrin. The first significant fall in concentration for albumin (+ $p<0.01$) and transferrin (*$p<0.001$) is shown.

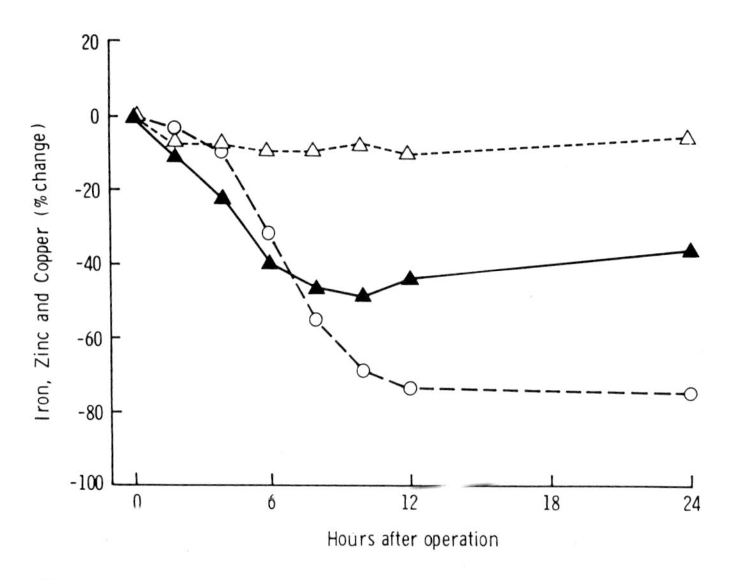

Fig. 5. Early changes in plasma iron, zinc and copper after hysterectomy. Figures are given as % change from baseline at zero time. △, Copper; ▲, zinc; ○, iron. The earliest significant fall in plasma zinc concentration occurred at 2 h ($p<0.05$) and that of iron at 6 h ($p<0.001$). See text for description of results.

The concentration of iron and zinc in plasma also decreases after operation (Fig. 5). It has been suggested that the fall in serum iron seen after injury is caused by a release of lactoferrin from neutrophils (Van Snick et al., 1974; Klempner et al., 1978) which complexes with iron from transferrin and this lactoferrin-iron complex is then taken up by the liver, spleen and reticuloendothelial system (Bennett and Kokocinski, 1979; Markowetz et al., 1979). However, it is possible that this pathway of iron exchange is more important at the site of injury, as it has been shown that the transfer of iron from transferrin to lactoferrin occurs optimally under acidic conditions which will exist at the site of injury and not in the general circulation (Van Snick et al., 1974). Much the greater proportion of the serum iron is bound to transferrin (Bothwell et al., 1979) and about 65% of plasma zinc is bound to albumin and 35% to macroglobulin (Giroux, 1975). The amount of zinc bound to α_2-M is constant, at about 1 mole zinc/mole of the protein and this zinc is metabolically unavailable (Song and Adham, 1979; Foote and Delves, 1984). Our observations (Figs. 6 and 7) are consistent with the initial fall in both metals being due to a decrease in the concentrations of their transport proteins, transferrin and albumin – due to an early increase in microvascular permeability (cf. p. 256). If the concentrations of iron and zinc are corrected for such changes in their transport proteins transferrin and albumin, by plotting the ratio of the metal to its transport protein it is seen that specific changes in the metals do not start until 6 h have elaps-

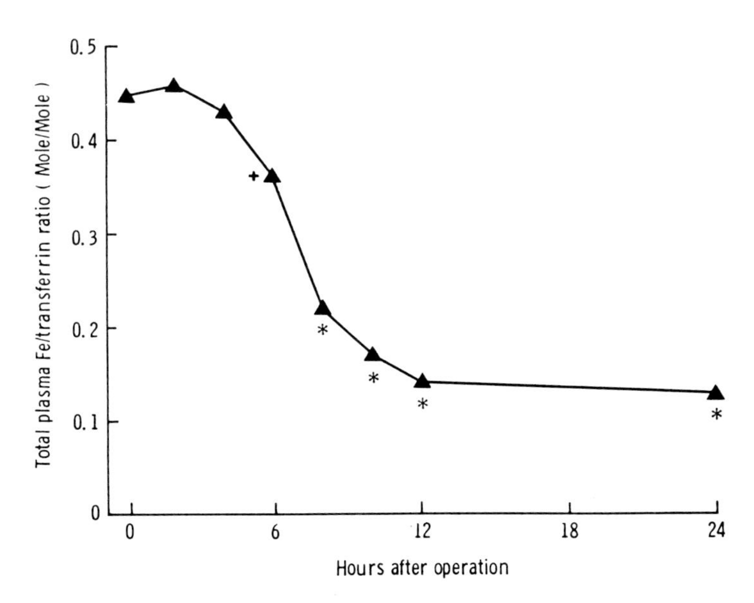

Fig. 6. Total plasma iron: transferrin ratio after hysterectomy. Significant differences from baseline levels at zero time are shown: + $p<0.05$; *$p<0.001$. Refer to text for explanation.

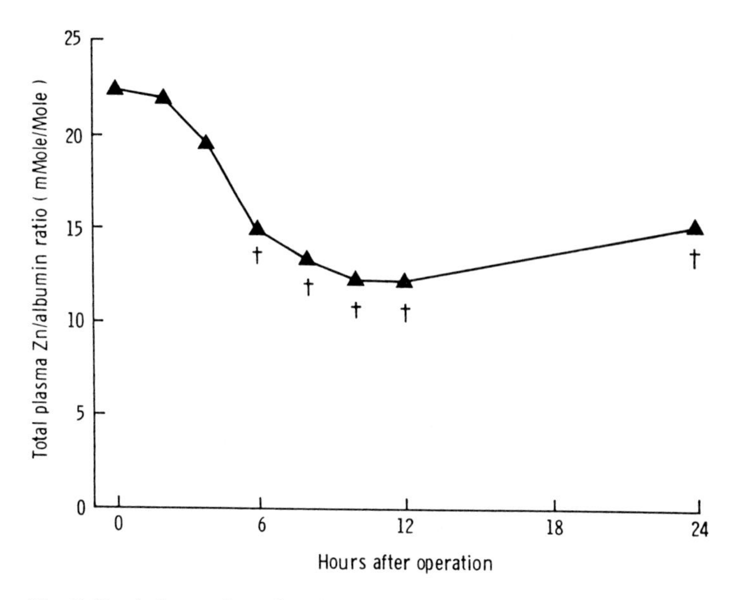

Fig. 7. Total plasma zinc: albumin ratio after hysterectomy. Significant differences from baseline levels at zero time are shown: $p<0.001$. Refer to text for explanation.

ed after the initial tissue damage. It has been shown that after interleukin-1 (IL-1) injection, iron and zinc accumulates predominantly in the liver (Pekarek et al., 1972) and it would appear that the liver takes up both metals after a delay similar to that seen with CRP, α_1-AGP and fibrinogen. In the case of iron, ferritin may be involved because its content in the liver has been found to be increased a few hours after injury (Konijn and Hershko, 1977) (cf. Chapter 13, Section 6).

The concentration of iron and zinc remains decreased for several days and the normal diurnal variation seen in these metals (Bothwell, 1979; McMillan and Rowe, 1982) disappears for at least 5 days. It is possible therefore that the stimulus for the increase in AP-proteins and hepatic uptake of zinc and iron could be the same, with both events occurring 6 h after the initial stimulus.

It is likely that the prompt fall in albumin and other plasma proteins after injury is due to a rapid general increase in microvascular permeability to albumin and to other plasma proteins of similar size (Fleck and Clark, 1980; Raines et al., 1984; Fleck, 1985). A measure of the permeability of small blood vessels to albumin can be obtained by determining the Transcapillary Escape Rate (TER) (Parving and Rasmussen, 1973), which is readily done by following the disappearance of radioiodinated albumin from the blood in the first hour after its injection intravenously. The TER has been found to be increased by more than 300% in septic shock and by up to 200% within the first 4 h of cardiac surgery (Raines et al., 1984).

Since the normal TER is 5% per hour (i.e. 120% per day) and the rates of synthesis and catabolism and albumin are only 10% per day, even large changes in the rates of synthesis and catabolism will have no measurable effect on the concentration of albumin in plasma within 6–12 h, whereas small percentage changes in microvascular permeability (measured as TER) could produce rapid and significant changes in the concentration of albumin in plasma (see Fleck, 1985). The rapidity of the increase in microvascular permeability is consistent with mechanisms (Hurley, 1983) different from those responsible for the initiation of the AP-response.

Speculation as to whether decreases in synthesis rates could account for the fall in the concentration of albumin in plasma is unneccessary when this phenomenon can be explained by existing knowledge of the changes in vascular permeability after injury. In our recent studies of the changes in serum iron after operation, referred to above, we found that although the serum iron began to fall immediately during operation, the iron transferrin ratio did not begin to fall until 4–6 h after the beginning of the operation (Fig. 6). Our experimental evidence is compatible with the fall in iron: transferrin ratio being due to hepatic uptake of iron triggered by the action of IL-1 on the liver, whereas the early fall in transferrin, which precedes this, is likely to be due to a prompt increase in microvascular permeability, since the molecular weight of transferrin and albumin are fairly close. The differences in the timing of the events (that is the prompt beginning of the fall in transferrin and the change in the ratio of iron to transferrin) render it unlikely that they are both mediated by IL-1.

There is now a problem of nomenclature. Should proteins such as albumin and transferrin, the concentration of which decreases after injury be referred to as 'negative AP-proteins' or 'synthesis inhibited AP-proteins'? Unfortunately, the meaning of the first phrase is ambiguous. Thus the use of either term is unsatisfactory (cf. p. 139).

We have argued above (and elsewhere, Myers et al., 1984) that the rapid fall in the concentrations of albumin and transferrin after trauma is unlikely to be mediated by IL-1. In addition the concentrations of albumin (and transferrin) remain low for many days after simple surgical procedures, and there is evidence that the ratio of extravascular to intravascular albumin is increased for up to one week after operation (Davies et al., 1962; Ballantyne and Fleck, 1973), which implies that the increase in vascular permeability may last for several days after injury. However, while there is tentative evidence that mediators such as histamine, or bradykinin could be responsible for the prompt increase in permeability (Hurley, 1983) the mechanism which leads to the prolonged increase is obscure.

Simple arithmetic applied to the rates of synthesis, breakdown and extravascular exchange of albumin indicates that changes in fractional rate of synthesis (i.e. the fraction of the intravascular pool synthesised each day – 10%) cannot account for the rapid decrease in albumin concentration. The synthesis rate of albumin after trauma has been measured directly in experimental animals and man. In rats, albumin synthesis is increased after laparotomy (Caine, 1971) whereas in rabbits compared with pair-fed controls it was decreased (Ballantyne et al., 1973). The rabbits, as do patients, had a considerably depressed food intake immediately after operation. Although the synthesis rate of albumin may be decreased after surgery, it may increase in response to intravenous amino acids (Skillman et al., 1967).

The above points lead us to consider that the term 'synthesis inhibited AP-proteins' would not be satisfactory because inhibition of synthesis is clearly not the only mechanism for the prompt fall shown by all plasma proteins (except α_2-M) after trauma, and to favour retention of the descriptively accurate term 'negative AP-proteins' (Lebreton et al., 1979) despite the possibility that the mechanisms of the early and prolonged decrease in the concentrations of these proteins might not involve IL-1.

Although much of the evidence concerning an increase in microvascular permeability comes from the use of labelled albumin, there is evidence that other proteins are affected by the same mechanism (cf. Chapter 20). The consequences of increased microvascular permeability are apparent for both AP-proteins and many others. We have shown that α_1-AGP and fibrinogen fall prior to elevation of AP-proteins (cf. Fig. 3) and Lamy et al. (1978) have shown a fall in α_1-AT prior to an increase). Also, as shown here, the concentration of transferrin falls after injury. Johannsson et al. (1972) found a decrease in prealbumin, transferrin, α-lipoprotein prothrombin, IgG, IgA and IgM following myocardial infarction.

It may be noted that in contrast with zinc and iron, copper falls only slightly after operation (Fig. 5). This small initial fall occurred presumably due to loss to the extravascular space of its binding protein ceruloplasmin, but as ceruloplasmin is an AP-protein, the concentration of copper is seen to rise significantly within a few days.

3. *The AP-response in various diseases in man*

There is widespread agreement that CRP is a very sensitive marker for the AP-response (see Figs. 1 and 2). Because increased levels are readily measured immunologically (e.g. by nephelometry) CRP is increasingly becoming used as a marker for inflammation (Kenny et al., 1981; Pepys, 1981a; Gewurz et al., 1982; Laurent, 1982; Vanlente, 1982; Carr, 1983). Indeed CRP has been again advocated as a replacement for the ESR as a general screening test for illness (Kenny et al., 1981; Moodley, 1981; Mallya et al., 1982; Sliwinski et al., 1983). Numerous factors determine the ESR, increases being caused by increased fibrinogen, decreased albumin, and increased immunoglobulins, altered red cells and plasma viscosity. Thus, during inflammation the ESR is strongly influenced by fibrinogen and other AP-proteins and therefore responds more slowly than CRP alone. Also the ESR has been found to have a diurnal variation which could lead to confusion in the monitoring of disease (Mallya et al., 1982).

It has been shown that anti-inflammatory steroids do not cause a direct decrease in CRP concentrations by suppression of synthesis and any decrease in CRP with the use of drugs is due to anti-inflammatory actions on the site of tissue damage (Stollerman et al., 1954; McConkey et al., 1979).

It has been suggested that another AP-protein – serum amyloid A component (SAA) – could be used as the main marker for the AP-response (Chambers et al., 1983) because it may change from its normal level of about 10 mg/l to up to 2,000 mg/l during inflammation (Pepys, 1981b). Changes in SAA generally correlate well with those of CRP, and increases in the concentration of SAA have been observed in conditions in which the concentration of CRP is not elevated, e.g. systemic lupus erythematosus (SLE) (Pepys and Baltz 1983). However, the present method for determining SAA is much less convenient than that for CRP and in consequence there have been considerably fewer observations on the changes in SAA in disease than of CRP (Pepys and Baltz, 1983).

It is accepted that the immediate stimulus for the AP-protein response is IL-1 which in turn is produced by macrophages responding to products of tissue damage or foreign material such as bacterial endotoxin (Dinarello, 1984). Since the causes of tissue damage are multitude, CRP is not pathognomic; it cannot be used as a specific diagnostic tool and its application is limited by its non-specificity. Nevertheless, measurement of CRP in a patient's plasma can provide information useful to the clinician.

3.1. Post-operative care and monitoring progress of illness

By far the major usefulness of measurements of CRP is that they constitute a sensitive indicator of inflammation. Serial measurements of CRP can give an indication of the patient's response to therapy and whether complications are developing. The use of CRP to monitor complications, especially infection, after surgery has been proposed (Fischer et al., 1976; Pantano et al., 1980; Ghoneim et al., 1982). In the absence of infection, CRP peaks at about 2 days after surgery and gradually returns to normal (Colley et al., 1983). If, however, complications develop such as infection, the CRP will either remain elevated or a secondary increase will occur. CRP has also been used to monitor the development of complications in patients with measles. Hirsch et al. (1983) and Griffin et al. (1983) found that the increases in CRP coincides with the appearance of the rash, then returns to normal. If complications such as pneumonia or encephalitis develop, a secondary elevation is seen.

Recently the value of using CRP or SAA to monitor the progress of patients who have received renal allografts or bone marrow transplants has been investigated. The surgical procedures of transplanting a kidney stimulate, in common with other surgical operations, prompt and moderate increases in CRP (Freed et al., 1984) and in SAA (Maury et al., 1984) with peak values within 3 days (CRP) or 2 days (SAA) and which decline rapidly towards the pre-operation values in the absence of episodes of rejection or infection.

In the case of CRP (Freed et al., 1984), out of 38 transplant cases there were seven instances when the CRP concentration declined after the operation then increased about 2-fold without other evidence of inflammation, rejection or infection. CRP was elevated in 89% of all episodes of rejection and these authors concluded that using a 2-fold increase in CRP as the criterion in the period 1 – 4 weeks after transplantation, prediction of almost 70% of episodes of rejection was possible from 1 to 12 days before the clinical diagnosis was made. The results with SAA were consistent with those of CRP. The predictive value for graft rejection of an SAA concentration of more than 100 mg/l was 88% and in eight of 15 examples of rejection the rise in SAA preceded clinical signs of rejection by an average of 2 – 5 days (Maury et al., 1984).

After bone marrow transplantation there seems to be little, if any, transient rise in CRP. Otherwise the general effects of graft versus host disease (GVHD) and infection on CRP concentration are similar to those after renal transplantation. In a study of 68 patients, 32 of whom had episodes of GVHD uncomplicated by infection within 3 months of the transplant, the CRP concentration increased but reached a value above 40 mg/l on only three occasions, whereas in patients with bacterial infections the median concentration of CRP was more than 100 mg/l (Walker et al., 1984a). Fungal or viral infections produced only a slight increase in CRP, if any, and in the later post-transplant period did not stimulate an increase in CRP concentration (Walker et al., 1984b).

In an earlier study of bone marrow transplant patients, CRP was found to attain a median concentration close to 90 mg/l and in one case 200 mg/l after uncomplicated GVHD (Rowe et al., 1984) whereas in contrast to the results of Walker et al. (1984a) the response of CRP to bacterial infection was slightly less than this. Despite these apparent contradictions there is clearly agreement that measurement of CRP concentration after bone marrow transplant can provide valuable indications of the presence of complications, whether GVHD or infection, and of the efficacy of therapy; a conclusion which also applies to the value of measuring CRP or SAA after renal transplantation.

Serial CRP estimations have been reported to be useful in the management of patients with acute pancreatitis, ulcerative colitis, chorioamnionitis, polyarteritis, rheumatoid arthritis and chemotherapy after operation because the concentration of CRP decreases rapidly if therapy is efficacious (Amos et al., 1977; Buckell et al., 1979; Evans et al., 1980; Ghoneim et al., 1982; Hind et al., 1984; Mayer et al., 1984).

The value of sequential measurements of CRP over several days in the management of patients with fractured skulls has been reported. Goester et al. (1981) found that if the CRP peaked at 2 days then returned to normal it could be assumed that the skull trauma was uncomplicated. However, in patients with complications, CRP values remained elevated for several days.

Although CRP may not be greatly elevated in some chronic diseases such as SLE, further increases in either CRP or SAA can be used as indicators of infection (Bertouch et al., 1983). Evidently an AP-response can still occur even in patients with SLE. However, CRP concentrations alone should not be relied on and should be accompanied by tests such as white blood cell counts. It has been suggested that α_1-AGP could be used as an index of chronic inflammation but as the concentration of this protein can be influenced by steroids (Laurell et al., 1968) which are often given in these conditions, this is not to be recommended.

A good correlation between CRP and the area, depth and severity of burns was found by Kohn (1961) who suggested that CRP can be used to reflect the clinical state of the patient and may be of value in prognosis. Others, (Laurent and Marichy, 1982) favour the use of the CRP/C4 ratio as an indicator of infection in the burns patient. CRP is high in these patients due to the tissue damage and it is suggested that CRP should not be used alone as a marker. However, in sepsis there is a consumption of C4 with a resulting increase of the CRP/C4 ratio. The use of ratios has also been suggested by Kohn et al. (1978). They used the CRP/prealbumin ratio in burns patients and in patients with ulcerative colitis. It was found that the ratio was positively correlated with severity of illness. The CRP/prealbumin ratio has also been used in the prognosis of advanced bowel cancer by Milano et al. (1978). Although the use of such ratios is of interest, care is required in their interpretation. Each protein has its own characteristic pattern during the AP-response so that ratios of proteins will also have different values during the progress of a single AP-response such as after an uncomplicated surgical operation (see Fig.2). That is, the ratio of any two proteins may also rise and fall during the course of a single AP-response.

3.2. The AP-response after myocardial infarction

It has been shown that after a single uncomplicated myocardial infarction (MI), the concentration of CRP rises rapidly and falls in a smooth exponential curve, in a pattern similar to that seen post-operatively (Johannsson et al., 1972; De Beer et al.,

1982; Laurent, 1982). As was mentioned earlier (p. 251) there is disagreement on whether CRP can be used to indicate the magnitude of the infarction (Kushner et al., 1978; Laurent, 1982) or not (Johannsson et al., 1972). In the latter study it was shown that the rises in the AP-proteins did not correlate with the enzyme changes. However, Smith et al. (1977) did find a correlation between AP-proteins and the activity of the enzyme β-hydroxybutyrate dehydrogenase in plasma. De Beer et al. (1982) and Chappelle et al. (1980) found that the activity of lactate dehydrogenase (LDH) is a better predictor of the short-term evolution of MI than creatine kinase (CK) activity in plasma, and Chappelle et al. (1981) also found that the rise in α_1-AGP measured at 24 h correlated with the LDH peak level. It has been suggested that after myocardial infarction plasma zinc levels were correlated with the incidence of further complications and were related to the patterns of cardiac enzyme levels (e.g. CK and LDH) but could not be used as a prognostic marker (Lekakis and Kalofoutis, 1980). However, there is a marked diurnal variation in the concentration of zinc in plasma (Sugarman, 1983) which complicates its use as a marker of inflammation.

Measurements of α_1-AGP levels after acute MI seem to be of little diagnostic value as the peak in concentration occurs too late (Chappelle et al., 1981), but in a 6-month study it was found that the concentration of α_1-AGP on day 8 after admission was significantly higher in non-survivors. Daily measurements of CRP may be useful in monitoring for possible complications after MI (Johannsson, 1972; De Beer et al., 1982). The authors also found that angina did not cause a rise in CRP and suggested that if the concentration of CRP is elevated in the absence of an elevation of myocardial CK (in the first 15 h) then the cause of the chest pain is neither MI nor angina. After exercise, the activity of serum CK is elevated and so too is the white blood count. This has caused Sylven et al. (1983) to cast doubt on the specificity of cardiac enzymes in the differential diagnosis of chest pain in marathon runners. The problem of timing of blood samples after an injury is illustrated by the suggestion that CRP could be used to determine whether marathon runners with chest pain had had a MI (Pepys and Hind, 1984). The evidence of the 6 h delay following the onset of tissue damage and the beginning of the rise in CRP and other AP-plasma proteins described above indicates that determination of CRP or other AP-proteins before this period has elapsed may lead to the erroneous conclusion that no tissue damage, i.e. no MI, had occurred. It is our view that the general evidence indicates that to be of value the CRP should be measured at least 12 h after the onset of pain, with preferably follow-up measurements at 24 and 48 h.

3.3. The AP-response in pelvic disease

In the differential diagnosis of pelvic diseases, Angerman et al. (1979) showed that, on the basis of presence or absence of elevated CRP two categories could be distinguished:

(1) Inflammatory type, in which CRP was elevated. This includes pelvic infection, tubo-ovarian abscess, and
(2) non-inflammatory type in which there is no elevation of CRP. This group includes benign ovarian cysts, uterine pregnancies, uterine leiomyoma.

An interesting observation is that no elevation of CRP has been found in patients with cystitis whereas CRP was elevated in all patients with pyelonephritis in whom it was measured (Jodal et al., 1975).

3.4. Organic versus mental disease

The present concept of the mechanism of the AP-response is that it is mediated by IL-1 produced by macrophages in response to tissue or organic damage due to any cause including infection or to the presence of 'foreign' material such as bacterial endotoxin. Since one of the most sensitive indicators of the response in man is the concentration of CRP in plasma (Belfrage, 1963), it should be possible to use its measurement to distinguish whether organic or mental illness is present. Consistent with this, preliminary studies indicate that stress alone, unaccompanied by tissue damage does not promote an AP-response (Gedeon, 1981).

Similarly it would be inconsistent with the hypothesis were neuroses or mental illness to lead to elevation of CRP. Thus while increased CRP or SAA indicates a source of inflammation, the absence of abnormally increased CRP should not be taken as an indicator of the absence of organic disease, as CRP can approach to, or be at normal levels in the quiescent phases of many rheumatoid diseases, and as mentioned above it may not be elevated in some patients with acute appendicitis, primary biliary cirrhosis, hepatitis, SLE and in the early stages of cancer (Morley and Kushner, 1982; Bertouch et al., 1983; Marchand et al., 1983). Also, CRP levels may not rise in certain endocrine disorders.

3.5. The AP-response in pregnancy and parturition

The presence or absence of CRP as an indicator of bacterial infection has been used in mothers with premature rupture of the membranes (Evans et al., 1980). Hawrylshyn et al. (1983) found that CRP was a better indicator of infection than white cell count (WBC), neutrophils and ESR. In this situation the obstetrician can decide whether to delay parturition if there is no infection or induce delivery if there is. However, Farb et al. (1983) found that CRP was not a good indicator of amnionitis and suggested that CRP should be used in conjunction with other signs and symptoms suggestive of chorioamnionitis, rather than as a pathognomic test. The concentration of α_1-AGP does not change throughout pregnancy (Chu et al., 1981) or may fall slightly (Laurell, 1968), and increased levels could be used as a marker of inflammation.

The concentrations of other AP-proteins increase during gestation: fibrinogen by 58%; caeruloplasmin by 130%; and α_1-AT by 85% (Ganrot, 1972). It has been suggested that this pattern could be caused in part by changes in hormone concentration as a similar pattern is seen after administration of combined estrogen-progestin contraceptives (Laurell et al., 1968). It is apparent therefore that the absolute concentrations of these proteins cannot be relied on as indicators of inflammation in pregnancy. Kindmark (1972) found that the concentration of CRP could rise in pregnancy, but only to levels of less than 5 mg/l and did not approach levels seen during inflammation. The evidence would appear to be that CRP is the best indicator of the AP-response during gestation.

3.6. *The AP-response in the neonate*

After a normal birth, there is no increase in the concentration of CRP in the neonate (Collet-Cassart et al., 1983; Speer et al., 1983). However, it has been shown that CRP, ESR, and WBC are good indicators of infection in full-term neonates with CRP being the best (Iwaszko-Krawczuk, 1974; Philip, 1979; Sabel and Wadsworth, 1979; Speer et al., 1983). In a clinical trial, CRP has been used as an indicator of the efficacy of antibiotic therapy in neonatal sepsis (Alt et al., 1983).

In contrast with CRP, haptoglobin and α_1-AT do not appear to increase during neonatal sepsis (Speer et al., 1983), even where CRP is elevated to levels seen in an adult AP-response, which indicates an inability of the neonate to produce the complete AP-response of the adult. This is consistent with Laurell's (1968) investigation that in premature babies, the concentrations of α_1-AT and ceruloplasmin were very low, suggesting immaturity of the mechanisms of synthesising these proteins in the fetal liver.

It has been shown that CRP is not always present in the premature or small-for-date newborns with bacterial infections and they may be more at risk from infection because of their lack of acute phase and related immune responses (Iwaszko-Krawczuk, 1974; Philip, 1979; Speer et al., 1983). This failure of the AP-response in premature and small-for-date neonates could be due to lack of endogenous pyrogen (EP)/IL-1 production which appears to be inhibited before birth (Dinarello et al., 1981). It would seem to be hazardous at present to rely on CRP alone as an indicator of neonatal sepsis because immature neonates may lack the fully developed mechanism for producing the adult AP-response.

The use of serum levels of α_1-AGP as an indicator of infection in the neonate has been suggested because the capacity to synthesise this protein develops early (Bienvenu et al., 1981). Indeed in an AP-response in neonates, the concentration of α_1-AGP may reach levels comparable with those seen in the AP-response of an adult. Elevations in the concentration of α_1-AGP were found in 80% of infants with bacterial infections, and constituted a more sensitive index of the disease than increases in fibrinogen and IgM. In contrast, α_1-AGP was not a useful indicator of viral or parasite infections. The decision as to whether a rise in α_1-AGP indicates the presence of infection is complicated by the fact that during the neonate's first week, increased levels of α_1-AGP are to be expected, presumably due to development of the neonatal liver.

3.7. Viral, fungal and parasitological disease

The presence or absence of elevated CRP can contribute in the diagnosis of other disease states where confusion or uncertainty could arise. The most useful hypothesis would seem to be that CRP will not be elevated at all or will show only a slight increase during a viral infection unless tissue damage resulting from this infection is also present (Belfrage, 1963). Thus Corrall et al. (1981) have suggested that if increased CRP is present in the serum or cerebrospinal fluid (CSF) in meningitis, then the cause is more likely to be bacterial and that absence of elevated CRP could indicate a viral cause. In this differentiation of viral from bacterial meningitis, CRP was more sensitive than CSF leucocyte count, CSF polymorphonuclear count, CSF glucose, CSF total protein and gram staining of the CSF. Serum CRP was increased above 10 mg/l in all cases of bacterial meningitis investigated and only 6% of non-bacterial type (Corrall et al., 1981; Peltola, 1982; Clark and Cost, 1983). However, Philip and Baker (1983) found that the CRP levels in the CSF could not be used to distinguish between infants with viral or bacterial meningitis, probably because of the absence of correlation between CSF and serum CRP. Clarke and Cost (1983) recommended that only the serum CRP values should be measured and that these values could differentiate between viral and bacterial meningitis in children. On the other hand, Benjamin et al. (1984) found that CRP could not be relied on in the differential diagnosis of the cause of meningitis. Salonen and Vaheri (1981) measured CRP in several types of acute viral infections and concluded that viral infections can indeed cause elevations in the concentration of CRP. Evidently a small increase in CRP concentration is not by itself sufficient evidence to exclude the possibility of viral meningitis.

The presence or absence of elevated CRP in serum has also been used in the differentiation between bacterial tracheitis and viral laryngotracheitis (Shabino, 1983). Again, CRP will only be elevated in the bacterial type, and only present in the viral type if inflammation and tissue damage occurs. Thus estimation of CRP after 6 h but within 24 h may assist in the early differentiation between viral and bacterial infection and thus allow prompt specific treatment of the infection before the results of conventional microbiological studies become available.

There is very little data on the effects of parasitological infections and increases of the AP-proteins. It is known that an AP-response occurs in the acute stage of malaria (Klainer et al., 1968; Murphy et al., 1971), but this may be caused by the tissue damage in the host at this stage rather than by the presence of the parasite itself (cf. bacterial endotoxin). Clearly more research in this area is required.

3.8. The AP-proteins in cancer

There has been considerable interest in the AP-response in patients with cancer (Cooper and Stone, 1979; Raynes and Cooper, 1983). Early studies on cancer patients showed that advanced cancer was accompanied by a rise in the α-globulins (Winzler, 1953), but because these and the other AP-proteins respond to such a wide variety of stimuli, it is self-evident that any changes seen in these proteins are non-specific. Although it has been reported that CRP can inhibit lung metastases (Deodhar et al., 1982) and promote tumoricidal activity of the macrophage (Barna et al., 1984), it is generally agreed that CRP is not a satisfactory tumour marker. It appears that CRP is only elevated in conditions where the tumour load is large enough to cause tissue damage (Drahovsky et al., 1981) and it is this tissue damage which stimulates the liver through mediators such as IL-1 to increase synthesis of the AP-proteins. These authors also found that CRP correlated with the staging of the progress of the disease which they based on the TNM classification (tumour-node-metastases) and with active disease. However, the CRP could not indicate the presence or absence of metastases. CRP has been advocated for use in conjunction with tumour cancer markers (Coombes et al., 1977) the most widely used of which include α-fetoprotein (AFP), human chorionic gonadotrophin (HCG), carcino-embryonic antigen (CEA), and placental alkaline phosphatase (PAP). It should be emphasised that these are not AP-proteins but specific products of cells which have reverted due to derepression of normally repressed genes, which are reproducing abnormally, and in an uncontrolled fashion. When CRP was added to the battery of tests, the sum of total positive tests increased considerably. Drahovsky et al. (1981) suggested that although CRP cannot be used as a cancer marker, the levels can be an indication of the disease activity and could be used to monitor infection in immunosuppressed patients. In support of this concept, Seal et al. (1976) found, in a study of prostatic cancer, that pre-treatment elevations of fibrinogen were correlated with increased proportions of deaths from all causes. The elevations in fibrinogen were related to severity of the illness thus indicating the patients at risk. Similar results were found in a later study by Seal et al. (1978) of haptoglobin in prostatic cancer. Haptoglobin showed elevations in stages III and IV, with stage IV having higher levels than stage III. The pre-treatment values correlated with death rates. The AP-proteins in both studies did not serve as markers of the presence of tumours as such, but were used as non-specific markers of tumour load. Following studies on breast cancer, Coombes et al. (1977) concluded that CRP and α_1-AGP may be useful in the monitoring of the disease, but that the levels of these AP-proteins are primarily related to the tumour load. Minton and Bianco (1974) reported that raised α-globulins were present in 32 out of 39 advanced breast cancers which had spread beyond local nodes. Despite this, Cowen et al. (1978) concluded from their studies that α_1-AT had no value in the monitoring of carcinoma of the breast.

In advanced large bowel cancer, Milano et al. (1978) found that a fall in prealbumin and coincidental rise of CRP, in the absence of infection, indicated terminal stages of the illness. However, these changes of concentration of prealbumin and CRP are also seen in benign disease of the bowel, such as ulcerative colitis, Crohns disease and most other non-intestinal inflammations. CRP, α_1-AGP and α_1-ACh have been used for long-term monitoring of cancer of the bladder, higher levels being found in advanced cancers with a higher tumour load (Bastable et al., 1979). Serial measurements of haptoglobin can give good indications of active disease in malignant ovarian cancers (Meuller et al., 1971). Primary tumours of less than 6 cm diametre produced no rise in haptoglobin, whereas patients with tumours of over 6 cm in diameter did have elevations of haptoglobin. Thus the elevations seen were related to tumour load and not specifically to the type of tumour. After successful treatment the concentration of haptoglobin returned to normal, but remained elevated or rose again if the tumour recurred or spread.

Elevations in α_1-AT have been found in women with a positive cervical smear and α_1-AT has been suggested as a sensitive marker of microinvasion of the cervix (Latner et al., 1976).

It appears, therefore, that AP-proteins are not specific markers in cancer but may have a role in monitoring the development of the tumour, especially the primary and related tissue damage, and as indicators of the efficacy of therapy (Fish et al., 1982).

4. *Summary and conclusions*

Of the many AP-proteins in man, CRP and SAA protein are the most sensitive indicators of the response. At present CRP can be determined readily by a variety of methods whereas, the assay of SAA is more tedious and the serum may require pretreatment (Pepys, 1981b). This is why CRP is most often used as an indicator of tissue damage or infection. Increased concentrations of the AP-proteins therefore indicate non-specifically the presence of inflammation, tissue damage or endotoxin. Thus the basic clinical use of measurements of CRP or SAA is to monitor the progress of a disease and its response to therapy. The value of this is becoming increasingly recognised in diseases such as rheumatoid arthritis, Crohn's disease and many other chronic diseases, in monitoring the success of allografts such as bone marrow and kidney, and in indicating the development of complications after burning injury, MI and other acute conditions.

Early suggestions that an increase in plasma CRP indicated that the cause of infection was bacterial rather than viral have not been fully substantiated. It would be interesting to examine whether a viral disease per se can promote an AP-response or whether it is cell damage caused by the virus that promotes the response.

In the neonate, the response patterns differ from those seen in the adult because the mechanisms of synthesis and control of the plasma concentration of specific proteins do not develop simultaneously.

It may be that the measurement of CRP will replace that of ESR because of its greater specificity and promptness of the response.

References

Allison, A.C. and Rees, W. (1957) Br. Med. J. *2,* 1137 – 1143.

Alt, R., Irazuzta, J., Erny, P., Messer, J., Monteil, H., Minck, R. and Willard, D. (1983) J. Antimicrobial. Chemother. *11,* (Suppl. C.), 51 – 55.

Amos, R.S., Constable, T.J., Crockson, R.A., Crockson, A.P. and McConkey, B. (1977) Br. Med. J. *1,* 195 – 197.

Angerman, N.S., Maravec, W.D. and Hajj, S. (1979) Surg. Forum *30,* 478 – 479.

Aronsen, K.F., Ekelund, G., Kindemark, C.O. and Laurell, C.B. (1972) Scand. J. Clin. Lab. Invest. *29* (Suppl. 124), 127 – 136.

Ballantyne, F.C. and Fleck, A. (1973) Clin. Chim. Acta *46,* 139 – 146.

Ballantyne, F.C., Tilstone, W.J. and Fleck, A. (1973) Br. J. Exp. Pathol. *54,* 409 – 415.

Barna, B.P., Deodhar, S.D., Gautam, S., Yen-Lieberman, B. and Roberts, D. (1984) Cancer Res. *44,* 305 – 311.

Bastable, J.R.G., Richards, B., Haworth, S. and Cooper, E.H. (1979) Br. J. Urol. *51,* 283 – 289.

Belfrage, S. (1963) Acta Med. Scand. (Suppl. 395), 1 – 169.

Benjamin, D.R., Opheim, K.E. and Brewer, L. (1984) Am. J. Clin. Pathol. *81,* 779 – 782.

Bennett, R.M. and Kokocinski, T. (1979) Clin. Sci. *57,* 453 – 460.

Bertouch, J.V., Roberts-Thompson, P.J., Feng, P.H. and Bradley, J. (1983) Ann. Rheum. Dis. *42,* 655 – 659.

Bienvenu, J., San, L., Bienvenu, F., Lahet, C., Divry, P., Cotte, J. and Bethenod, M. (1981) Clin. Chem. *27,* 721 – 726.

Bostian, K.A., Blackburn, B.S., Wannemacher, Jr., R.J., McGann, V.G., Beisel, W.R. and Dupont, H.L. (1976) J. Lab. Clin. Med. *87,* 577 – 585.

Bothwell, T.H., Charlton, R.W., Cook, J.D. and Finch, C.A. (1979) in Iron Metabolism In Man, pp. 284 – 310, Blackwell Scientific Publications, Oxford, U.K.

Buckell, N.A., Lennard-Jones, J.E., Hernandez, M.A., Kohn, J., Riches, P.G. and Wadsworth, J. (1979) Gut *20,* 22 – 27.
Carr, W.P. (1983) Clin. Rheum. Dis. *9,* 227 – 239.
Caine, S. (1971) Partial hepatectomy and plasma protein synthesis, Ph.D. Thesis, University of Glasgow.
Chambers, R.E., MacFarlane, D.G., Whicher, J.T. and Dieppe, P.A. (1983) Ann. Rheum. Dis. *42,* 665 – 668.
Chapelle, J.P., Albert, A., Heusghem, C., Smeets, J.P. and Kulbertus, H.E. (1980) Clin. Chim. Acta *106,* 29 – 38.
Chapelle, J.P., Albert, A., Smeets, J.P., Heusghem, C. and Kulbertus, H.E. (1981) Clin. Chim. Acta *115,* 199 – 209.
Chu, C.Y.T., Singla, V.P., Wang, H.P., Sweet, B. and Lai, L.T.Y. (1981) Clin. Chim. Acta *112,* 235 – 240.
Clarke, D. and Cost, K. (1983) J. Paediatrics *102,* 718 – 720.
Collet-Cassart, D., Maraschal, J.C., Sindic, C.J.M., Tomosi, J.P. and Masson, P.L. (1983) Clin. Chem. *29,* 1127 – 1131.
Colley, C.M., Fleck, A., Goode, A.W., Muller, B.R. and Myers, M.A. (1983) J. Clin. Pathol. *36,* 203 – 207.
Coombes, R.L., Powels, T.J. and Neville, A.M. (1977) Proc. R. Soc. Med. *70,* 843 – 845.
Cooper, E.H. and Stone, J. (1979) Adv. Cancer Res. *30,* 1 – 44.
Corrall, C.J., Pepple, J.M., Moxon, E.R. and Hughes, W.T. (1981) J. Pediatrics *99,* 365 – 369.
Cowan, D.M., Searle, F., Milford-Ward, A., Benson, G.A., Smiddy, F.G., Eaves, G. and Cooper, E.H. (1978) Eur. J. Cancer *14,* 885 – 894.
Davies, J.W.L., Ricketts, C.R. and Bull, J.P. (1962) Clin. Sci. *23,* 411 – 423.
De Beer, F.C., Hind, C.R.K., Fox, K.M., Allan, R.M., Maseri, A. and Pepys, M.B. (1982) Br. Heart J. *47,* 239 – 243.
Deodhar, S.D., James, K., Chiang, T., Edinger, M. and Barna, B.P. (1982) Cancer Res. *42,* 5084 – 5088.
Dinarello, C.A. (1984) Rev. Infect. Dis. *6,* 51 – 95.
Dinarello, C.A., Shparber, M., Kent, E.F. and Wolff, S.M. (1981) J. Infect. Dis. *144,* 337 – 343.
Dominioni, L., Dionigi, R. and Cividini, F. (1980) Eur. Surg. Res. *12* (Suppl. 1), 133.
Drahovsky, D., Dunzendorfer, U., Ziegenhagen, G., Drahovsky, M. and Kellen, J.A. (1981) Oncology *38,* 286 – 291.
Evans, M.I., Hajj, S.N., Dero, L.D., Angerman, N.S. and Moawad, A.H. (1980) Am. J. Obstet. Gynecol. *138,* 648 – 652.
Farb, H.F., Arnesen, M., Geistler, P. and Knox, G.E. (1983) Obstet. Gynaecol. *62,* 49 – 51.
Fischer, C.L., Gill, C., Forrester, M.G. and Nakumura, R. (1976) Am. J. Clin. Pathol. *66,* 840 – 846.
Fish, R.G., Yap, A.K.L. and James, K. (1982) Clin. Biochem. *15,* 4 – 8.
Fleck, A. (1984) Ann. Clin. Biochem., in press.
Fleck, A. and Clark, B. (1980) Adv. Physiol. Sci. *26,* 81 – 89.
Foote, J.W. and Delves, H.T. (1984) J. Clin. Pathol. *37,* 1050 – 1054.
Freed, B., Walsh, A., Pietrocola, D., MacDowell, R., Laffin, R. and Lempert, N. (1984) Transplantation *37,* 215 – 219.
Ganrot, P.O. (1972) Scand. J. Clin. Lab. Invest. *29* (Suppl. 124), 83 – 88.
Gedeon, G.S. (1981) Protein metabolism and catecholamine response in trauma and injury, Ph.D. Thesis, University of Glasgow.
Gewurz, H., Mold, C., Siegel, J. and Fiedel, B. (1982) Adv. Int. Med. *27,* 345 – 372.
Ghoneim, A.T.M., McGoldrick, J. and Ionescu, M.I. (1982) Ann. Thorac. Surg. *34,* 166 – 175.
Giroux, E.L. (1975) Biochem. Med. *12,* 258 – 266.
Goester, C., Ferard, G., Klumpp, T. and Metais, P. (1981) Clin. Chim. Acta *117,* 43 – 51.

Griffin, D.E., Hirsch, R.L., Johnson, R.T., De Soriano, I.L., Roedenbeck, S. and Vaisberg, A. (1983) Infect. Immun. *41*, 861 – 864.

Hawrylshyn, P., Bernstein, P., Milligan, J.E., Soldin, S., Pollard, A. and Papsin, F.R. (1983) Am. J. Obstet. Gynecol. *147*, 240 – 247.

Hedlund, P. (1961) Acta Med. Scand. *35* (Suppl.), 361.

Hind, C.R.K., Savage, C.O., Winearls, C.G. and Pepys, M.B. (1984) Br. Med. J. *288*, 1027 – 1030.

Hirsch, R.L., Johnson, R.T., Lindo De Soriano, I., Roedenbeck, S. and Vaisberg, A. (1983) Infect. Immun. *41*, 861 – 864.

Hurley, J.V. (1983) Acute Inflammation, 2nd ed., Churchill Livingstone, Edinburgh – London – Melbourne – New York.

Iwaszko-Krawczuk, W. (1974) Acta Paediatr. Acad. Sci. Hung. *15*, 115 – 118.

Jodal, U., Lindberg, V. and Lincoln, K. (1975) Acta Paediatr. Scand. *64*, 201.

Johansson, B.G., Kindmark, C.O., Trell, E.Y. and Wollheim, F.A. (1972) Scand. J. Clin. Lab. Invest. *29* (Suppl. 124), 117 – 126.

Kenny, M.W., Worthington, D.J., Stuart, J., Davies, A.J., Farr, M., Davey, P.G. and Chughtai, M.A. (1981) Clin. Lab. Haematol. *3*, 299 – 305.

Kindmark, C.O. (1972) Scand. J. Clin. Lab. Invest. *29*, 407 – 411.

Klainer, A.S., Clyde, D.F., Bartelloni, P.J. and Beisel W.R. (1968) J. Lab. Clin. Med. *72*, 794 – 802.

Klempner, M.S., Dinarello, C.A. and Gallin, J.I. (1978) J. Clin. Invest. *61*, 1330 – 1336.

Kohn, J. (1961) In Protides of the Biological Fluids, Proceedings of the 8th Coloqui. Bruges 1960 (Peeters, H., ed.), pp. 315 – 318, Elsevier, Amsterdam.

Kohn, J., Hernandez, M. and Riches, P. (1978) Ricerca Clin. Lab. *8* (Suppl. 1), 61 – 70.

Konijn, A.M. and Hershko, C. (1977) Br. J. Haematol. *37*, 7 – 16.

Kostiala, I. (1984) J. Infect. *8*, 212 – 221.

Kushner, I., Broder, M.L. and Karp, D. (1978) J. Clin. Invest. *61*, 235 – 242.

Lamy, Y., Ibrahim, S., Lomanto, C. and Dombrowiecki, A. (1978) Clin. Chim. Acta *89*, 387 – 391.

Latner, A.L., Turner, G.A. and Lamin, M.M. (1976) Oncology *33*, 12 – 14.

Laurell, C.B. (1968) Scand. J. Clin. Lab. Invest. *21*, 136 – 138.

Laurell, C.B., Kullander, S. and Thorell, J. (1968) Scand. J. Clin. Lab. Invest. *21*, 337 – 343.

Laurent, P. (1982) in Marker Proteins in Inflammation (Allen, R.C., Bienvenu, J., Laurent, P. and Suskind, R.M., eds.), pp. 69 – 88, Walter de Gruyter, Berlin and New York.

Laurent, P. and Marichy, J. (1982) in Marker Proteins in Inflammation (Allen, R.C., Bienvenu, J., Laurent P. and Suskind, R.M. eds.), pp. 497 – 502, Walter de Gruyter, Berlin and New York.

Lebreton, J.P., Joisel, F., Raoult, J.P., Lannuzel, B., Rogez, J.P. and Humbert, G. (1979) J. Clin. Invest. *64*, 1118 – 1129.

Lekakis, J. and Kalofoutis, A. (1980) Clin. Chem. *26*, 1660 – 1661.

Mallya, R.K., Mace, B.E.W., De Beer, F.C. and Pepys, M.B. (1982) Lancet. *1*, 389 – 390.

Marchand, A., Van Lenk, P. and Galen, R.S. (1983) Am. J. Clin. Pathol. *80*, 369 – 374.

Markowetz, B., Van Snick, J.L. and Masson, P.L. (1979) Thorax. *34*, 209 – 212.

Maury, C.P.J., Teppo, A.M., Ahonen, J. and Willerbrand, E.V. (1984) Br. Med. J. *288*, 360 – 361.

Mayer, A.D., McMahon, M.J., Bowen, M. and Cooper, E.H. (1984) J. Clin. Pathol. *37*, 207 – 212.

McConkey, B., Davies, P., Crockson, R.A., Crockson, A.P. and Butler, M. (1979) Ann. Rheum. Dis. *38*, 141 – 144.

McMillan, E.M. and Rowe, D.J.F. (1982) Clin. Exp. Dermatol. *7*, 629 – 632.

Milano, G., Cooper, E.H., Gollgher, J.C., Giles, G.R. and Munro Neville, A. (1978) J. Natl. Cancer Inst. *61*, 687 – 691.

Minton, J.P. and Bianco, M.A. (1974) Arch. Surg. *109*, 238 – 240.

Moodley, G.P. (1981) S. Afr. Med. J. *60*, 545 – 547.

Morley, J.J. and Kushner, I. (1982) Ann. N.Y. Acad. Sci. *389,* 406–418.
Mueller, W.K., Handshumacher, R. and Wade, M.E. (1971) Obstet. Gynaecol. *38,* 427–435.
Murphy, S.G., Klainer, A.S. and Clyde, D.F. (1971) J. Lab. Clin. Med. *79,* 55–61.
Myers, M.A., Fleck, A., Sampson, B., Colley, C.M., Bent, J. and Hall, G. (1984) J. Clin. Pathol. *37,* 862–866.
Owen, J.A. (1967) Adv. Clin. Chem. *9,* 1–41.
Pantano, E., Pisani, M. and deJaco, M. (1980) Ricera Clin. Lab. *10,* 281–284.
Parving, H.H. and Rasmussen, S.M. (1973) Scand. J. Lab. Invest. *32,* 81–87.
Pekarek, R.S., Wannemacher, R.W. and Beisel, W.R. (1972) Proc. Soc. Exp. Biol. Med. *140,* 685–688.
Peltola, H.O. (1982) Lancet, May 1, 980–982.
Pepys, M.B. (1981a) Lancet *i,* 653–657.
Pepys, M.B. (1981b) in Clinics in Immunology and Allergy (Holborow, E.J., ed.), pp. 77–102, W.B. Saunders, London–Philadelphia–Toronto.
Pepys, M.B. and Baltz, M.L. (1983) in Advances in Immunology (Dixon, F.J. and Kunkel, H.G., eds.), Vol. 34, pp. 141–213, Academic Press, New York.
Pepys, M.B. and Hind, C.R.K. (1984) Lancet *i,* 278.
Philip, A.G.S. (1979) Acta Paediat. Scand. *68,* 481–483.
Philip, A.G.S. and Baker, C.J. (1983) J. Paediatr. *102,* 715–718.
Raines, G., Fleck, A., Ledingham, I.McA., Wallace, P. and Colman, K.C. (1984) Scot. Med. J. *29,* 45.
Raynes, J.G. and Cooper, E.H. (1983) J. Clin. Pathol. *36,* 798–803.
Rowe, I.F., Worsley, A.M., Donnelly, Y.P., Catovsky, D., Goldman, J.M., Galton, D.A.G. and Pepys, M.B. (1984) J. Clin. Pathol. *37,* 263–267.
Sabel, K.G. and Wadsworth, C.H. (1979) Acta Paediatr. Scand. *68,* 825–831.
Salonen, E-M. and Vaheri, A. (1981) J. Med. Virol. *8,* 161–167.
Seal, U.S., Doe, R.P., Byar, D.P. and Corle, M.S. (1976) Cancer *38,* 1108–1117.
Seal, U.S., Doe, R.P., Byar, D.P. and Corle, D.K. (1978) Cancer *42,* 1720–1729.
Shabino, C.L. (1983) J. Paediatr. *103,* 1010.
Skillman, J.J., Awwad, H.K. and Moore, F.D. (1967) Surg. Gynecol. Obstet. *125,* 983–996.
Sliwinski, A.J., Weber, L.D. and Nashel, D.J. (1983) Arch. Pathol. Lab. Med. *107,* 387–388.
Smith, S.J., Bos, G., Esseveld, M.R., Van Eijk, H.G. and Gerbrandy, J. (1977) Clin. Chim. Acta *81,* 75–85.
Song, M.K. and Adham, N.F. (1979) Clin. Chim. Acta *99,* 13–21.
Speer, C., Bruns, A. and Gahr, M. (1983) Acta Paediatr. Scand. *72,* 679–683.
Stollerman, G.H., Glick, S.K. and Anderson, H.C. (1954) Proc. Soc. Exp. Biol. Med. *87,* 241–245.
Sugarman, B. (1983) Rev. Infect. Dis. *5,* 137–147.
Sunderman, Jr., F.W. (1964) Am. J. Clin. Pathol. *42,* 1–21.
Sylven, J.C., Jansson, E., Brandt, S. and Kallner, A. (1983) Lancet *ii,* 1505.
Tillet, W.S. and Francis, J.T. (1930) J. Exp. Med. *52,* 561–571.
Vanlente, F. (1982) Hum. Pathol. *13,* 1061–1062.
Van Snick, J.L., Masson, P.L. and Heremans, J.F. (1974) J. Exp. Med. *140,* 1068–1084.
Walker, S.A., Rogers, T.R., Riches, G.P., White, S. and Hobbs, J.R. (1984a) J. Clin. Pathol. *37,* 1018–1021.
Walker, S.A., Riches, P.G., Rogers, T.R., White, S. and Hobbs, J.R. (1984b) J. Clin. Pathol. *37,* 1022–1026.
Werner, M. (1969) Clin. Chim. Acta *25,* 299–305.
Werner, M. and Odenthal, D. (1967) J. Lab. Clin. Med. *70,* 302–310.
Winzler, R.J. (1953) Adv. Cancer Res. *1,* 506–549.
Yocum, R.S. and Doerner, A.A. (1957) AMA. Arch. Int. Med. *99,* 74–81.

Gordon/Koj (eds.) The acute-phase response to injury and infection.

CHAPTER 22

Acute-phase proteins in chronic inflammation

J.D. SIPE

1. Introduction

While it is thought that the acute inflammatory response can cease and disappear without a trace, chronic inflammation is characterized by persistence of the irritating agent or by repeated recurrence of inflammation episodes. As indicated in Table 1, a single event of cell necrosis or tissue injury sets into motion a sequence of events requiring 48 – 72 h to be completed. One of the earliest events, occurring within the first 3 h (Sipe et al., 1979) is macrophage production of IL-1. Inflammatory agents used experimentally in the AP-response such as turpentine oil and endotoxin are capable of perturbing the macrophage cell membrane (Dinarello, 1984a) resulting in IL-1 output, as is the process of phagocytosis that begins shortly after tissue injury and cell necrosis. As a result of the IL-1 release which follows (cf. p. 33), the hypothalamus, liver, muscle, connective tissue and fibroblasts are all stimulated in a number of different ways. If a second and then a third inflammatory episode is initiated before the first has come to its natural conclusion, then the origin of some of the abnormalities associated with chronic inflammation may be explained. Two marked features of chronic inflammation are proliferation of connective tissue cells and of the extracellular matrix consisting of collagens, fibronectin and other structural glycoproteins. Occasionally, as a consequence of incomplete proteolysis of the AP-reactant SAA, reactive systemic amyloidosis can occur secondary to chronic inflammatory diseases such as rheumatoid arthritis or recurrent acute inflammatory disease such as familial Mediterranean fever (FMF) (reviewed by Glen-

TABLE 1
Times after injury at which maximum responses occur

Stage	AP-response	Time of maximum appearance after initiating stimulus
1	SAA inducer/IL-1	90 min
2	Interferon	4 h
3	CSF	6 h
4	Neutrophil response	6 – 24 h
5	SAA response	16 – 24 h
6	Macrophage response	24 – 72 h

ner, 1980; Skinner and Cohen, 1983). This chapter will discuss human chronic inflammatory disease, particularly rheumatoid arthritis, and animal models of chronic inflammation. It will describe studies which indicate that there is an altered pattern of acute phase and other plasma proteins in the host after repeated inflammatory episodes or after chronic inflammation. It will describe the role of AP-proteins serum amyloid A (SAA) and serum amyloid P (SAP) in the pathogenesis of secondary amyloidosis. Finally, it will present some of the issues to be considered in future decisions on therapeutic enhancement of the AP-response.

2. *Chronic inflammation in human rheumatoid arthritis*

Rheumatoid arthritis as defined by Billingham (1983) is a disease of articulating joints in which the cartilage and bone is slowly eroded away by a proliferating invasive connective tissue, called pannus, which is derived from the synovial membrane and its microvasculature (cf. p. 56). The disease is characterized by elevated concentrations of AP-proteins and thus an elevated ESR Furthermore, AP-proteins such as SAA, CRP, fibrinogen as well as IL-1 are found in synovial fluid (J. Sipe, unpublished observations; Bodel and Hollingsworth, 1968; Fontana et al., 1982; Wood et al., 1983).

It has been recognized that bacterial products are capable of inducing chronic synovitis (Hadler and Granovetter, 1978). Bacterial products such as endotoxin are, of course, potent stimulators of IL-1 production, and it has been postulated that lipopolysaccharides (LPS) may remain as part of the persistent bacterial debris in joints (Goldenberg and Rice, 1984). Thus, the presence of LPS or other nonviable bacterial components may set into motion many of the sequelae of chronic arthritis (Goldenberg, 1983).

First the effects of IL-1 on joint structure will be considered. IL-1 has been shown to stimulate secretion of collagenase and prostaglandins from synoviocytes (Dayer et al., 1980; Mizel et al., 1981). Furthermore, IL-1 can stimulate articular chondrocytes to secrete proteases and prostaglandins (Desmukh-Phadke et al., 1980; McGuire et al., 1982).

In addition, IL-1 has been shown to stimulate proliferation of fibroblasts and possibly of synoviocytes and chondrocytes (Postlethwaite et al., 1982; Schmidt et al., 1982), although a report from another laboratory (Estes et al., 1984) showed the macrophage-derived growth factor for fibroblasts to be different from IL-1. Furthermore, IL-1 has been shown to stimulate bone resorption (Gowen et al., 1983). IL-1 activity has been found in joint fluid from patients with osteoarthritis as well as rheumatoid arthritis (Wood et al., 1983), supporting the concept that this activity is derived from a cell in the joint rather than by ultrafiltration from serum. Therefore, it seems quite possible that the erosion of bone and cartilage associated with arthritic disease may arise from IL-1/catabolin (Saklatvala and Dingle, 1980; Saklatvala et al., 1984) released from the synovial tissue itself. IL-1 has been demonstrated to stimulate the cellular responses which lead to connective tissue destruction. These include release of collagenase, plasminogen activators, proteoglycans, glycosaminoglycans, prostaglandins and calcium.

As well as its effects on joint architecture, IL-1 exhibits systemic effects which will contribute to the other sequelae of rheumatoid arthritis. These include increased AP-protein synthesis and thus, an elevated ESR, neutrophil infiltration resulting from neutrophilia and increased chemotaxis of neutrophils, increased IgG secretion into synovial fluid and a high proportion of stable erythrocyte rosetting T cells (Wood et al., 1983). For the many effects of IL-1 on the immune system cf. Chapter 7.

3. Animal models of chronic inflammation

3.1. Arthritis

The pathogenesis of adjuvant arthritis has not been completely defined, although it undoubtedly involves macrophage activation and T-lymphocyte proliferation. Thus IL-1 almost certainly must play a major role. That there is a requirement for T lymphocytes is demonstrated by the observation that polyarthritis does not develop in athymic rats or mice (Kohashi et al., 1982; Billingham, 1983). There is a migration of activated macrophages and lymphocytes from lymph nodes to lesion sites during the second phase (Table 2). Immunity to the arthritogen is considered to be an exacerbating, but a secondary phenomenon.

It would seem that an animal model of rheumatoid arthritis that accurately reflects the time course and pathology of the human disease has not yet been found.

TABLE 2
Some features distinguishing distinct phases of adjuvant disease[a]. After Baumgartner et al. (1974)

Phase	Approximate time (days after adjuvant inoculation)	General characteristics	Local inflammation[b]	Splenomegaly	Impaired DMA	Distal foot swelling	Hypoalbuminemia	Thiol depression[c]
(a)	1 – 4	Acute local inflammation; systemic effect (liver)	+ +	0	+ (0)	0	+ (0)	(±)
(b)	7 – 12	Remission of acute inflammation; prearthritis	+	0	+	0	+ + (±)	0
(c)	12 – 28	Chronic inflammation with periarthritis	+ + +	+	+ +	+	+ +	+ +
(d)	21 – onwards	Residual systemic inflammation; osteogenic activity	+	+	+	+ +	+	+ +
(e)	35 onwards (indefinitely)	Permanent articular deformity, minimal (burnt-out) inflammation	0	0	↓ 0	+(+)	0	↓ 0

[a] Characteristic responses are those following tail/rear paw injection of an arthritogenic adjuvant. Alternative responses following ear inoculation of adjuvant are shown in parentheses.
[b] At site of adjuvant inoculation.
[c] Thiol titer of plasma proteins.

In all models so far investigated, only the ESR and CRP concentrations were found to correlate with radiological progression, which is considered the best observational basis on which to assess progression of rheumatoid arthritis (as reviewed by Billingham, 1983).

The most frequently used animal model of arthritis is adjuvant-induced arthritis in the rat (Stoerk et al., 1954; Pearson, 1956). This model has been used extensively for drug screening and has proven successful in identifying drug treatment that has been effective in the acute inflammatory stage of the disease. Unfortunately it is of limited use in screening for agents which may alter the fundamental progression of the disease. The disease is induced by injection of dispersed mycobacteria into the foot-pad or base of the tail. The rats are then monitored for at least 35 days. Baumgartner et al. (1974) have described five phases of adjuvant arthritis according to change in plasma protein concentrations, serum thiols, hepatic drug metabolism and paw size (Table 2). Billingham (1983) comments that monitoring of biochemical changes is more useful than following changes in paw size. These changes are compared in three different strains of rat in Fig. 1. Certain other agents, especially type II collagen, which is the phenotypic collagen of articular cartilage, have also been used to induce arthritis. Such experiments have shown that regardless of the inducing agent, the early stages of the disease are very similar (reviewed by Billingham, 1983).

The rat seems uniquely susceptible to induction of adjuvant arthritis and attempts to establish the disease in hamster, mouse, gerbil, guinea pig, rabbit, pig, dog and chicken have generally been unsuccessful (reviewed by Billingham, 1983). Furthermore, there is a significant strain variation with respect to incidence and severity of disease in rats (cf. Fig. 1). Mice, however, are susceptible to muramyldipeptide (MDP) induced arthritis (Kohashi et al., 1982), and to a complex immunization involving bovine serum albumin in complete Freunds adjuvant supplemented with extra mycobacteria and *Bordetella pertussis* organisms (Brackertz et al., 1977). Rabbit has been used in studies of gonococcal arthritis (Goldenberg and Rice, 1984), and antigen (fibrin) induced monoarticular arthritis (Dumonde and Glynn, 1962).

Adjuvant arthritis in rats has remained a consistently popular method for screening of anti-inflammatory agents. However, the adjuvant disease in rats is quite distinct from human rheumatoid arthritis in that all phases of adjuvant arthritis are inhibited by nonsteroidal anti-inflammatory drugs whereas in the human disease, symptoms, but not progression of the disease are affected. Furthermore, drugs which do induce clinical remission of human rheumatoid arthritis (gold, penicillamine and chloroquine) have little effect on adjuvant arthritis in rat (Billingham, 1983).

However, 2-(4-*p*-chlorophenyl benzyloxy)-2-methylpropionic acid-ICI55897, Clozic which like gold and penicillamine lowers the concentration of AP-proteins, the ESR and which slows the erosive progression of human rheumatoid arthritis, also inhibits progression of adjuvant arthritis in rats (Billingham, 1983). The mechanism by which Clozic works is not known, however, like gold and penicillamine, it is remission-inducing, but not anti-symptomatic.

Thus, while rat may not be ideally suited as a model which parallels the course

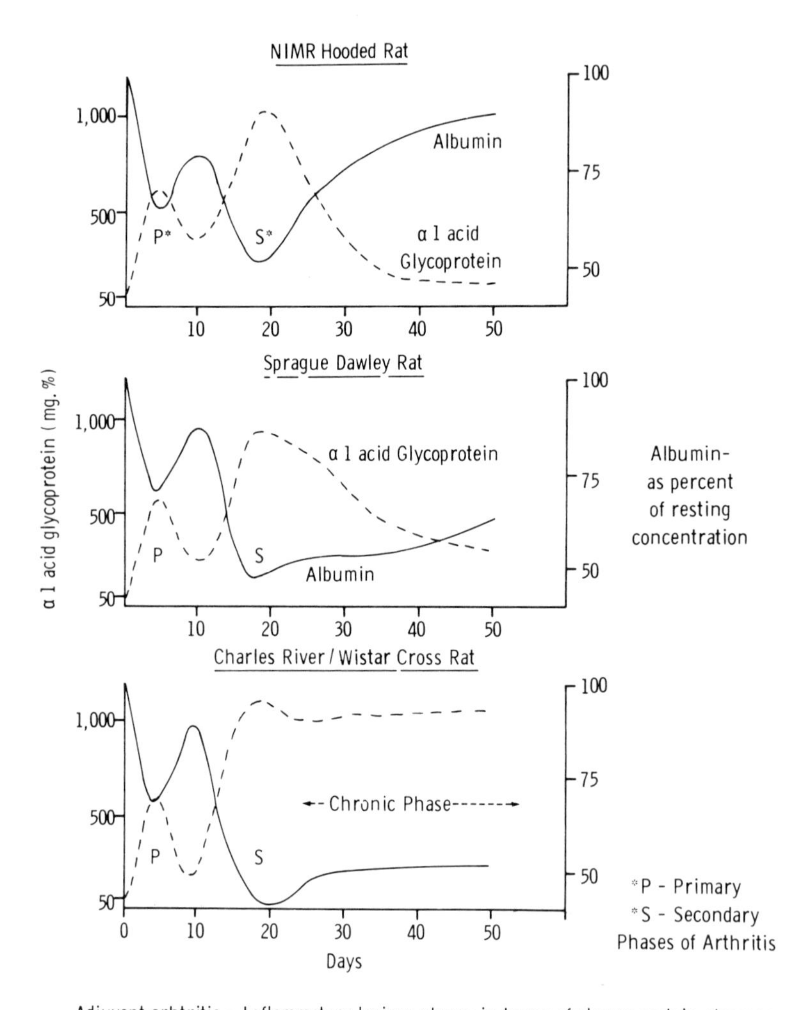

Fig. 1. Adjuvant arthritis in three rat strains. Variation is demonstrated in terms of the AP-proteins (α_1-AGP), and albumin levels in the serum. Arthritis subsided rapidly in the NIMR (National Institute for Medical Research. U.K.) hooded rat strain after day 20, as seen by the return of α_1-AGP and albumin to normal serum levels. In contrast the arthritis persisted for many weeks in the Charles River/Wistar cross strain, as shown by the abnormal serum protein levels of α_1-AGP and albumin.

of human disease, study of its inflammatory response may, because of its innate differences, yield important clues about species specific responses to inflammation.

3.2. Amyloidosis

Amyloidosis is the generic term given to a group of diseases characterized by the extracellular deposition of insoluble protein fibrils derived from plasma protein precursors by incomplete proteolysis. Amyloid fibrils are characterized by a β-pleated sheet conformation which is responsible for Congo red staining and birefringence of amyloid deposits (Glenner, 1980; Skinner and Cohen, 1983).

At present, three systemic forms of amyloidosis are recognized each of which has a distinct serum protein precursor. The primary or myeloma-associated forms of amyloid are made up of immunoglobulin light chains or portions thereof. Another form of amyloidosis which occurs where there is an underlying infection or inflammation (secondary amyloidosis) is characterized by deposition of amyloid A (AA) protein, believed to be derived from SAA by proteolytic removal of a carboxyl terminal peptide (Benditt et al., 1982). However, amyloid deposition does not necessarily result from the elevation of SAA. A third type of systemic amyloid is the hereditary type found in familial amyloid polyneuropathy. This autosomal dominantly transmitted disease is characterized by deposition of the negative AP-plasma protein prealbumin as insoluble β-pleated sheet fibrils. In addition to the systemic amyloidoses, there are several localized forms in which fibrils of prealbumin, and other as yet uncharacterized proteins are deposited (Cohen et al., 1983).

All amyloid deposits are defined by their fibrillar appearance, by their ability to bind the planar hydrophobic dye Congo red and when so stained to exhibit apple green birefringence and by their characteristic cross-β X-ray diffraction pattern.

The availability of animal models for acquired (secondary) systemic amyloidosis has greatly advanced our understanding of this form of the disease. The mouse has been most frequently employed for such studies (reviewed by Kisilevsky, 1983). Repeated daily injections of casein (Shirahama and Cohen, 1974) have been used in many laboratories to study the pathogenesis of amyloidosis. Recently, the accelerated mouse model using amyloid enhancing factor (AEF) has been used. This model has the effect of eliminating the lengthy and somewhat unpredictably preamyloid phase (Sipe et al., 1978). AEF administration has the effect of synchronizing the animals so that all lay down amyloid fibrils in spleen 48 h after the simultaneous administration of AEF and an inflammatory stimulus such as subcutaneous $AgNO_3$ injection.

One very active area of current investigation is to define the molecular and cellular factors involved in AA fibril deposition. The efforts of Axelrad et al. (1982)

to develop an experimental model of accelerated amyloidosis led to isolation of AEF. AEF is a high molecular weight glycoprotein extracted from tissues of mice that are either on the verge of or are actively undergoing amyloid deposition. AEF appears to be a product of chronic inflammation. A single inflammatory episode does not result in either AEF formation or amyloid deposition (Deal et al., 1982). AEF formation always precedes amyloid deposition by one or two days (Kisilevsky, 1983). Like other strains of mice, the LPS nonresponder C3H/HeJ mice develop amyloidosis when a potent inflammatory stimulus such as injection of silver nitrate is administered simultaneously to intravenous AEF injection. However, when AEF was administered to C3H/HeJ mice along with LPS-induced IL-1, there was no amyloid deposition although SAA was significantly elevated (C.L. Deal, J.D. Sipe, A.S. Cohen, unpublished observations). This suggests that another product of chronic inflammation is involved in AA fibril formation, perhaps an enzyme at the surface of mononuclear cells that is involved in the degradation of SAA (stage 1, Table 1 and Fig. 2).

4. AP-plasma proteins in different species: effects of chronic inflammation

The rheumatic diseases are characterized by the appearance in serum of new proteins of plasma cell origin such as rheumatoid factor and antinuclear antibodies.

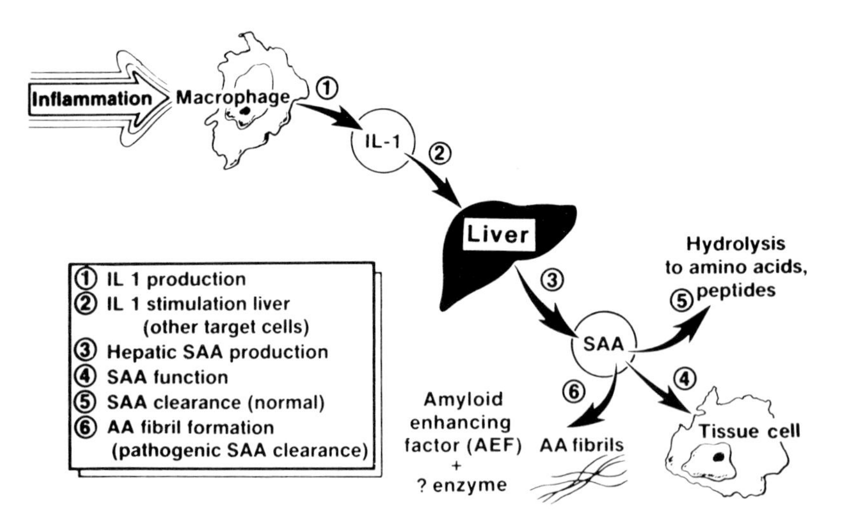

Fig. 2. Outline of SAA metabolism. SAA was identified during studies of inflammation-associated amyloidosis. The function of SAA is largely unknown although its role as AA precursor suggests that after SAA arrives at tissues (4), and is not cleared normally (5) then insoluble AA fibrils will accumulate, (6).

While serum concentrations of CRP, SAA, SAP and the AP-glycoproteins are generally elevated in chronic or recurrent acute inflammatory episodes, they almost never approach the values seen in acute episodes. In fact, tachaphylaxis is observed upon repeated inflammatory stimulation of rabbits and mice (Macintyre et al., 1982; Brandwein et al., 1983) (cf. Chapters 4 and 18).

The mouse model which is so useful for study of SAA synthesis and metabolism in acute and chronic inflammation is limited by the lack of a significant CRP-response. This has led many laboratories (reviewed by Sipe, 1985) to use SAP, a structurally related but antigenically distinct pentaxin, as an analog or a substitute for CRP. Following appropriate stimulation, 2 to 20-fold elevations in mouse SAP concentration have been recorded (Skinner et al., 1982).

Human SAP is not a significant AP-reactant. However, slightly higher levels have been found in males than in females (reviewed by Sipe, 1985). In Syrian hamsters, an SAP-like protein is under hormonal regulation. Amyloid P component (APC) is a pentaxin that is intimately associated with all amyloid deposits except those occurring in certain types of cerebral amyloid. APC is apparently identical to SAP and APC/SAP binding to amyloid fibrils is mediated by calcium (Skinner et al., 1982; Skinner and Cohen, 1983). The half-life of SAP is in the order of hours, thus a rapid and continuous production is required to maintain (in man) levels of 50 – 79 μg/ml. APC/SAP is usually present in amyloidotic tissues at levels ranging from 10 to 15% of the isolated fibril weight (Skinner et al., 1982). It remains to be determined whether SAP plays a role in the pathogenesis of amyloidosis. However, it is not an integral part of the fibril structure.

A thorough study of the AP-response in rats repeatedly injected with turpentine (Weimer and Humelbaugh, 1967) suggested that fibrinogen and complement levels were lower following the third challenge than the first, although the inducible α_2-AP globulin was greater following oil than the first. This early study suggests diversity in the regulation of AP-protein synthesis.

Also important is species variation in AP-protein profiles (Kushner, 1982; Chapter 12). Man and rabbit (Anders et al., 1977) are similar in that both SAA and CRP are inducible proteins, whereas mouse exhibits an inducible SAA fluctuation similar to man and rabbit. However, mouse does not have a significant CRP response although it does exhibit an alteration in SAP concentration. Rat exhibits two pentaxins, CRP and SAP both of which are constitutive proteins, both are stimulated to a limited extent during the AP-response. The rat is unique in its resistance to inflammation associated amyloidosis, and it is quite possible that the rat does not make SAA. Certainly the rat exhibits unique AP-proteins in that the α_1-AP-globulin appears to be the major AP-protein (Koj, 1974).

The rat is very resistant to endotoxin relative to other species. As shown in Fig. 3, the response in this species, as indicated by increased plasma concentration of a

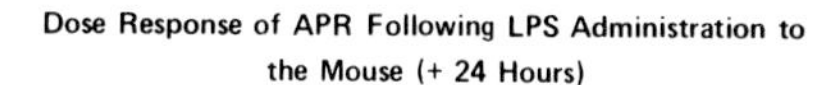

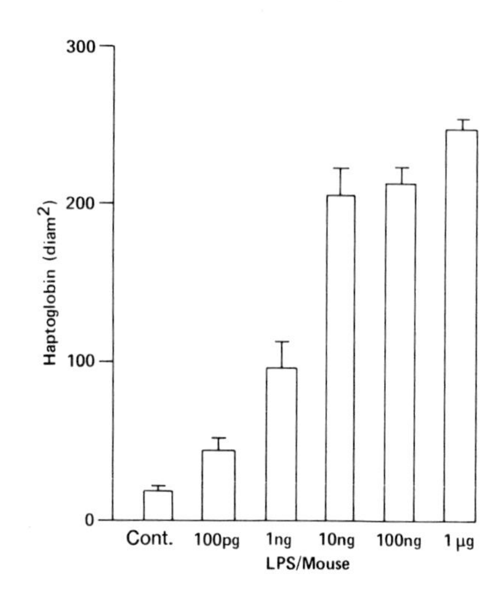

Dose Response of APR Following LPS Administration to
the Rat (+48 Hours)

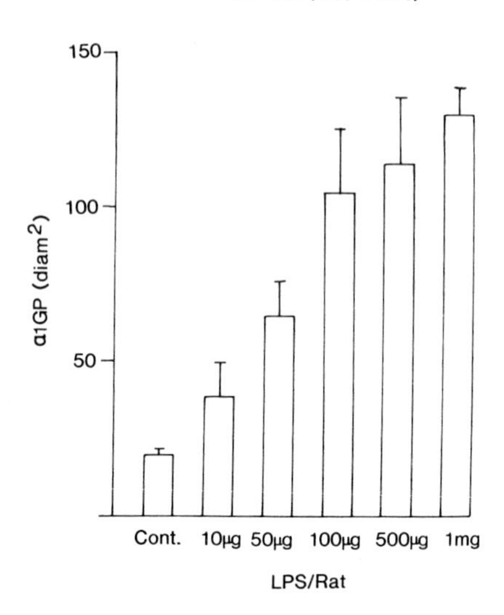

Fig. 3. Dose response to LPS (*Salmonella typhimurium* – Sigma) in the rat and mouse, given as a single treatment (intraperitoneally). AP-proteins were determined by radial immunodiffusion at the time of maximal response. Allowing for body weight (mice ~ 25 g; rat ~ 200 g) the difference in sensitivity between the species is several thousandfold: the minimum level for stimulation of the Alderley Park albino mouse is 4 ng/kg, whereas in the Alderley Park albino rat, the value is 50 µg/kg. (Cont. = the control value for the serum concentration of the proteins in unstimulated animals.)

major AP-protein α_1-AGP AP protein, compared with a similar AP-response in mice, required the injection of more than 1000 times as much LPS (M.E. Billingham, personal communication).

5. *Therapeutic intervention in chronic inflammation*

In addition to the dramatic increases and decreases which occur within the first 24 h of trauma and rapidly progressing infections, lesser but elevated concentration of AP-reactants can be measured after several days or weeks of more persistent disease processes. In addition to rheumatoid arthritis, granulomatous bowel disease, hypersensitivity, cancer and low grade infection are marked by lesser elevations of the AP-proteins.

Furthermore, there are secondary changes which occur weeks after the disappearance of clinical signs of illness, such as anemia, hypergammaglobulinemia and changes in the T-lymphocyte population (Dinarello, 1984b).

As was mentioned in the introductory paragraph of this chapter, timing is of the essence if intervention in IL-1 mediated host responses is to be attempted. The sequence of events following an inflammatory stimulus includes IL-1 production as an early and essential step. However, IL-1 has both stimulatory and inhibitory effects and thus superimposed inflammatory events become inextricably complicated if each episode is not allowed to reach its natural conclusion.

In the case of secondary amyloidosis, a long-lived product (AEF) of chronic inflammation can arise in conjunction with additional inflammatory episodes and can lead to AEF amyloid fibril deposition. It is possible that if IL-1 action on SAA synthesis could be blocked, further amyloid deposits would not develop. However, since IL-1 can have both beneficial and harmful effects, the history of an inflammatory process as well as its current status would need to be considered before a decision to intervene could be made.

References

Anders, R.F., Natvig, J.B., Sletten, K., Husby, G. and Nordstoga, K. (1977) J. Immunol. *118,* 229 – 234.

Axelrad, M.A., Kisilevsky, R., Willmer, J., Chen, S.J. and Skinner, M. (1982) Lab. Invest. *47,* 139 – 146.

Baumgartner, W.A., Back, F.W.J., Lorber, A., Pearson, C.M. and Whitehouse, M.W. (1974) Proc. Soc. Exp. Med. *145,* 625 – 630.

Benditt, E.P., Hoffman, J.S., Eriksen, N., Parmelee, D.C. and Walsh, K.A. (1982) Ann. N.Y. Acad. Sci. *389,* 183 – 189.

Billingham, M.E. (1983) Pharmacol. Ther. *21,* 389 – 428.

Bodel, P.T. and Hollingsworth, J.W. (1968) Br. J. Exp. Pathol. *49,* 11 – 19.

Brackertz, D., Mitchell, G.F. and MacKay, I.R. (1977) Arthritis Rheum. *20,* 841 – 850.

Brandwein, S.B., Sipe, J.D., Skinner, M. and Cohen, A.S. (1983) Fed. Proc. *42,* 3406.

Cohen, A.S., Shirahama, T., Sipe, J.D. and Skinner, M. (1983) Lab. Invest. *48,* 1 – 3.

Dayer, J.M., Goldring, S.R., Robinson, D.R. and Krane, S.M. (1980) in Collagenase in Normal and Pathological Connective Tissues (Woolley, D.E. and Evanson, J.M., eds.), pp. 83 – 104, Wiley and Sons, New York.

Deal, C.L., Sipe, J.D., Tatsuta, E., Skinner, M. and Cohen, A.S. (1982) Ann. N.Y. Acad. Sci. *389,* 439 – 441.

Desmukh-Phadke, K., Nanda, S. and Lee, K. (1980) Eur. J. Biochem. *104,* 175 – 180.

Dinarello, C.A. (1984a) Rev. Infect. Dis. *6,* 51 – 95.

Dinarello, C.A. (1984b) Surv. Immunol. *3,* 29 – 33.

Dumonde, D.C. and Glynn, L.E. (1962) Br. J. Exp. Pathol. *43,* 373 – 383.

Estes, J.E., Pledger, W.J. and Gillespie, G.Y. (1984) J. Leukocyte Biol. *35,* 115 – 129.

Fontana, A., Hengartner, H., Weber, E., Fehr, K., Grob, P.J. and Cohen, G. (1982) Rheumatol. Int. *2,* 49 – 53.

Glenner, C.G. (1980) N. Engl. J. Med. *302,* 1283 – 1292, 1333 – 1343.

Goldenberg, D.L. (1983) Am. J. Med. *74,* 925 – 928.

Goldenberg, D.L. and Rice, P.A. (1984) Prog. Clin. Rheum. *1,* 179 – 201.

Gowen, M., Woods, D.D., Ihrie, E.J., McGuire, M.K.B. and Russell, R.G.G. (1983) Nature (London) *306,* 378 – 380.
Hadler, N.M. and Granovetter, D.A. (1978) Semin. Arthritis Rheum. *8,* 1 – 16.
Kisilevsky, R. (1983) Lab. Invest. *49,* 381 – 390.
Kohashi, O., Aihara, K., Ozawa, A., Kotani, S. and Azcima, I. (1982) Lab. Invest. *47,* 27 – 36.
Koj, A. (1974) in Structure and Function of Plasma Proteins (Allison, A.C., ed.), Vol. 1, pp. 73 – 131, Plenum Press, New York.
Kushner, I. (1982) Ann. N.Y. Acad. Sci. *389,* 39 – 48.
Macintyre, S.S., Schultz, D. and Kushner, I. (1982) Ann. NY Acad. Sci. *389,* 76 – 87.
McGuire, M.K.B., Wood, D.D., Meats, J.R., Ebsworth, N.M., Gowen, M., Beresford, J., Gallagher, J.A. and Russell, R.G.G. (1982) Calcif. Tissue Int. *34,* S10.
Mizel, S.B., Dayer, J., Krane, S.M. and Mergenhagen, S.E. (1981) Proc. Natl. Acad. Sci. USA *78,* 2474 – 2477.
Pearson, C.M. (1956) Proc. Soc. Exp. Biol. Med. *91,* 96 – 101.
Postlethwaite, A.E., Lachman, L.B., Mainardi, C.L. and Kang, A.H. (1982) *157,* 801 – 806.
Saklatvala, J. and Dingle, J.T. (1980) Biochem. Biophys. Res. Commun. *96,* 1225 – 1231.
Saklatvala, J., Pilsworth, L.M.C., Sarsfield, S.J., Gavrilovic, J. and Heath, J.K. (1984) Biochem. J. *224,* 461 – 466.
Schmidt, J.A., Mizel, S.B., Cohen, D. and Green, I. (1982) J. Immunol. *128,* 2177 – 2182.
Shirahama, T. and Cohen, A.S. (1974) J. Exp. Med. *140,* 1102 – 1107.
Sipe, J.D., Vogel, S.N., Ryan, J.L., McAdam, K.P.W.J. and Rosenstreich, D.L. (1979) J. Exp. Med. *150,* 597 – 606.
Sipe, J.D., McAdam, K.P.W.J. and Uchino, F. (1978) Lab. Invest. *38,* 110 – 114.
Sipe, J.D. (1985) Lymphokines, Vol. 12, in press.
Skinner, M. and Cohen, A.S. (1983) in Connective Tissue Disease (Wagner, B.M., Fleischmajer, R. and Kaufman, N. eds.), pp. 97 – 119, Williams and Wilkins, Baltimore.
Skinner, M., Sipe, J.D., Yood, T., Shirahama, T. and Cohen, A.S. (1982) Ann. N.Y. Acad. Sci. *389,* 190 – 198.
Stoerk, H.C., Bielinski, T.C. and Budzilovich, T. (1954) Am. J. Pathol. *30,* 616.
Weimer, H.E. and Humelbaugh, C. (1967) Can. J. Physiol. Pharmacol. *45,* 241 – 247.
Wood, D.D., Ihrie, J., Dinarello, C.A. and Cohen, P.L. (1983) Arthritis Rheum. *26,* 975 – 983.

PART V

Assays of monokines

Gordon/Koj (eds.) The acute-phase response to injury and infection.

CHAPTER 23

Assays of monokines

A.H. GORDON

1. Comparison of various assays of IL-1

Following recognition of the important role of interleukin 1 (IL-1) as the primary mediator of the acute-phase response (AP-response), a greatly increased need has arisen for suitable means for its assay. Ideally such an assay would combine specificity, sensitivity, rapidity, operational convenience and the possibility of handling many samples at a time and most important a sufficient degree of accuracy. Needless to say, economy may sometimes be an overriding consideration. As shown in Table 1, a number of approaches have been tried. These may be divided into those where IL-1 has an effect which can be measured as such, e.g. induction of fever or increased incorporation of [^{3}H]thymidine into cells and others in which addition of IL-1 leads to production of something, e.g. prostaglandin E_2 (PGE_2) or collagenase which can then itself be estimated. Though convenient, this division is only of limited validity because IL-2 is an essential intermediate in the thymocyte costimulator assay and could, at least in theory, be estimated. However, except for fever in rabbits and the mouse thymocyte costimulator assay, all the rest are at a preliminary stage and remain to be investigated in depth.

TABLE 1
Bioassays, including cell culture assays for IL-1

		Accuracy	Sensitivity	Specificity	Sample economy	Time required	No. of samples/ assay	Ref.
In vivo responses								
Fever	Rabbits[a]	★★	★★	★★★	★	★★	★	Allison et al. (1973)
	Mice		★★	★★★	★★★	★★★	★	Bodel and Miller (1978)
								Gordon and Parker (1980)
Plasma constituents	AP-protein	ND	★	★★★	★	★	★	Sztein et al. (1981)
								Kampschmidt et al. (1983)
	zinc	ND	★★	★★	★	★★	★	Kampschmidt and Upchurch (1970)
	iron	ND	★★	★★	★	★★	★	Kampschmidt et al. (1983)
Cell culture								
Thymocyte costimulator assay		★★	★★★	★★	★★★	★	★★★	Mizel et al. (1978)
Fibroblasts	[^{3}H]thymidine incorporation	★★	★★★		★★★	★	★★★	Sellers and Reynolds (1977)
	collagenase		★					Jaffe and Behrmann (1977)
	PGE_2	★★★						
	[^{3}H]thymidine incorporation							

Chondrocytes[b]								
	collagenase		★					
	PGE_2		★★★					
Hepatocytes	AP-protein		★★	★★★	★★	★	★★	Koj et al. (1984)
Leucocytes	chemotaxis	★	★★	ND	★★	★★★	★★★	Werner et al. (1980)
Organ culture								
Cartilage, chondroitin sulphate release		★★★	★★★	★	★★★	★	★★★	Saklatvala et al. (1983)

On the basis of the limited data available the usefullness of these assays may be ranked as indicated:
★★★ Good, relatively rapid.
★★ Medium.
★ Poor.
[a] log dose/response ratio available.
[b] Includes synovial cells.

2. *Assays based on fever*

2.1. *Fever in rabbits*

At least three rabbits are required for each dose. Before use, the animals must be acclimatised to accustom them to their surroundings. This requires 1 – 2 weeks in a quiet room maintained at an even temperature. For the assay the rabbits should be lightly restrained in head stocks and equipped for measurement of rectal temperature with thermistors inserted to a depth of at least 10 cm. Temperature should be recorded at intervals of only a few minutes. Any animal showing a variation of more than 0.2°C during 30 min before injection must be excluded from the assay.

Increases in temperature during 90 min post-injection constitute the basis of the assay. Two assays per day can be carried out but the second injection must be at least 3½ h after the end of the first period of measurement and should only be attempted provided the response in the first assay has not exceeded 0.6°C. If this schedule is employed the rabbits cannot be used for more than 4 days (Allison et al., 1973). The same authors also noted that mean responses in the afternoons were lower than those occurring in the mornings.

If it is suspected that there may be an antigenic substance present with the IL-1 then the rabbits should not be used more than once. However, immune fevers are not expected before 5 to 7 days (Dinarello, 1982).

As shown in Fig. 1, the temperature response to endogenous pyrogen (EP) is rapid and easily distinguishable from that given by endotoxin. Both the area under the curve and the maximum rise in temperature during 90 min after injection have been shown to be linearly related to the log of the injected dose (Allison et al., 1973).

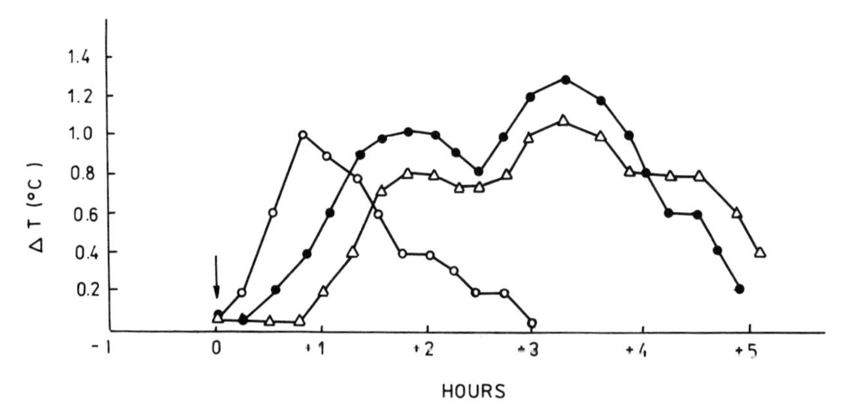

Fig. 1. Mean fever in rabbits injected with various pyrogens. (○—○), EP/IL-1; (●—●), 5 ng E. Coli endotoxin; (△—△), human serum albumin in rabbits sensitized to this protein. (Dinarello, 1982.)

Probably the most serious disadvantage of this assay is that the immune response caused by giving an antigenic substance more than once to the same animal may delay, interfere with, or even abolish the rise in temperature. Complete confidence that this has not occurred can only be achieved by single use only of each rabbit. A less satisfactory alternative is the use of rabbits rendered refractory by previous injection of a large dose of endotoxin. Unfortunately however, such animals show tachyphyllaxis. One millilitre of typhoid vaccine has been used for this purpose (Allison et al., 1973). A much more serious difficulty is interindividual differences in temperature response which means that the assay cannot be expected to detect much less than 2-fold changes in EP/IL-1 concentration.

The fever response in rabbits to bacterial endotoxin is the basis of the British Pharmacopoeia test for the substance, for full details of the conditions required cf. British Pharmacopoeia (1980). As with EP/IL-1 there is severe variability. This has been investigated in detail in an international collaborative study (WHO ECBS unpublished working document No BS/83) in 22 countries which showed that the threshold doses of a variety of endotoxins causing a 0.6°C rise in temperature in rabbits varied widely but all fell within one order of magnitude. This compared favourably with a much greater variability obtained using the Limulus lysate assay for the same preparations.

2.2. Fever in mice

The use of mice for assay of EP/IL-1 was introduced by Bodel and Miller (1976). Initially EP/IL-1 from mouse peritoneal macrophages was used. Similar results were obtained with supernatants obtained by incubation of human blood leucocytes with heat-killed staphylococci (Bodel and Miller, 1977, 1978).

The sensitivity of mice to both homologous and human EP/IL-1 was demonstrated when fever occurred, after injection into a single mouse of supernatant, from as few as 5×10^3 appropriately stimulated blood monocytes. Direct comparison of the fever response of rabbits to the same material indicated that mice are approximately 100 times more sensitive to EP/IL-1 than are rabbits. This statement does not take into account the relative sizes of mice and rabbits, which also differ by about the same factor. Thus, if expressed on a dose per animal weight basis, the mouse is no more than equally sensitive to EP/IL-1 than is the rabbit. Despite this similarity the use of mice for EP/IL-1 assay must be considered as potentially very advantageous because so many more mouse responses compared with rabbit responses can be obtained for each unit of EP/IL-1.

Although the usual reaction of mice to EP/IL-1 is hyperthermia, a small proportion respond with a rapid fall in temperature. Fortunately this hypothermic response to EP/IL-1 becomes vanishingly rare if the mice are warmed before the assay or retained throughout at a temperature of 35°C. Using a group of 20 Swiss Webster mice under these conditions, Bodel and Miller (1978) obtained a mean maximum increase of 0.45 ± 0.05°C. Alternatively the mice can be prewarmed for 1 h at 37 – 39°C and then used for the assay in a room at ambient temperature. Working thus with groups of eight mice per dose, Gordon and Limaos (1979) and Gordon and Parker (1980), were able to establish log dose-temperature increase relationships for supernatants obtained from human leukocytes and also for pyrogen R obtained from the same source (cf. Chapter 3, Section 4.5). Apparently the strain of mice employed is not of critical importance because Parkes mice were used for the latter assays and another strain obtained locally in Brazil have also been successfully used.

3. *Assays using cells or tissues in culture*

3.1. *Thymocyte costimulator assay*

Mouse thymocytes maintained in culture, in presence of PHA, will in presence of IL-1 take up [^{3}H]thymidine proportionately to the amount of IL-1 present. A general description of how and why this comes about has been attempted in Chapter 7, Section 3.1. It remains to outline the conditions necessary for successful employment of murine thymocytes for this purpose.

The mice should be at the weanling stage, male or female weighing no more than 20 g. A number of strains have proved suitable including C57/Black 6, Parkes and C_3H/HeJs. These latter are advantageous because of their limited sensitivity to endotoxin and should therefore be chosen if endotoxin is likely to be present. The thymus is removed aseptically; then on teasing with a probe the gland will be found to disintegrate easily. After this has been done the cells are suspended in RPMI 1640 medium containing 5% foetal calf serum (FCS), 5×10^{-5} M mercaptoethanol, fresh 2 mM glutamine and penicillin plus streptomycin or other suitable antibiotics. After standing for 10 min to allow sedimentation under gravity of cell debris, the cells are spun down and resuspended in fresh medium at a concentration of at least 1.5×10^7 cells per millilitre.

Incubation is most conveniently conducted on a cell culture plate designed to take 96 samples each of 0.2 ml. Before transfer of the cell suspension to the plate, a small part is retained as a control in absence of PHA. To the remainder is added sufficient PHA to give a final concentration of 1 – 2 μg/ml. Next, the plate is loaded with the samples to be assayed, usually in triplicate and only in sufficient wells to allow dilutions to be made by transfer into other wells pre-loaded with medium. Finally aliquots of cells with and without PHA are added to the wells in sufficient amounts to give between 5×10^5 and 1.5×10^6 cells per well. Alternatively the cells may be put into the plate before the addition of the samples. This has the advantage that cell attachment may be inhibited in presence of the samples. After 2 – 3 days incubation in a 5% CO_2 incubator [^{3}H]thymidine is added, 0.5 μc per well, and after a further 5 – 18 h free [^{3}H]thymidine is removed from the cells after these have been sucked from the microwells using an automatic cell harvester. After the cells had undergone multiple washing, on the special glass fibre paper appropriate for the cell harvester and drying, discs of paper carrying cells from and to the 96 wells are dried, transferred into vials with scintillation fluid and estimated for [^{3}H]thymidine.

Dilutions of a standard solution of IL-1 are incorporated and the units found in the unknown are obtained by comparison of the 50% of maximum response values for the standard and the unknown. For these to be obtained, percentages of maximum response are plotted against the logs of the dilutions of the standard and the unknown (cf. Fig. 1 in Mizel, 1980).

With slight variations of conditions, this assay has yielded reproducible results in a large number of laboratories. Horse serum was used as an alternative to FCS by Mizel and Mizel (1981). Increased incorporation of [^{3}H]thymidine at 39°C instead of 37°C was reported by Duff and Durum (1982). As mentioned above, an important limitation of the usefulness of this assay is its lack of specificity (cf. Chapter 7, Section 3.1). A less important difficulty may occur if inhibitory substances are present in samples. This is the main reason for the use of sample dilutions as described above. Under optimum conditions increases of as much as 10 to 20-fold over the PHA control have been achieved. Finally the remarkable sensitivity of this assay must be emphasised. As little as 10^{-5} of the amount of IL-1 sufficient to produce detectable fever in a rabbit can cause significant enhancement of [^{3}H]thymidine incorporation (Rosenwasser and Dinarello, 1981), thus the assay has sensitivity comparable to that of radioimmunoassay (cf. Section 4.1).

3.2. Inhibitors of EP/IL-1

3.2.1. Presence of inhibitors in plasma

Because the thymocyte costimulator assay is normally carried out in 96 well cell culture plates, sequential dilutions of the samples under test are simple to perform. The importance of this procedure becomes obvious when relative crude preparations of IL-1 are under examination. In such cases sequential dilution often leads to greatly increased incorporation of [^{3}H]thymidine. As shown in Table 2, 256-fold dilution of serum from mice previously injected I.V. with *Proprionibacterium acnes* was required to give the maximum response (Wood et al., 1983). In this experiment interference by IL-2 was shown to be minimal by the use of C57 BL/6 nu/nu (nude) mice in place of C57 BL/6 mice. It is concluded that when thus stimulated, both IL-1 and an inhibitor of thymocyte proliferation are released into the plasma.

TABLE 2
LAF activity in serum from *P. acnes*-primed mice after injection with lipopolysaccharide (LPS)

Sample dilution	Thymocyte proliferation (cpm)			
	P. acnes + LPS	*P. acnes* + saline	Control + LPS	PU-5-IR supernatant
1:16	25 ± 7	20 ± 8	32 ± 4	4820 ± 20
1:32	813 ± 131	22 ± 8	61 ± 2	4237 ± 24
1:64	2047 ± 191	32 ± 8	128 ± 6	3475 ± 318
1:128	3756 ± 458	41 ± 3	142 ± 35	1912 ± 167
1:256	4258 ± 46	63 ± 15	161 ± 36	1304 ± 4
1:512	3938 ± 55	56 ± 4	120 ± 4	512 ± 30
1:1024	2798 ± 242	71 ± 1	84 ± 12	288 ± 44
1:2048	1367 ± 455	72 ± 10	87 ± 22	134 ± 25
1:4096	626 ± 91	76 ± 11	86 ± 4	128 ± 16

Activity in cells plus con A alone was 99 ± 8.
Data are given as mean counts per minute (cpm) ± S.E. of duplicate cultures.
CBA mice were injected I.V. with P. acnes (0.5 mg) and challenged 2 weeks later with LPS-W (20 μg/mouse) or saline by the same route. From Wood et al. (1983).

Evidently the IL-1, rather than the inhibitor, had the greater effect on the thymocytes since dilution led to loss of inhibition.

Several years previously the presence of an inhibitor of the thymocyte costimulator assay in serum from human volunteers with fever was reported by Dinarello et al. (1981). In this work all the serum samples were pretreated with an immunoadsorbant for IL-1 and only tested after elution in a manner which would have released IL-1 if it had been present. Unexpectedly, because the presence of IL-1 had been anticipated, the costimulator assay showed only inhibition. This finding suggests either, sufficient homology between IL-1 and an inhibitor to explain binding of both by the antiserum, or the existence of IL-1 and the inhibitor as a complex in the plasma. In more recent work, using a different method, it has been possible to assay the amount of IL-1 in human plasma in various disease states (Cannon and Dinarello, 1984). Interestingly in virus infections little or no increase in IL-1 over basal levels could be detected. Whether this finding may be the result of exceptionally raised levels of IL-1 inhibitor in such infections remains to be ascertained.

3.2.2. Presence of IL-1 inhibitors in human urine

As occurred with human plasma, a search for IL-1 in urine of febrile patients did not succeed but did reveal the existence of an inhibitor (Laio et al., 1984). Preliminary purification by precipitation with ammonium sulphate and chromatography on Sephacryl S-200, led to the identification of material eluting in the range of 20 – 40 kDa, with strong IL-1 inhibitory power in the thymocyte costimulator assay. As shown in Fig. 2, competitive antagonism occurred with murine IL-1 derived from P388/D_1 cells. Because inhibition of the thymocyte costimulator assay can be brought about by SAA, Liao et al. (1984) suggested the possibility that the inhibitor which they had identified in human urine, might be a fragment of this protein. They suggested this hypothesis because the urine inhibitor itself is too small to be serum amyloid A (SAA) or C-reactive protein (CRP) which is another plasma protein with an immunomodulatory effect. As SAA and CRP are both AP-proteins, their synthesis is stimulated by IL-1. Inhibition of the effectiveness of IL-1 by a fragment of such an AP-protein, if shown to occur, would be of special interest because of the possible provision of negative feedback and thus limitation of the consequences of IL-1 production in an inflammatory state.

Confirmation that inhibitors of IL-1 can be found in human urine has been obtained using urine from normal individuals (Kimball et al., 1984). The presence of IL-1 inhibitory activity was made apparent in two ways. Firstly, using the thymocyte costimulator assay, sequential dilutions of concentrates prepared from certain, but not all, normal urines led to greatly increased activities. Secondly, additions of such urine concentrates to IL-1 led to inhibition of the IL-1 costimulator activity. These

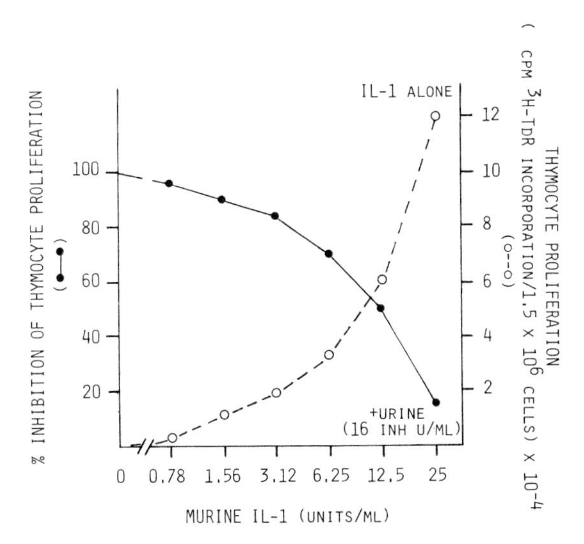

Fig. 2. Competitive antagonism between IL-1 and the urine inhibitor (INH) on mouse thymocyte proliferation. Mouse thymocytes were cultured in presence of increasing concentrations of murine IL-1 and a fixed amount of sterile, dialysed febrile urine with known inhibitory activity. Results represent the arithmetic means of triplicate cultures. (Liao et al., 1984.)

results were explicable assuming not only the presence of inhibitory activity against IL-1 in the thymocyte costimulator assay but also the presence of IL-1 itself in the normal urines. The presence of IL-1 in concentrates of urine made by ultrafiltration with an Amicon PM 10 membrane was confirmed by means of the fibroblast proliferation assay, described by Schmidt et al. (1982), using human early passage dermal fibroblasts. With this assay there was strong stimulation but no evidence of inhibition. As epidermal growth factor (EGF) is known to affect fibroblasts in these conditions, and the presence of this mediator is to be expected in urine, care was taken to distinguish its effect from that of the IL-1.

Gel filtration on S.200 of the concentrated urine revealed the presence of at least three molecular species smaller than 15 kDa. However, as a 60 kDa IL-1 species was also present, the urinary forms of IL-1 could not be characterised only as proteolytic fragments. As shown in Table 3, assay of both 4 and 2 kDa pools from the S.200 column by both thymocyte costimulation and by fibroblast proliferation were carried out. Most interestingly, the 4 kDa pool was completely inactive in the latter assay. Both results of Liao et al. (1984) and Kimball et al. (1984) have been described in some detail because the clear demonstration that there are IL-1 inhibitors in urine must underline that in any assay of IL-1 the possible presence of specific inhibitors must always be taken into account.

3.3. Cartilage resorption assay

Cartilage explants in organ culture will resorb their matrix when variously stimulated, e.g. by vitamin A (Fell and Mellanby, 1952), or by products of synovium

TABLE 3
Estimation of IL-1 activity[a]

Sample	Assay system	
	Thymocyte activation (U/ml)	Fibroblast proliferation (U/ml)
IL-1[b]	150	7000
Urine[c]	0.8 – 80	0.15 – 2.5
2 kDa pool[d]	0.4	13
4 kDa pool[d]	0.27	<1

[a] Results of probit analyses on serially diluted urine samples compared to a standard of IL-1 preparation and expressed in U/ml (cf. Section 3.1).

[b] Culture supernatant of silica-activated human peripheral blood monocytes.

[c] Range of fibroblast proliferation responses obtained for 12 normal urine samples tested. Only the six (of 12) thymocyte assay values that provided significant activation were included.

[d] Units/ml in the original urine. From Kimball et al. (1984).

(Fell and Jubb, 1977). The relevance of this effect to the properties of IL-1 became apparent as a result of the purification of catabolin, a 21 kDa protein obtained from both pig synovium and also from pig leucocytes (Saklatvala et al., 1983). As described in Chapter 10, Section 3, this protein shares many properties with human and murine IL-1, including the power to stimulate incorporation of [^{3}H]thymidine in the co-stimulator assay. A comparison of the activity of highly purified catabolin from both pig synovium and pig leucocytes in respect to cartilage resorption and the thymocyte costimulator assay has been made by Saklatvala et al. (1984). Interestingly, both catabolins were active at lower concentrations using cartilage than in the thymocyte assay. In the former assay catabolin gave a dose-response curve at between 2 and 20 pM, whereas an order of magnitude more was required for a comparable result in the thymocyte assay. The concentrations of pig catabolin required for thymocytes corresponds well with the concentration of highly purified murine IL-1 used in the same assay by Mizel and Mizel (1981). However, it remains to be seen whether the same ratio of activities will be found for IL-1 obtained from other species on tissues.

The assay is most conveniently conducted using 2 mm diameter × 1 mm thick discs of bovine nasal cartilage placed in the wells of a cell culture plate in suitable volumes of Dulbecco's modification of Eagles medium containing 5% heat inactivated normal sheep serum and the test sample. After 3 days at 37°C in 5% CO_2, 95% air, the medium is estimated for chondroitin sulphate using dimethyl methylene blue (Saklatvala et al., 1983). In these conditions there is no evident collagenase loss from the cartilage perhaps because collagenase, if it is produced, is in a latent form. The amount of chondroitin released has been found to increase approximately linearly with the amount of catabolin added. Important advantages of this assay are that it utilises a direct product of catabolin/IL-1 action and that radioactivity is not required.

3.4. Collagenase assay for IL-1

As IL-1 has been shown to bring about release of collagenase from synovial cells (Dayer et al., 1976) from chondrocytes (Gowen et al., 1984) and from muscle (Baracos et al., 1983), estimation of this action could, under appropriate circumstances, constitute an indirect assay for IL-1. Digestion of [^{14}C]acetylated rat skin collagen fibrils is often used for this purpose (Sellers and Reynolds, 1977). Pretreatment of samples for 10 min with trypsin at 100 μg/ml and 37°C followed by soya bean trypsin inhibitor is recommended in order to activate the collagenase and at the same time destroy collagenase inhibitors.

3.5. Assay of hepatocyte stimulating factor (HSF)/IL-1 in hepatocyte cultures

Recently, primary cultures of rat and mouse hepatocytes have been widely used for studying the mechanisms of increased synthesis of AP-proteins (reviews in Chapter 17). The amounts of HSF/IL-1 which can be measured by means of such hepatocyte cultures is 2 to 3 orders of magnitude lower than those required for in vivo assays based on determinations of AP-proteins. In addition, up to 100 samples can be tested simultaneously in the in vitro assay. Use of hepatocyte cultures has made possible testing of individual samples obtained by gel chromatography or chromatofocusing and has resulted in the separation of the lymphocyte activating factor (LAF) and HSF activities present in various cytokine preparations (cf. Chapter 17). So far two quantitative HSF assays with rat hepatocyte cultures have been described in detail: enzyme-linked immunosorbent assay of fibrinogen by Ritchie and Fuller (1981), and electroimmunoassay of albumin and α_2-macroglobulin (α_2-M) by Koj et al. (1984). In both assays hepatocytes were isolated from the liver of adult healthy rats by perfusion with collagenase using the established procedure (Sweeney et al., 1978). Only preparations of hepatocytes containing more than 85% viable cells, as judged by trypan blue exclusion, were used for the cultures.

3.5.1. HSF assay of Ritchie and Fuller (1981)

Approximately 3×10^6 hepatocytes suspended in modified Williams' E medium were plated in 35 mm dishes. After 30–45 min of incubation required for cell attachment the supernatants were aspirated and the cell monolayers were maintained in 1 ml of medium which was changed daily. The material undergoing test (or saline in the control) was added to the culture with each change of medium. After 2 days of culture the final supernatants were used for ELISA-based quantitative determination of rat fibrinogen as described by Kwan et al. (1977). The absorbances obtained from these assays were then converted into μg/ml of fibrinogen. Fibrinogen concentrations relative to cell protein present in each dish were expressed as μg fibrinogen secreted/mg protein/24 h. The amount of fibrinogen secreted by control cultures thus expressed was in the range of 1 to 2 μg/ml, while in HSF-treated cultures it reached 6 μg/ml. Fibrinogen secretion was found to be related to HSF in a dose-dependent manner (Fig. 3). The concentration of HSF in the culture media necessary to achieve a 50% maximal fibrinogen response was taken to represent one unit of HSF activity. Although the authors did not report on the accuracy and reproducibility of the bioassay, the method has been successfully used in further studies (for reference see Chapter 17).

3.5.2. HSF assay of Koj et al. (1984b)

Hepatocytes were suspended at a concentration of 2×10^6 cells per ml in complete

Williams' E medium enriched in heparin, insulin, dexamethasone, FCS and a mixture of antibiotics. Hepatocyte monolayers were obtained by pipetting 0.1 ml cell suspension (2×10^5 cells) into flat bottom wells (approximately 17mm in diameter) of Linbro plates coated with collagen and containing 0.3 ml of medium. After 2 h at 37°C in 5% CO_2, the unattached cells were aspirated and fresh 0.2 ml portions of the medium added followed by 50 μl of phosphate buffered saline (PBS) or serial dilutions of the material being tested (expressed as μl of undiluted preparation). The plates were incubated at 37°C in 5% CO_2 with daily changes of the media, each time PBS or the preparation under test were introduced. The samples of media collected after the last day of culture (i.e. 48 – 72 h period) were used for determination of rat albumin and rat α_2-M by rocket immunoelectrophoresis according to Weeke (1973). The heights of the rockets were used for estimation of albumin and α_2-M concentration by reference to protein standards (cf. Chapter 17, Fig. 4). The results were expressed in μg of albumin and α_2-M secreted into the medium by 2×10^5 hepatocytes during 24 h incubation, or as the ratio of α_2-M:albumin concentration in the medium of each hepatocyte culture (cf. Chapter 17, Fig. 5). As discussed in Chapter 17, there are several advantages in using α_2-M:albumin ratios although it is still uncertain whether the reduced synthesis of albumin and increased synthesis of α_2-M by rat hepatocytes are caused by a single component in the cytokine preparations.

Since different batches of hepatocytes vary in respect to basal synthesis of albumin and α_2-M, and also in their responses to HSF/IL-1, a standardisation procedure is required. This can be done either by incorporation of an arbitrary standard cytokine preparation into each assay, or by construction of an experimental curve based on the maximum hepatocyte response, such as that used by Ritchie and Fuller (1981) Fig. 3. Alternatively, a simpler approach is to use an arbitrary unit of HSF activity defined as the amount of cytokine required to increase the α_2-M:albumin ratio from 0.3 to 0.4. In this region linear dose-response relationships were observed for almost all the human cytokine preparations which were investigated (cf. Chapter 17, Fig. 5 and Koj et al., 1984). HSF activity can most easily be calculated from the log-log plot of α_2-M:albumin ratio versus the consecutive dilutions of the preparation under test. The reciprocal value thus obtained will be related to the concentration of the active component in a cytokine preparation.

As shown in Table 4, human cytokine preparations tested on three different hepatocyte cultures gave values ranging from 286 to 312 U/ml (mean 300 ± 11 S.D.). These results show the inherent variability of this bioassay and its quantitative limitations. The method appears to be well suited for screening assays and comparison of HSF activity in various cytokine preparations. Also it has proved to be useful in evaluating the efficiency of consecutive purification steps of human HSF (Koj, Gauldie and Sauder, unpublished observations).

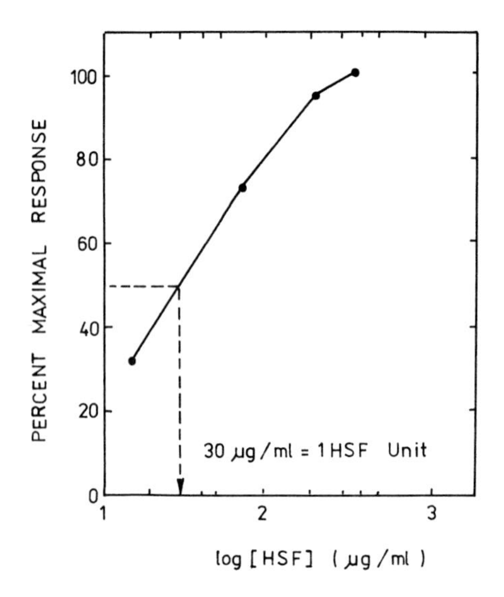

Fig. 3. HSF/IL-1 dose-response curve. Rat hepatocytes were cultured for 24 h in the presence of FCS and 40 nM dexamethasone. The cells were then washed and fresh medium added. Increasing amounts of HSF/IL-1 were then added to each pair of dishes. Incubation was allowed to proceed for an additional 24 h at which time medium was collected for fibrinogen assay by the method of Lowry and the cells for measuring total protein. (After Ritchie and Fuller, 1981.)

TABLE 4
Comparison of HSF/IL-1 activity in a single preparation of COLO-16 cytokine purified on Sephadex G-100 and tested in three different hepatocyte cultures (A, B and C)

Experiment	Treatment of the culture	Albumin		α_2-M		α_2-M / Albumin	HSF (U/ml)
		µg	%	µg	%		
A	PBS	3.31	100	0.63	100	0.19	312
	Cytokine	1.34	40.5	1.23	195	0.92	
B	PBS	3.12	100	0.47	100	0.15	286
	Cytokine	1.20	38.4	1.30	276	1.08	
C	PBS	4.30	100	0.49	100	0.11	303
	Cytokine	1.75	40.7	1.60	326	0.91	

Five dilutions of cytokine were tested in each experiment to establish the dose-response curves for albumin and α_2-M and calculate HSF units (amount of cytokine required to increase α_2-M:albumin ratio from 0.3 to 0.4). Here are shown only the results obtained with the highest cytokine concentration (50 µl) and corresponding control (PBS-treated) cultures. For comparison the results found in control cultures were expressed as 100%.

3.6. Potential usefulness of fibroblasts for assay of IL-1

Evidence of several kinds that IL-1 can stimulate metabolic changes in fibroblasts is now available. These include the production of prostaglandins and collagenase (Mizel et al., 1981) by synovial fibroblasts and uptake of [^{3}H]thymidine (Luger et al., 1982; Schmidt et al., 1982).

Further evidence as to how IL-1 acts to bring about increased uptake of [^{3}H]thymidine has become available (Ester et al., 1984). When added to quiescent fibroblasts, IL-1 was inactive but together with platelet-derived growth factors (PDGF) it was able to initiate proliferation. The presence of traces of PDGF in serum would doubtless explain the findings of both Schmidt et al. (1982) and Luger et al. (1982). As mentioned on p. 42, Estes et al. (1984) found that both P388/D_1 cells and another macrophage-like murine cell line RAW 246.1 can produce not only IL-1, but also another mediator, macrophage-derived competence factor (MDCF), which is able to stimulate fibroblasts. This factor of 56 kDa could easily be distinguished from IL-1 and was assayed by incorporation of [^{3}H]thymidine in BALB/C 3T3 cells, followed by autoradiography.

In addition to the above effects induced by IL-1, it has been claimed that fibroblasts of several kinds can produce IL-1, e.g. dermal fibroblasts (Iribe et al., 1983) and 3T3 fibroblasts (Okai et al., 1982). Whether the same fibroblast cell lines both act as producers of, and can be stimulated by, IL-1 is not yet known. In view of the generally inhibitory effect of PGE_2 on cell growth it is perhaps surprising that its production and the stimulation of [^{3}H]thymidine uptake can take place simultaneously. However, three dermal fibroblast cell lines obtained originally from adult humans, CRL 1445, 1224 and 1222 from the American Type Culture Collection, have given satisfactory results in an assay using uptake of [^{3}H]thymidine (Schmidt et al., 1982). Certain other fibroblasts have not behaved in the same way (J. Saklatvala, personal communication).

3.7. Chemotaxis as an assay for IL-1

Because IL-1 has been shown to possess chemotactic properties for both macrophages and neutrophils, its presence can be thus revealed. A 48 well chamber suitable for this purpose has been described by Werner et al. (1980). Migration takes place through 5 μm holes in a nucleopore filter sheet into chambers containing the attractant. After 2 h incubation, the sheet is removed from between the upper and lower plates, scraped twice and fixed in methanol. The cells still present on the sheet are then counted which can be facilitated by means of a photomicroscope and an image analyser (Sauder et al., 1984). Obviously the chief disadvantage of this method is its lack of specificity.

4. Possible use of AP-response for assay of IL-1

Although it has been firmly established that injection of IL-1/EP/leucocyte endogenous mediator (LEM) into rats (Eddington et al., 1972), rabbits and mice, leads to increased concentration of AP-proteins and the responses in the three species have been compared (Kampschmidt et al., 1983), none of these responses has been investigated in sufficient detail to provide a secure basis for an assay. Earlier work was mainly concerned with CRP, fibrinogen and α_2-M. The increases in concentration of the former two of these proteins did not exceed 5-fold. Later with the isolation of human and murine SAA (Sipe et al., 1976; Sztein et al. 1981) and the development of radioimmunoassays for these proteins, much greater responses were found to occur. Thus increases of murine SAA concentration reaching 35 μg/ml (20-fold increases) were obtained in C3H/HeJ mice by injection of murine IL-1 obtained from the P388/D_1 murine macrophage cell line. As shown in Fig. 4, the observed SAA concentration and the log of the IL-1 dose were roughly propor-

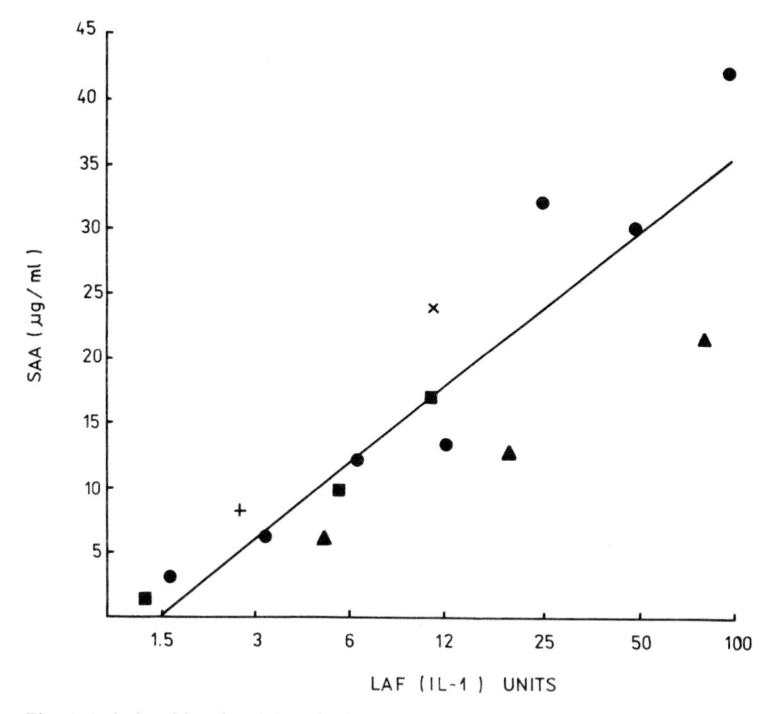

Fig. 4. Relationship of activity of injected LAF/IL-1 and SAA response. Five different preparations of murine LAF/IL-1, containing supernatants, were injected at various concentrations into C3H/HeJ mice and SAA concentrations in serum measured 18 h later. Each different symbol represents a distinct LAF/IL-1 preparation, tested at various dilutions, 5 mice/dilution. (Sztein et al., 1981.)

tional. However, it is important to note that approximately 100 times more IL-1 was found to be necessary to elicit the SAA response in vivo compared with that needed for a response in the thymocyte co-stimulator assay in culture (Sztein et al., 1981). Although the in vivo AP-protein response to IL-1 requires much larger amounts of IL-1 than are necessary, either for the thymocyte costimulator assay or for radioimmunoassay, the specificity of the former assay makes it potentially of great value. Two types of specific action have been demonstrated for rabbit IL-1 obtained from peritoneal macrophages (Kampschmidt et al., 1983). The IL-1, before testing, had been purified and shown to consist of two components differing in p*I*. The material with p*I* 7.0 when injected into rabbits led to increased concentrations of fibrinogen whereas that with lower p*I*, 4.5 – 5.0 had no such effect (cf. Chapter 17, Fig. 1). The same material showed species specificity in so far as increased concentration of fibrinogen was produced in mice but not in rats. In contrast, the p*I* 7 material was effective in stimulating an increase of plasma fibrinogen in both species. Evidently these findings have potential usefulness in future studies on the multiple components now included under the general name of IL-1. Possibly also they may be applicable in studies of individual molecular species of IL-1 in respect to active sites responsible for particular biological effects.

4.1. Radioimmunoassay for IL-1

Production of antisera against IL-1 has proved to be difficult (Dinarello et al., 1977a). In fact, 7 monthly injections of human EP/IL-1 with 25 rabbit pyrogenic doses per rabbit were required before antibodies capable of neutralising pyrogenicity were obtained. The initial immunisation was in Freund's complete adjuvant and was followed by EP/IL-1 in the incomplete adjuvant. The antiserum thus obtained was shown to be species specific in that it failed to bind EP/IL-1 from rabbit and guinea pig peritoneal macrophages and from monkey monocytes under conditions in which human EP/IL-1 was completely removed. With the availability of more highly purified IL-1 obtained from cultures of mouse P388/D_1 cells, antibodies in a goat have been obtained in a rather shorter period; 6 weekly injections of IL-1α in Freund's complete adjuvant were used (Mizel et al., 1983). The antibodies thus obtained were found to recognise human IL-1 as well as all seven molecular variants of mouse IL-1 but to be less effective against heterospecific IL-1. As yet, no report of the use of goat antisera for RIA of IL-1 has appeared.

No fully satisfactory reason for the apparent poor antigenicity of EP/IL-1 has yet been advanced. Possibly its having a highly conserved structure, with little difference between species may be important in this respect. Evidently the recent studies on antisera to muramyldipeptide against which antibody able to bind purified IL-1 have been produced may prove highly relevant (cf. Chapter 7, Section 5).

The antibody to human EP/IL-1 produced as described above proved to be of high affinity since as little as 0.003 ml was sufficient to neutralise one rabbit pyrogen dose. After binding to Sepharose, its value as an aid to purification of EP/IL-1 has been amply demonstrated (Dinarello et al., 1982). On the other hand, it proved to be unsuitable for radioimmunoassay of EP/IL-1. Thus 25% of ^{125}I-radiolabelled EP/IL-1 (Bolton and Hunter, 1973) remained bound in a solid phase RIA (Dinarello et al., 1977b). This was reduced to 5% in presence of one rabbit pyrogen dose of EP/IL-1. At present RIA is not routinely used for the assay of EP/IL-1.

4.2. *Assay for PGE_2 produced by IL-1*

As PGE_2 is formed under the same circumstances in which collagenase is released, estimation of PGE_2 can also be used as an indirect assay for IL-1. Assuming the availability of appropriate cells as PGE_2 producers, when stimulated by IL-1, this method may be advantageous. Although lacking in specificity, as many stimuli may lead to PGE_2 formation, it is highly sensitive if RIA is used. An RIA method has been described by Jaffe and Behrman (1979).

References

Allison, E.S., Cranston, W.I., Duff, G.W., Luff, R.H. and Rawlins, M.D. (1973) Clin. Sci. Mol. Med. *45,* 449 – 459.

Baracos, V., Rodemann, H.P., Dinarello, C.A. and Goldberg, A.L. (1983) N. Engl. J. Med. *308,* 553 – 558.

Bodel, P. and Miller, H. (1976) Proc. Soc. Exp. Biol. Med. *151,* 93 – 96.

Bodel, P. and Miller, H. (1977) J. Exp. Med. *145,* 607 – 617.

Bodel, P. and Miller, H. (1978) Inflammation *3,* 103 – 110.

Bolton, A.E. and Hunter, W.M. (1973) Biochem. J. *133,* 529 – 539.

Cannon, J.C. and Dinarello, C.A. (1984) Fed. Proc. *43,* 462.

Dayer, J.M., Krane, S.M., Russell, G.G. and Robinson, D.R. (1976) Proc. Natl. Acad. Sci. USA *73,* 945 – 959.

Dinarello, C.A. (1982) Hormone Drugs Proc. FDA-USP Workshop. Published by US Pharmacological Convention Inc., pp. 36 – 47.

Dinarello, C.A., Renfer, L. and Wolff, S.M. (1977a) J. Clin. Invest. *60,* 465 – 472.

Dinarello, C.A., Renfer, L. and Wolff, S.M. (1977b) Proc. Natl. Acad. Sci. USA *74,* 4624 – 4627.

Dinarello, C.A., Rosenwasser, L.J. and Wolff, S.M. (1981) J. Immunol. *127,* 2517 – 2519.

Dinarello, C.A., Bendtzen, K. and Wolff, S.M. (1982) Inflammation *6,* 63 – 78.

Duff, G.W. and Durum, S.K. (1982) Yale J. Biol. Med. *55,* 437 – 442.

Eddington, C.L., Upchurch, H.F. and Kampschmidt, R.F. (1972) Proc. Soc. Exp. Biol. Med. *139,* 565 – 569.

Estes, J.E., Pledger, W.J. and Gillespie, G.Y. (1984) J. Leukocyte Biol. *35,* 115 – 129.

Fell, H.B. and Jubb, R.W. (1977) Arthritis Rheum. *20,* 1359 – 1371.

Fell, H.B. and Mellanby, E. (1952) J. Physiol. (London) *116,* 320 – 349.

Gordon, A.H. and Limaos, E.A. (1979) Br. J. Exp. Pathol. *60,* 441 – 446.
Gordon, A.H. and Parker, I.D. (1980) Br. J. Exp. Pathol. *61,* 534 – 539.
Gowen, M., Wood, D.D., Ihrie, E.J., Meats, J.E. and Russell, R.G.G. (1984) Biochim. Biophys. Acta *797,* 186 – 193.
Iribe, H., Koga, T., Kotani, S., Kusumoto, S. and Tetsuo, S. (1983) J. Exp. Med. *157,* 2190 – 2195.
Jaffe, B.M. and Behrmann, H.R. (1979) in Methods of Hormone Radioimmunoassay (Jaffe, B.M. and Behrmann, H.R., eds.), pp. 19 – 42, Academic Press, New York.
Kampschmidt, R.F., Upchurch, H.F. and Worthington, M.L. (1983) Infect. Immun. *41,* 6 – 10.
Kimball, E.S., Pickeral, S.F., Oppenheim, J.J. and Rossio, J.L. (1984) J. Immunol. *133,* 256 – 260.
Koj, A., Gauldie, J., Regoeczi, E., Sauder, D.N. and Sweeney, G.D. (1984) Biochem. J., *224,* 505 – 514.
Koj, A., Gauldie, J., Regoeczi, E. and Sauder, D.N. (1985) J. Immunol. Methods *76,* 317 – 328.
Kwan, S.W., Fuller, G.M., Kroutter, M.A., van Bavel, J.H. and Goldblum, R.M. (1977) Anal. Biochem. *83,* 589 – 596.
Liao, Z., Grimshaw, R.S. and Rosenstreich, D.L. (1984) J. Exp. Med. *159,* 126 – 136.
Mizel, S.B. (1980) Mol. Immunol. *17,* 571 – 577.
Mizel, S.B. and Mizel, D. (1981) J. Immunol. *126,* 834 – 837.
Mizel, S.B., Dukavich, M. and Rothstein, J. (1983) J. Immunol. *131,* 1834 – 1837.
Mizel, S.B., Dayer, J.M., Krane, S.M. and Mergenhagen, S.E. (1981) Proc. Natl. Acad. Sci. USA *78,* 2475 – 2477.
Okai, Y., Tashiro, H. and Yamashita, U. (1982) FEBS Lett. *142,* 93 – 95.
Ritchie, D.G. and Fuller, G.M. (1981) Inflammation *5,* 287 – 299.
Rosenwasser, L. and Dinarello, C.A. (1981) Cell. Immunol. *63,* 134 – 142.
Saklatvala, J., Pilsworth, L.M.C., Sarsfield, S.J., Gavrilovic, J. and Heath, J.K. (1984) Biochem. J. *224,* 461 – 466.
Saklatvala, J., Curry, V.A. and Sarsfield, S.J. (1983) Biochem. J. *215,* 385 – 392.
Sauder, D.N., Mounessa, N.L., Katz, S.I., Dinarello, C.A. and Gallin, J.I. (1984) J. Immunol. *132,* 828 – 832.
Schmidt, J.A., Mizel, S.B., Cohen, D. and Green, I. (1982) J. Immunol. *128,* 2177 – 2182.
Sellers, A. and Reynolds, J.J. (1977) Biochem. J. *167,* 353 – 360.
Sipe, J.D., Ignaczak, T.F., Pollock, P.S. and Glenner, G.G. (1976) J. Immunol. *116,* 1151 – 1156.
Sipe, J.D., Johns, M., DeMaria, A., Cohen, A.S. and McCabe, W.R. (1984) Fed. Proc. *43,* 1542.
Sweeney, G.D., Jones, K.G. and Knestynski, F. (1978) J. Lab. Clin. Med. *91,* 444 – 454.
Sztein, M.B., Vogel, S.N., Sipe, J.D., Murphy, P.A., Mizel, S.B., Oppenheim, J.J. and Rosenstreich, D.L. (1981) Cell. Immunol. *63,* 164 – 176.
Weeke, B. (1973) Scand. J. Immunol. *2* (Suppl. 1), 37 – 46.
Werner, F., Goodwin, R.H. and Leonard, E.J. (1980) J. Immunol. Methods *33,* 239 – 247.
Wood, P.R., Audrus, L. and Clark, J.A. (1983) Immunology *50,* 637 – 644.

PART VI

Assay of acute-phase proteins

Gordon/Koj (eds.) The acute-phase response to injury and infection.

CHAPTER 24

General methods applied to assay of acute-phase proteins in plasma, body fluids and tissue cultures

A. KOJ

The technical details necessary for quantitative assays for particular proteins are available in numerous textbooks, research monographs and experimental papers. Many of these techniques can be used directly for measuring changes in concentration of acute-phase proteins (AP-proteins) during the acute-phase response (AP-response), or for detecting unusual protein variants in the plasma. The purpose of the brief review presented below is to aid in the selection of an appropriate technique (cf. Table 1). To evaluate injury-induced changes in the blood level of a protein, the range of 'normal' values should be obtained in each laboratory. Apart from selecting a proper control group in respect of age, sex, etc., one must bear in mind the phenotypic differences known to affect certain AP-proteins, such as human haptoglobin or α_1-proteinase inhibitor (α_1-PI). The data obtained should be evaluated statistically and cautiously analysed, especially if they concern several proteins and include variables. Cooper and Stone (1979) in their exhaustive review on AP-proteins in cancer recommended three multivariate methods: discriminant analysis, principal component analysis and cluster analysis. By using these methods the authors were able to distinguish some recognizable patterns in AP-protein profiles in patients with neoplastic diseases. Clinical aspects of the assay of AP-proteins are discussed at length in Chapter 21.

1. Electrophoretic techniques

Single stage polyacrylamide gel electrophoresis in cylindrical or slab gels (Gordon,

TABLE 1
Comparison of quantitative immunological methods in assays of certain AP-proteins

Method	Lower limit of working range (μg/ml)	Protein assayed	Precision or accuracy	Simplicity	Rapidity	Sample economy	No. of samples per assay	Ref.
Radial immuno-diffusion	5 – 50	α_1-PI α_1-AGP Hp	+	+ + +	+	+ + +	+ +	Smith et al. (1977)
	2 – 10	CRP						Kushner et al. (1978)
Electro-immuno-assay	2 – 10	α_1-AGP α_2-M Alb	+ +	+ +	+ +	+ + +	+ +	Koj et al. (1984)
		SAP						Pepys (1981)
Nephelo-metry	4 – 16	α_2-APl	+ + +	+	+ + +	+ + +	+ + +	Schmitz-Huebner et al. (1980)
	5 – 20	SAP						Gertz et al. (1983)
RIA	1 – 5	SAA	+ + +	+	+ + +	+ + +	+ + +	Sipe et al. (1976)
ELISA	0.05 – 0.1	Fib	+ + +	+	+ + +	+ + +	+ + +	Kwan et al. (1977)

Lower limit of working range also gives information about sensitivity of the assay which depends largely on individual properties of an antigen and the avidity of its antibody. Specificity of all assays is very high, providing the monospecific antisera are available. For definitions of sensitivity, precision, accuracy and specificity see Skelley et al. (1973). Only relative evaluation of the methods is given and the sign + + + stands for the highest precision, simplicity of procedure, rapidity of assay, sample economy and ability to process a large number of samples in one assay.

1969) permits detection of gross changes in the patterns of plasma proteins occurring as part of the AP-response but is unsuitable for a more detailed analysis. Due to the efforts of O'Farrell (1975) and Anderson and Anderson (1977) a high-resolution two-dimensional gel electrophoresis of plasma proteins became available. In its original version this system involved isoelectric focusing under dissociating conditions (9 mol of urea and 20 ml of a non-ionic detergent per liter) in the first dimension and sodium dodecyl sulphate (SDS) slab gel electrophoresis in the second dimension. As modified by Anderson and Anderson (1977), serum or plasma is mixed with a dissociation buffer containing SDS and mercaptoethanol and heated for 5 min at 95°C; the SDS separates from the proteins during isoelectric focusing with urea and Nonidet P-40 but is reintroduced before electrophoresis in the second dimension. The inclusion of SDS and mercaptoethanol ensures that individual polypeptide subunits are separated. Finally the proteins are visualized by staining of the gel with Coomassie Brilliant Blue R-250 which enables detection of as little as 0.1 μg of protein (O'Farrell, 1975). The sensitivity can be increased by at least one order of magnitude if a silver stain is used (Tracy et al., 1982a). Identification of protein spots is based on the use of protein standards, or antisera of known specificity. Anderson and Anderson (1977) were able to identify over 30 human plasma proteins on such two-dimensional gel maps and to detect simultaneously numerous genetic variants differing in charge or size. By employing this technique Baumann et al. (1984a) carried out genetic analysis of α_1-acid glycoprotein (α_1-AGP) and serum amyloid A protein (SAA) produced by various strains of mice (Chapter 18, Fig. 3). Resolution and reproducibility of the separations can be improved by initial removal from plasma of albumin (by adsorption on immobilized Cibacron Blue) and IgG (by adsorption on immobilized Protein A), and by minimising protein losses during gel equilibration between development in the first and second dimensions (Tracy et al., 1982a, b). The method is specially advantageous for phenotyping and for simultaneous investigation of microheterogeneity of several plasma proteins. On the other hand, for quantitation of individual proteins it is much less satisfactory even when based on scanning of a gel photograph with a suitably calibrated computer-guided densitometer (Tracy et al., 1982a).

Similar two-dimensional gel electrophoresis permits rather good estimation of relative synthesis rates of individual proteins in bacterial cultures (O'Farrell, 1975) or tissue cultures (Ivarie and O'Farrell, 1978) after these have been incubated with labelled amino acids. On the basis of an autoradiograph, appropriate spots are cut from the dry gel and their radioactivities are determined. Neighbouring areas not containing detectable spots serve as background. The results are often expressed in values relative to the total counts of protein-bound radioactivity applied to the gels (Baumann et al., 1983, 1984b; Chapter 17, Fig. 3). Variable protein losses during equilibration of the focusing gel in the SDS buffer are a possible source of error

(Tracy et al., 1982b). As yet the full potential of two-dimensional slab gel electrophoresis for studies of AP-proteins has not been explored.

2. *Direct immunological techniques*

If an antiserum of sufficient specificity and avidity is available assay of a particular AP-protein may be carried out by either one of several convenient methods (Table 1). Of these radial immunodiffusion of Mancini (1965) in various modifications is probably the easiest and most economical. The protein antigen is placed in a well cut in a thin layer of agarose containing the relevant monospecific antiserum. After incubation for 1 or 2 days a characteristic precipitation ring is formed. The diameter of this ring is proportional to the log concentration of the antigen. For human plasma proteins ready-to-use immunodiffusion plates are commercially available. Faster and more accurate results are obtained by the electroimmunoassay method introduced by Laurell (1966) now often known as rocket immunoelectrophoresis due to the characteristic shape of the immunoprecipitation lines. An excellent description of quantitative immunoelectrophoresis including electroimmunoassay is given by Weeke (1973a, b). The appropriately diluted samples to be compared are applied in circular wells side by side on glass plates usually 80 × 100 × 1.5 mm in size. Electrophoresis is performed in 1% agarose gel (usually 1 – 2 mm thick) containing a suitable proportion of the antiserum. Sometimes it is advantageous to use a mixture of two monospecific antisera in order to assay two AP-proteins simultaneously (Chapter 17, Fig. 4). After the separation the excess of antibodies and other unbound proteins are removed from the gel by soaking it in water, pressing under filter paper, drying and staining with Coomassie Brilliant Blue R-250. The position of the protein antigen is identified by the rocket-shaped precipitate. Quantitation can be achieved by measuring the heights of the rockets and comparison with those of the standards run in parallel wells. In the case of broad, blunt rockets often observed with proteins of low electrophoretic mobility (transferrin, fibrinogen) measurement of the peak area may give more accurate results. With a precipitating antibody of high avidity as little as 10 ng of antigen can be accurately determined. Assuming that the well contains 5 μl of the solution under investigation this gives sensitivity of 2 μg of protein per millilitre. Rocket immunoelectrophoresis has been used for the measurement of changes in concentration of 10 plasma proteins in rats during acute inflammation or during hepatoma growth (Koj et al., 1982), and also for estimation of synthesis and secretion of 11 plasma proteins by rat hepatocyte cultures stimulated with human cytokines (Koj et al., 1984). Thus this method can be recommended as a convenient, sensitive, versatile and reproducible technique.

Because simultaneous analysis by electroimmunoassay of large numbers of samples requires much time and skill, automated immunonephelometric techniques are now gaining wide acceptance. The well known principle of light scattering by immune complexes was first used by Schultze and Schwick (1959) for assay of plasma proteins in a manual nephelometer. Today a whole range of sophisticated laser nephelometers with internal microprocessors are routinely used both in research laboratories and clinical practice (Ritchie, 1974; Walker and Gauldie, 1978; Hoffken and Schmidt, 1981; Colley et al., 1983; Gertz et al., 1983). In all these techniques serial dilutions of standards and the samples to be tested are mixed with monospecific antiserum in the presence of dextran or polyethylene glycol. These substances serve to accelerate the antigen-antibody reaction rates, increase the light scattering at equilibrium and shift the equilibrium toward higher antigen concentration (Walker and Gauldie, 1978). A disadvantage is that rather large quantities of monospecific antisera are required for the assay. After a preset time of incubation, amount of scattered light given by the sample is measured and compared with the standard curve obtained for purified antigens or pooled normal plasma. In practice the continuous-flow nephelometric procedure gives precision better than 5% (coefficient of variation) within each batch, and approximately 10% between batches. However, the precision is often less at low concentration of antigen, e.g. for human C-reactive protein (CRP) below 10 μg/ml (Colley et al., 1983). The accuracy of the method is good and when its results are compared with those obtained by other established procedures, such as radial immunodiffusion or rocket immunoelectrophoresis, the coefficient of regression is usually near to 0.95 (Schmitz-Huebner et al., 1980; Gertz et al., 1983). Precision of measurements of murine serum amyloid P protein (SAP) was found to be greater with rate nephelometry than with either immunoelectrophoresis or radioimmunoassay (Gertz et al., 1984). As summarized by Gertz et al. (1983) 'the speed, technical ease, and automated operation of a rate nephelometer make this quantitatively sensitive system feasible for widespread use for human SAP measurements'. Indeed, laser nephelometry is the method of choice for measuring the early time course of the AP-protein response in several diseases (Colley et al., 1983).

Radial immunodiffusion, rocket immunoelectrophoresis and immunonephelometry are suitable for handling numerous samples but usually a single antigen is measured (except in some cases of electroimmunoassay: cf. de Beer et al., 1982; Koj et al., 1984); moreover, protein variants remain undetected. On the other hand, by the method of two-dimensional immunoelectrophoresis (crossed immunoelectrophoresis), as originally proposed by Laurell (1965) and modified by Clarke and Freeman (1968), simultaneous quantitative assay of many proteins is possible. A detailed description of how to obtain a high degree of resolution by this procedure is given by Weeke (1973c). At first the protein mixture is separated by electrophoresis in barbital buffer pH 8.6 in a strip of 1% agarose gel near to one edge of a glass plate. Then 1% agarose containing antibodies is poured onto the remaining part of the glass plate and electrophoresis is run perpendicularly to the first direction. Finally, the gel plate is soaked, dried and stained as described for rocket immunoelectrophoresis. The protein peaks are much broader than during rocket immunoelectrophoresis and in some cases show asymmetry which may indicate the existence of protein variants differing in charge (cf. Chapter 16, Fig. 1 and Chapter 17, Fig. 2). The absolute amount of a particular protein in the mixture can be estimated from the area under the precipitate peak in comparison with the peak area given by a standard subjected to crossed immunoelectrophoresis on a separate plate. Alternatively, the peak area in the acute-phase serum can be compared with that given by normal serum and expressed as per cent change. Ruhenstroth-Bauer and co-workers were able to estimate relative changes in concentration of several proteins in the serum of rats injected with endotoxin (Abd-el-Fattah et al., 1976) or with leucocyte endogenous mediator LEM/IL-1 preparation (Scherer and Ruhenstroth-Bauer, 1977). The accuracy and reproducibility of this method are rather limited, and moreover the method is time-consuming and unsuitable for large number of samples. On the other hand, it can be very useful for screening assays of many plasma proteins, or for assessing the so-called ‘plasma profile’ during the AP-response.

Certain molecular variants of glycoproteins occurring in the plasma can be detected by a technique known as affinity immunoelectrophoresis which has been developed by Bøg-Hansen et al. (1975, 1983). In this method a lectin, such as Con A, is incorporated in the agarose gel used in the first dimension, or in an intermediate gel as used in certain modifications of the technique. Proteins showing affinity for the lectin are retarded during the affinity electrophoresis stage and show up as additional peaks during immunoelectrophoresis in the second dimension (cf. Chapter 16, Fig. 2). A suitable range of concentration of Con A in agarose gel is from 0.1 to 1.0 mg/ml. In addition, Ca, Mn and Mg ions are added to the buffer in order to facilitate interaction of glycoproteins with the lectin. This method has permitted detection of the variants of α_1-AGP (Nicollet et al., 1981) and of α_1-antichymotrypsin, (α_1-ACh) (Bowen et al., 1982) in plasmas of patients with inflammatory diseases, and variants of α_1-AGP and α_1-AP-globulin in plasma of rats with acute inflammation or with an implanted tumour (Koj et al., 1982, and Chapter 16). Affinity electrophoresis is very valuable in the studies of AP-proteins although reproducibility remains a serious problem. Thus results may vary depending on the batch of lectin or small changes of technique during the electrophoretic separation.

3. *Indirect immunological techniques: radioimmunoassay and enzyme-linked immunoassay*

Various modifications of immunodiffusion, immunoelectrophoresis or immunonephelometry are suitable for routine measurements of proteins at concentrations above 10 μg/ml, although in certain cases of electroimmunoassay as little as 2 μg/ml can be determined (Pepys, 1979; Koj et al., 1984; Chapters 17 and 23). Higher sensitivity enabling accurate measurements of proteins in the nanogram/ml range is offered by radioimmunoassay (RIA) and enzyme-linked immunosorbent assay (ELISA).

The technique of RIA was first developed for estimation of hormone concentration based on the original observation of Yalow and Berson that plasma levels of antibodies to insulin could be quantitated by their ability to bind ^{131}I-labelled hormone (for references see Odell and Daughaday, 1971; Skelley et al., 1973). In RIA unknown concentration of antigen may be determined due to the fact that the non-labelled molecules in the sample under test compete with radiolabelled antigen for a limited number of antibody binding sites. The experimental conditions are chosen such that there is a fixed amount of antibody and a relative excess of antigen. A standard curve has to be prepared for each new batch of antiserum or labelled antigen. As for most of the quantitative immunochemical techniques monospecific antisera of strong affinity are required. Purified antigen has to be labelled to a high specific radioactivity, usually with ^{125}I, by one of the established procedures such as the chloramine-T method or by means of lactoperoxidase.

The separation of free and antibody-bound antigen, which is an essential step in RIA, can be done by removal of free antigen (e.g. by absorption on dextran-coated charcoal or ion-exchange resin) but more commonly is achieved by removal of the antigen-antibody complex. Unspecific methods, such as salting out or organic solvents used in early assays, have been almost completely replaced by specific procedures, such as the second antibody technique or absorption on insoluble protein A from *Staphylococcus aureus*. Comparison of the latter two methods in the assay of plasma ferritin is described by Gauldie et al. (1980). The need to improve the method used for removal of the antigen-antibody complex led to elaboration of a simplified technique now commonly used, i.e. solid phase RIA. In this method the appropriate antibody is either covalently attached to a solid support, or is absorbed onto the walls of test tubes or wells in a microtitration plate. After incubation the free antigen is decanted and the solid phase is washed a few times before counting.

In comparison with other methods of protein assay RIA has very high sensitivity, good precision, accuracy, reproducibility and specificity (Skelley et al., 1973) but it is a highly complex and expensive technique. It has been used in various modifications for measurements of AP-proteins occurring in the blood in trace amounts, such as murine SAA and SAP (Sipe et al., 1976, 1982; Sztein et al., 1981; Tatsuta et al., 1983), rabbit and human CRP (Claus et al., 1976; Macintyre et al., 1983) or rat α_2-macroglobulin (α_2-M) (Panrucker and Lorscheider, 1982). RIA may in future be the method of choice for measurement of AP-protein synthesis by cultures of hepatocytes (Macintyre et al., 1983; Tatsuta et al., 1983) or in subcellular fractions of the liver (Kudryk et al., 1982). RIA is particularly suitable for multisample operation but the short shelf-life of the reagents, the rather sophisticated and expensive equipment required and strict regulations concerning the use of radioisotopes have tended to exclude RIA from many small laboratories. These considerations have led to a search for alternative labels for antibodies and antigens, the most promising and versatile labels so far discovered are enzymes (for references see Voller et al., 1978; Saunders, 1979; Ngo and Lenhoff, 1982). These enzymes which are used both as markers and amplifiers should be easily available, suitable for linkage to proteins, should have high molecular activities and cheap substrates to give stable and readily measurable products. Fluorogenic methyl-umbelliferin substrates have been used to increase sensitivity, which may then exceed even that obtained in RIA. Unfortunately they are more expensive and require a special technique of measurement. The enzyme most commonly used in ELISA is alkaline phosphatase. Horseradish peroxidase and β-galactosidase have also been employed. The enzyme in question can be coupled to antigen or antibody using glutaraldehyde, periodate or other reagents.

Ngo and Lenhoff (1982) distinguish at least six categories of ELISA but of these two techniques are the most important. In competitive ELISA an enzyme-labelled ligand (L-E) competes with non-labelled ligand from the sample under test for a limited amount of immobilized antibody, after which the antibody-bound L-E is separated from the unbound L-E. Either fraction can then be assayed for enzyme activity. If the antibody-bound fraction is so assayed, as is commonly done, the enzyme activities obtained are inversely related to the concentrations of the tested ligands. The determination of the coloured product of the enzymic reaction is carried out either in a spectrophotometer microcuvette, or more often directly in the titration wells of a plastic plate using a special multiscan photometer. In the so-called 'sandwich-type' ELISA an excess of immobilized antibody is added to the sample to be tested and is followed by the enzyme-labelled antibody. The enzyme activity remaining associated with immobilized antibody will then be directly proportional to the amount of antigen in the test sample. Sandwich ELISA does not require purified antigens and does not involve competition between labelled and unlabelled antigens. It is also a more sensitive technique than competitive ELISA but it requires two incubations and two washing steps (Ngo and Lenhoff, 1982; Voller et al., 1978). A superb review of the art of performing solid phase enzyme immunoassay, including some experimental protocols, has been written by Saunders (1979).

So far ELISA techniques have been used only occasionally for assay of AP-proteins. In all cases alkaline phosphatase has been the marker enzyme. Wang and Chu (1979) measured α_1-AGP, and Harpel (1981) determined α_2-antiplasmin (α_2-APl) and antiplasmin-plasmin complexes in human plasma, while Lewin et al. (1983) used a similar approach for C1 inactivator-plasma kallikrein complexes, and Brower and Harpel (1983) for α_1-antitrypsin (α_1-AT) leucocyte elastase complexes. Robey et al. (1983) utilized ELISA to demonstrate immunologic cross-reactivity of dogfish CRP and rabbit CRP. Fuller and co-workers elaborated an ELISA competitive assay for quantitative determination of rat fibrinogen and albumin in primary hepatocyte cultures (Kwan et al., 1977; Rupp and Fuller, 1979; Ritchie and Fuller, 1981, 1983).

It is necessary to conclude that although RIA and ELISA offer unsurpassed sensitivity for the assay of proteins occurring in trace amounts, in many other cases the cheaper and simpler direct immunologic techniques are sufficient for determinations of AP-proteins.

References, p. 323.

Gordon/Koj (eds.) The acute-phase response to injury and infection.

CHAPTER 25

Specific methods of assay for certain acute-phase proteins

A. KOJ

1. Potential of affinity chromatography for assay of glycoproteins and pentraxins

Many AP-proteins are glycosylated and thus show variable degrees of affinity toward Con A or other lectins. Furthermore, the proportion of Con A-reactive constituents present in some microheterogenous glycoproteins increase in plasma during acute inflammation (cf. Chapter 16). Separation of unfractionated plasma on a Con A-Sepharose column does not provide sufficient information on the kinetics of the AP-response due to the presence of numerous 'neutral' proteins, the concentrations of which are not affected by inflammation. When, however, a [^{3}H]lysine containing perfusate from a rat liver was analyzed on a Con A-Sepharose column a good correlation ($r = 0.92$) was found between the radioactivities in the seromucoid fraction and in the glycoproteins eluted from the column with 0.1 M methyl-glucoside (Koj and Dubin, 1978).

Pentraxin proteins, i.e. CRP and SAP, bind to agarose and its derivatives in the presence of calcium ions; however, only CRP shows high affinity for phosphorylcholine and pneumococcal C-polysaccharide (CPS) (for references see Pepys and Baltz, 1983, and Chapter 14). Due to this fact CRP and SAP can be recovered from the plasma almost quantitatively and in a fairly pure state by sequential chromatography on agarose and phosphorylcholine-Sepharose or CPS-Sepharose (Baltz et al., 1982; de Beer et al., 1982), or AH-Sepharose and Sepharose CL-4B columns (Robey et al., 1983). Application of these techniques enabled detection and partial characterization of pentraxins in several vertebrate species (Pepys et al., 1978; Baltz et al., 1982; Robey et al., 1983). It is clear, however, that assay of pentraxins by affinity chromatography would be cumbersome and inaccurate hence various immunological techniques are used for this purpose (cf. Chapter 24).

2. *Assay of proteinase inhibitors*

Depending on species, several proteinase inhibitors present in plasma are classified as AP-proteins (Chapter 13). The majority inhibit trypsin so that changes in serum trypsin inhibitory capacity (STIC) are a rather sensitive indicator of the level of these inhibitors: microgram quantities of α_1-PI (α_1-AT) can be accurately determined using various modifications of a trypsin inhibition assay (Eriksson, 1965; Koj et al., 1972; Edy and Collen, 1977; Kueppers and Mills, 1983). STIC is often expressed in milligrams of trypsin inhibited by 1 ml of serum (Ihrig et al., 1975).

In such an assay approximately 5 μg of trypsin is exposed for 10 – 15 min to various dilutions of plasma. Then casein is added and incubation is continued for a fixed time (30 or 60 min). The reaction is stopped by addition of trichloroacetic acid and the filtrate absorbance at 280 nm measured. The amount of inhibitor is calculated assuming 1 : 1 molar ratio of proteinase : inhibitor in the inactive complex. The sensitivity of the assay can be increased by using azocasein, radiolabelled casein or certain synthetic substrates. Unfortunately the method suffers from very limited specificity because almost all proteinases are inhibited by more than one plasma inhibitor, with α_2-M showing the broadest range of action (cf. Chapter 13, Fig. 1). Since, however, proteinases bound to α_2-M preserve most of their activity with low molecular mass substrates, the contribution of α_2-M towards the inhibition of a given proteinase by plasma can be calculated (Starkey and Barrett, 1977).

An alternative approach proposed by Schapira et al. (1982) is based on the observation that methylamine selectively inactivates the proteinase-binding activity of α_2-M. Following methylamine treatment and using a specific substrate for kallikrein its inhibition by C1 inhibitor in plasma could be accurately measured. A good correlation (r = 0.90) between the level of C1-esterase inactivator (C1-INA) (in the range of 50 – 300 μg/ml plasma) determined by radial immunodiffusion and by functional assay based on inhibition of kallikrein was obtained.

An even simpler method for determination of C1-inhibitor activity in plasma has been proposed by Wiman and Nilsson (1983). They observed that C1-INA is the single effective inhibitor of C1-esterase. By using the C1s component of complement and specific substrate they successfully measured the amount of functional C1-INA in human plasma in the range of 50 – 500 μg/ml. When the results were compared with those obtained by electroimmunoassay the correlation coefficient was r = 0.96.

For this class of proteins assays based on inhibitory activity may have advantage over immunological methods in which detection of the antigen is used since the latter methods cannot distinguish between functional molecules and those occurring in complexes or rendered inactive by previous proteolytic attack. However, certain precautions must be taken for accurate determination of proteinase inhibitors in a functional assay:

(a) The amount of active enzyme in the employed preparations should be determined by active site titration (cf. Koj and Regoeczi, 1981);
(b) stoichiometry of the reaction between the enzyme and inhibitor should be established since in some cases only a proportion of the inhibitor molecules block the enzyme while the remainder are inactivated by proteolysis (Koj and Regoeczi, 1981);
(c) accurate results can be obtained only from data for partial enzyme inhibition since the lower part of the inactivation curve approaching total inhibition often deviates from rectilinearity due to dissociation of the enzyme-inhibitor complex (Bieth, 1980).

Employment of proteinases of narrow specificity and introduction of highly sensitive substrates may lead to the improvement of these functional assays for AP-proteins which are antiproteases.

3. Fibrinogen assay

Measurements of changes in plasma fibrinogen concentration provided in the past the main body of information on the kinetics of the AP-response (cf. Koj, 1974). Indeed, such measurements are still common in both clinical and research laboratories. Fibrinogen concentrations in plasma ranging between 1 and 10 mg/ml can be easily determined by simple and cheap techniques based on its precipitation with concentrated salts solutions or by heating followed by nephelometric and turbidimetric measurements. Several variations of this procedure have been compared by Stakenburg and Neumann (1977) and by Desvignes and Bonnet (1981). Excellent correlations ($r = 0.99$) between turbidimetric or nephelometric method on one hand and radial immunodiffusion or clotting technique on the other, were claimed.

All specific methods for fibrinogen estimation involve clotting of plasma with thrombin and subsequent isolation of the clot which can then be dissolved in NaOH for determination of protein by Lowry et al. procedure (Koj, 1974), or in alkaline urea solution for direct measurement of absorbance at 280 nm (Regoeczi, 1970). The clotting assays have the advantage of measuring biologically active fibrinogen but are too laborious for use on large numbers of samples. In special cases when functional fibrinogen needs to be estimated with high accuracy the isotope dilution method of Atencio et al. (1965) can be recommended. At low fibrinogen concentrations, i.e. below 50 μg/ml, immunological techniques are the method of choice (cf. Chapter 24).

4. Haptoglobin assay

The specific methods of haptoglobin determination are based on its interaction with haemoglobin. The Hp-Hb complex thus formed exhibits peroxidase activity and was used first by Jayle (1951) for quantitation of plasma haptoglobin. A more convenient procedure was described by Owen et al. (1960). An excess of methaemoglobin is added to a sample of serum (0.05 – 0.2 ml). A portion of the mixture (0.1 ml) is transferred to 5 ml guaiacol solution pH 4.0 followed by 1 ml of 0.05 M H_2O_2. After 8 min incubation at 25°C absorbance is measured at 470 nm and Hp concentration determined from a standard curve. The sensitivity of the assay is limited to 100 μg Hp per ml. The results are usually expressed as mg of haemoglobin binding capacity per 100 ml of serum.

A simple method for serum haptoglobin estimation described by Roy et al. (1969) is based on the observation that haptoglobin protects haemoglobin from acid denaturation. In this procedure a cyanmethaemoglobin reagent is mixed with the serum to be tested, the pH is adjusted to 3.7 and after 20 min absorbance at 407 nm is measured relative to a blank of the reagents only. The sensitivity of the assay is lower than that in the Owen method.

Another simple assay for haptoglobin was proposed by Calhoun and Englander (1979). It is based on the fact that the Soret peak absorbance at 415 – 430 nm of normal tetrameric deoxyhaemoglobin is about 20% higher than that of dimeric haemoglobin in the Hp-Hb complex. In this method a small aliquot of haemoglobin is added to a sample of plasma diluted 10 to 20-fold and absorbance at 415 nm is measured. Then oxyhaemoglobin is reduced with dithionite and absorbance read at 430 nm. From these readings the haptoglobin concentration can be calculated in the range of 10 – 100 mg haemoglobin-binding capacity per 100 ml of plasma.

All three methods briefly described above are suitable for measuring functional haptoglobin in plasma and body fluids at concentrations above 50 μg/ml. At lower concentrations immunologic techniques are indispensable.

5. Ceruloplasmin assay

Ceruloplasmin regarded as a weak AP-protein (cf. Chapter 12, Table 2) can be determined on the basis of its oxidase activity, usually with *p*-phenylene diamine (PPD) as substrate (for references see Sunderman and Nomoto, 1970; Koj, 1974). In the method described by Ravin (1961) 0.05 – 0.1 ml of serum is incubated at 37°C for 1 h with 4 ml 0.4 M acetate buffer pH 5.5 and 0.5 ml 0.5% PPD. The reaction is stopped with sodium azide, absorbance at 530 nm is measured and converted to microgram of ceruloplasmin using a calculated absorbance coefficient (Sunderman and Nomoto, 1970) which had been obtained with a purified standard of ceruloplasmin. If no ceruloplasmin standard is available the results may be expressed as units of enzyme activity of ceruloplasmin. A similar method was applied by Linder and Moor (1977) for measuring tissue ceruloplasmin; nonspecific substrate oxidation was prevented by 0.01 M EDTA.

The method based on PPD oxidation is rather sensitive and enables determination of 10 – 100 μg of ceruloplasmin with a good accuracy. For human serum the coefficient of variation within a single run was reported as 1.25%, while day-to-day variability amounted to barely 2.8% (Sunderman and Nomoto, 1970).

The presence of ceruloplasmin in particular fractions of serum separated by disc electrophoresis or immunoelectrophoresis can be demonstrated either by incubation of the gel with PPD solution, or with ferrous sulphate solution and differential staining (Schen and Rabinowitz, 1966).

References (Chapters 24 and 25)

Abd-el-Fattah, M., Scherer, R. and Ruhenstroth-Bauer, G. (1976) J. Molec. Med. *1*, 211 – 221.

Anderson, L. and Anderson, N.G. (1977) Proc. Natl. Acad. Sci. U.S.A., *74*, 5421 – 5425.

Atencio, A.C., Budrick, D.C. and Reeve, E.B. (1965) J. Lab. Clin. Med. *66*, 137 – 145.

Baltz, M.L., de Beer, F.C., Feinstein, A., Munn, E.A., Milstein, C.P., Fletcher, T.C., March, J.F., Taylor, J., Bruton, C., Clamp, J.R., Davies, A.J.S. and Pepys, M.B. (1982) Ann. N.Y. Acad. Sci. *389*, 49 – 73.

Baumann, H., Jahreis, G.P. and Gaines, K. (1983) J. Cell. Biol. *97*, 866 – 867.

Baumann, H., Held, W.A. and Berger, C. (1984a) J. Biol. Chem. *259*, 566 – 573.

Baumann, H., Jahreis, G.P., Sauder, D.N. and Koj, A. (1984b) J. Biol. Chem. *259*, 7331 – 7342.

Bieth, J. (1980) Bull. Eur. Physiopathol. Respir. *16*, (Suppl.), 183 – 195.

Bøg-Hansen, T.C., Bjerrum, O.J. and Ramlau, J. (1975) Scand. J. Immunol., *4* (Suppl. 2), 141 – 147.

Bøg-Hansen, T.C., Teisner, B. and Hau, J. (1983) in Modern Methods in Protein Chemistry (Tschesche, H., ed.), pp. 125 – 148, Walter de Gruyter and Co., Berlin and New York.

Bowen, M., Rayness, J.G. and Cooper, E.H. (1982) in Lectins – Biology, Biochemistry, Clinical Biochemistry (Bøg-Hansen, T.C., ed.), Vol. 2, pp. 403 – 411, Walter de Gruyter and Co., Berlin and New York.

Brower, M.S. and Harpel, P.C. (1983) Blood *61*, 842 – 849.

Calhoun, D.B. and Englander, S.W. (1979) Anal. Biochem. *99,* 421 – 426.
Clarke, M.H.G. and Freeman, T. (1968) Clin. Sci. *35,* 403 – 413.
Claus, D.R., Osmand, A.P. and Gewurz, H. (1976) J. Lab. Clin. Med. *87,* 120.
Colley, C.M., Fleck, A., Goode, A.W., Muller, B.R. and Myers, M.A. (1983) J. Clin. Pathol. *36,* 203 – 207.
Cooper, E.H. and Stone, J. (1979) Adv. Cancer Res. *30,* 1 – 43.
De Beer, F.C., Baltz, M.L., Munn, E.A., Feinstein, A., Taylor, J., Bruton, C., Clamp, J.R. and Pepys, M.B. (1982) Immunology *45,* 55 – 70.
Desvignes, P. and Bonnet, P. (1981) Clin. Chim. Acta *110,* 9 – 17.
Edy, J. and Collen, D. (1977) Biochim. Biophys. Acta *484,* 423 – 432.
Eriksson, S. (1965) Acta Med. Scand. *177,* (Suppl. 432), 1 – 85.
Gauldie, J., Tang, H.K., Corsini, A. and Walker, W.H.C. (1980) Clin. Chem. *26,* 37 – 40.
Gertz, M.A., Skinner, M., Cohen, A.S. and Kyle, R.A. (1983) J. Lab. Clin. Med. *102,* 773 – 778.
Gertz, M.A., Sipe, J.D., Skinner, M., Cohen, A.S. and Kyle, R.A. (1984) J. Immunol. Methods, *69,* 173 – 180.
Gordon, A.H. (1969) in Laboratory Techniques in Biochemistry and Molecular Biology (Work, T.S. and Work, E. eds.), Vol. 1, pp. 1 – 145, Elsevier/North-Holland, Amsterdam and London.
Harpel, P.C. (1981) J. Clin. Invest. *68,* 46 – 55.
Hoffken, K. and Schmidt, C.G. (1981) Methods Enzymol. *74,* 628.
Ihrig, J., Schwartz, H.J., Rynbrandt, D.J. and Kleinerman, J. (1975) Am. J. Clin. Pathol. *64,* 297 – 303.
Ivarie, R.D. and O'Farrell, P.H. (1978) Cell *13,* 41 – 55.
Jayle, M.F. (1951) Bull. Soc. Chim. Biol. *33,* 876.
Koj, A. (1974) in Structure and Function of Plasma Proteins (Allison, A.C., ed.), Vol. 1, pp.73 – 131, Plenum Press, London and New York.
Koj, A. and Dubin, A. (1978) Br. J. Exp. Pathol. *59,* 504 – 513.
Koj, A. and Regoeczi, E. (1981) Int. J. Peptide Protein Res. *17,* 519 – 526.
Koj, A., Chudzik, J., Pajdak, W. and Dubin, A. (1972) Biochim. Biophys. Acta *268,* 199 – 206.
Koj, A., Dubin, A., Kasperczyk, H., Bereta, J. and Gordon, A.H. (1982) Biochem. J. *206,* 545 – 553.
Koj, A., Gauldie, J., Regoeczi, E., Sauder, D.N. and Sweeney, G.D. (1984) Biochem. J., *224,* 505 – 514.
Kudryk, B., Redman, C.M. and Blomback, B. (1982) Biochim. Biophys. Acta *703,* 77 – 86.
Kueppers, F. and Mills, J. (1983) Science *219,* 182 – 184.
Kushner, I., Broder, M.L. and Karp, D. (1978) J. Clin. Invest. *61,* 235 – 242.
Kwan, S.W., Fuller, G.M., Krautter, M.A., van Bavel, J.H. and Goldblum, R.M. (1977) Anal. Biochem. *83,* 589 – 596.
Laurell, C.B. (1965) Anal. Biochem. *10,* 358 – 361.
Laurell, C.B. (1966) Anal. Biochem. *15,* 45 – 52.
Lewin, M.F., Kaplan, A.P. and Harpel, P.C. (1983) J. Biol. Chem. *258,* 6415 – 6421.
Linder, M.C. and Moor, J.R. (1977) Biochim. Biophys. Acta *499,* 329 – 336.
Macintyre, S.S., Schultz, D. and Kushner, I. (1983) Biochem. J. *210,* 707 – 715.
Mancini, G., Carbonara, A.C. and Heremans, J.F. (1965) Immunochemistry *2,* 235 – 254.
Ngo, T.T. and Lenhoff, H.M. (1982) Molec. Cell. Biochem. *44,* 3 – 12.
Nicollet, I., Lebreton, J.P., Fontaine, M. and Hiron, M. (1981) Biochim. Biophys. Acta *668,* 235 – 245.
Odell, W.D. and Daughaday, W.H. (eds.) (1971) Principles of Competitive Protein-Binding Assays, J.B. Lippincott and Co., Philadelphia.
O'Farrell, P.H. (1975) J. Biol. Chem. *250,* 4007 – 4021.
Owen, J.A., Better, F.C. and Hoban, J. (1960) J. Clin. Pathol. *13,* 163 – 167.
Panrucker, D.E. and Lorscheider, F.L. (1982) Biochim. Biophys. Acta *705,* 184 – 191.
Pepys, M.B. (1979) in Immunoassays for the 80s (Voller, A., Bartlett, A. and Bidwell, D., eds.), pp. 341 – 352, University Park Press, Baltimore.

Pepys, M.B. and Baltz, M.L. (1983) Adv. Immunol. *34,* 141 – 212.
Pepys, M.B., Dash, A.C., Fletcher, T.C., Richardson, N., Munn, E.A. and Feinstein, A. (1978) Nature (London) *273,* 168 – 170.
Ravin, H.A. (1961) J. Lab. Clin. Med. *58,* 161 – 164.
Regoeczi, E. (1970) Clin. Sci. *38,* 111 – 121.
Ritchie, D.G. and Fuller, G.M. (1981) Inflammation *5,* 275 – 287.
Ritchie, G.G. and Fuller, G.N. (1983) Ann. N.Y. Acad. Sci. *408,* 491 – 502.
Ritchie, R.F. (1974) Prot. Biol. Fluids *21,* 569 – 577.
Robey, T.A., Tanaka, T. and Liu, T.Y. (1983) J. Biol. Chem. *258,* 3889 – 3894.
Roy, R.B., Shaw, R.W. and Connell, G.E. (1969) J. Lab. Clin. Med. *74,* 698 – 704.
Rupp, R.G. and Fuller, G.M. (1979) Biochim. Biophys. Res. Commun. *88,* 327 – 334.
Saunders, G.C. (1979) in Laboratory and Research Methods in Biology and Medicine (Nakamura, R.M., Dito, W.R. and Tucker, E.S., III, eds.), Vol. 3, pp. 99 – 118, Alan R. Liss, Inc., New York.
Schapira, M., Silver, L.D., Scott, C.F. and Colman, R.W. (1982) Blood *59,* 719 – 724.
Schen, R.J. and Rabinowitz, M. (1966) Clin. Chem. Acta *13,* 537.
Scherer, R. and Ruhenstroth-Bauer, G. (1977) Naturwissenschaften *64,* 471 – 478.
Schmitz-Huebner, U., Nachbar, J. and Ashbeck, F. (1980) J. Clin. Chem. Clin. Biochem. *18,* 221 – 225.
Schultze, H.E. and Schwick, G. (1959) Clin. Chim. Acta *4,* 15 – 25.
Sipe, J.D., Ignaczak, T.F., Pollock, P.S. and Glenner, G.G. (1976) J. Immunol. *116,* 1151 – 1156.
Sipe, J.D., Vogel, S.N., Sztein, M.B., Skinner, M. and Cohen, A.S. (1982) Ann. N.Y. Acad. Sci. *389,* 137 – 150.
Skelley, D.S., Brown, L.P. and Besch, P.K. (1973) Clin. Chem. *19,* 146 – 186.
Smith, S.J., Bos, G., Esseveld, M.R., van Eijk, H.G. and Gerbrandy, J. (1977) Clin. Chim. Acta *81,* 75 – 85.
Stakenburg, J. and Neumann, H. (1977) Clin. Chim. Acta *80,* 141 – 149.
Starkey, P.M. and Barrett, A.J. (1977) in Proteinases in Mammalian Cells and Tissues (Barrett, A.J., ed.), pp. 663 – 696, Elsevier/North-Holland Publ. Co., Amsterdam – Oxford – New York.
Sunderman, F.W.Jr. and Nomoto, S. (1970) Clin. Chem. *16,* 903 – 910.
Sztein, M.B., Vogel, S.N., Sipe, J.D., Murphy, P.A., Mizel, S.B., Oppenheim, J.J. and Rosenstreich, D.L. (1981) Cell. Immunol. *63,* 164 – 176.
Tatsuta, E., Sipe, J.D., Shirahama, T., Skinner, M. and Cohen, A.S. (1983) J. Biol. Chem. *258,* 5414 – 5418.
Tracy, R.P., Currie, R.M. and Young, D.S. (1982a) Clin. Chem. *28,* 890 – 899.
Tracy, R.P., Currie, R.M. and Young, D.S. (1982b) Clin. Chem. *28,* 908 – 914.
Voller, A., Bertlett, A. and Bidwell, D.E. (1978) J. Clin. Pathol. *31,* 507 – 520.
Walker, W.H.C. and Gauldie, J.D. (1978) in Automated Immunoanalysis (Ritchie, R.F., ed.) Part 1, pp. 203 – 225, Marcel Dekker, New York.
Wang, H.P. and Chu, C.Y.T. (1979) Clin. Chem. *25,* 546 – 549.
Weeke, B. (1973a) Scand. J. Immunol. *2* (Suppl. 1), 15 – 35.
Weeke, B. (1973b) Scand. J. Immunol. *2* (Suppl. 1), 37 – 46.
Weeke, B. (1973c) Scand. J. Immunol. *2* (Suppl. 1), 47 – 56.
Wiman, B. and Nilsson, T. (1983) Clin. Chim. Acta *128,* 359 – 366.

Index